Applied Health Analytics and Informatics Using SAS®

Joseph M. Woodside

§.sas.

sas.com/books

Contents

About This Book

What Does This Book Cover?

Health Anamatics is formed from the intersection of data analytics and health informatics. There is significant demand to take advantage of increasing amounts of data by using analytics for insights and decision-making in healthcare. This comprehensive textbook includes data analytics and health informatics concepts along with applied experiential learning exercises and case studies using SAS Enterprise Miner in the healthcare industry setting. The intersection of distinct areas enables connections between data analytics, clinical informatics, and technical software to maximize learning outcomes.

Is This Book for You?

This textbook is intended for professionals, lifelong learners, senior-level undergraduates, and graduate-level students, it can be used for professional development courses, health informatics courses, health analytics courses, and specialized industry track courses.

What Are the Prerequisites for This Book?

An introductory statistics course and an introductory computer applications course are the recommended prerequisites for this book. Topics in an introductory statistics course might include descriptive statistics (frequency, central tendency, and variation) and inferential statistics (sampling, probability, correlation, and experimental design). Topics included in an introductory computer applications course might include computer hardware, productivity software (Microsoft Office, Excel, Word), data access and manipulation, and strategic use of technology.

What Should You Know about the Examples?

Experiential learning activities and applications are included in each chapter so that you can gain hands-on experience with SAS in various healthcare disciplines and in real-world settings. The practical nature of this book helps you to integrate healthcare, analytics, and informatics into health anamatics knowledge, skills, and abilities.

Software Used to Develop the Book's Content

SAS Enterprise Miner 14 is the graphical user interface (GUI) software for data mining and analytics.

Example Code and Data

You can access the example code and data for this book by linking to its author page at https://support.sas.com/woodside.

About the Author

Dr. Joseph M. Woodside is an Assistant Professor of Business Intelligence and Analytics at Stetson University teaching undergraduate, graduate, and executive courses on analytics, health informatics, business analysis, and information systems. He has been a SAS user for over ten years and is responsible for updating the analytics learning goals and course content for the SAS Joint Certificate Program. Before accepting the Business Intelligence and Analytics position at Stetson, Dr. Woodside worked with KePRO, a national healthcare management company, as the Vice President of Health Intelligence, with responsibility for healthcare applications, informatics, business intelligence, data analytics, customer relationship management, employee wellness online platforms, cloud-based systems deployment strategy, technology roadmaps, database management systems, multiple contract sites, and program management. Dr. Woodside previously held positions with Kaiser Permanente, with responsibility for HIPAA Electronic Data Interchange (EDI), national claims and electronic health record implementations, National Provider Identifiers, cost containment financial analytics, and various data analytic initiatives. Learn more about this author by visiting his author page at http://support.sas.com/woodside. There you can download free book excerpts, access example code and data, read the latest reviews, get updates, and more.

We Want to Hear from You

SAS Press books are written *by* SAS Users *for* SAS Users. We welcome your participation in their development and your feedback about SAS Press books that you are using. Please visit sas.com/books to do the following:

- Sign up to review a book.
- Recommend a topic.
- Request information on how to become a SAS Press author.
- Provide feedback on a book.

Do you have questions about a SAS Press book that you are reading? Contact the author through saspress@sas.com or https://support.sas.com/author_feedback.

SAS has many resources to help you find answers and expand your knowledge. If you need additional help, see our list of resources: http://sas.com/books.

Acknowledgments

I would like to thank the numerous individuals who have provided input and feedback in support of Health Anamatics. Thanks to my family members, editors, colleagues, leadership, and students in my previous healthcare and analytics coursework who have encouraged me to develop a customized textbook to maximize learning outcomes. This is an area of great interest to me. The efforts of the support team at SAS Press in preparing the manuscript copies and final textbook are greatly appreciated. I would like to provide individual appreciation to the following people:

SAS Press editor Lauree for the high level of personalized support and feedback throughout the publishing process.

The SAS technical reviewers Roy, Malorie, Catherine, Laurie and Jeremy for their valuable reviews and recommendations.

The SAS Press team of Julie, Stacey, and Sian for the topic design plan and publication opportunity.

The academic leadership team Wendy, Noel, Neal, Monica, and Yiorgos for their support of the interdisciplinary teacher-scholar role.

All my departmental colleagues Betty, Bill, Fred, John, Mahdu, Petros, Shahram, and school and university colleagues for their encouragement and contributions to my development.

My family members, parents, and Stephanie for their lifetime of care.

Chapter 1: Introduction

Introduction

Health Anamatics is formed from the intersection of data analytics and health informatics. Healthcare systems generate nearly 1/3 of the world's data, and healthcare stakeholders are promised a better world through data analytics and health informatics by eliminating medical errors, reducing re-admissions, providing evidence-based care, demonstrating quality outcomes, and adding cost-efficient care among others. Although healthcare has traditionally lagged behind other industries, the turning point is near with an increased focus across the healthcare sector by way of cost pressures, new technologies, population changes, and government initiatives. There is significant demand to take advantage of increasing amounts of data by using analytics for insights and decision making in healthcare. Healthcare costs keep rising and we can use our technology and analytics capabilities to help address these costs while also improving quality of care. It is our aim to use our knowledge for good and worthwhile causes.

Having conducted several health analytics and informatics related courses and professional education workshops, I have found a need for a comprehensive and current textbook that combines the applied analytics knowledge using SAS with the clinical healthcare informatics concepts. In addition to my ten years of healthcare industry experience, I have met with over 50 industry organizations and executives over the last several years to research relevant content, topics, and applications for health anamatics. This textbook provides a distinguishing feature as a holistic approach as shown in Figure 1.1.

Figure 1.1: Health Anamatics Textbook Distinguishing Approach

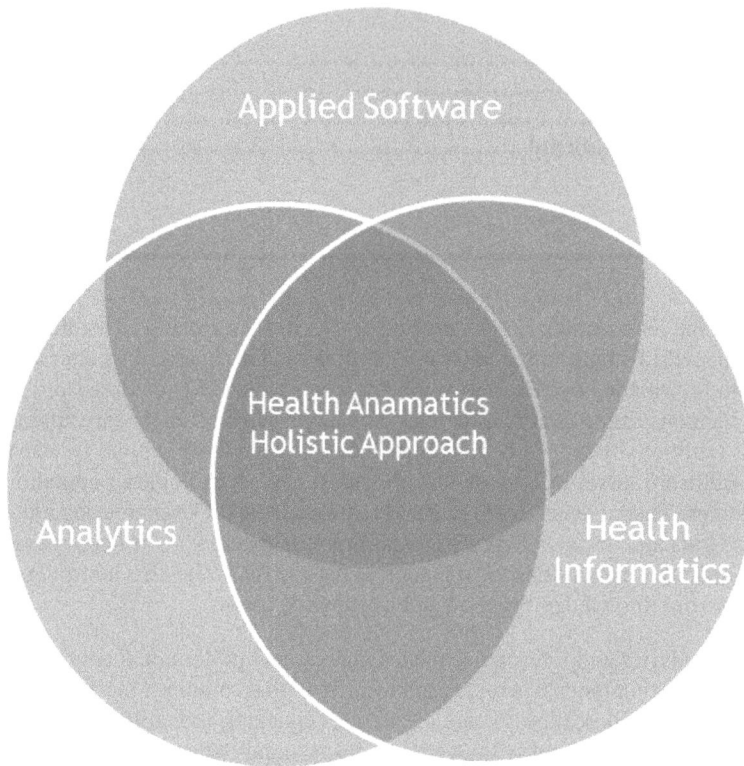

Related resources have a primary focus on clinical informatics, technical software, or analytics aspects exclusively, without a connection between all areas to integrate knowledge and maximize learning outcomes.

This textbook contains content and learning objectives, including data analytics and health informatics concepts along with applied experiential learning exercises and case studies using SAS Enterprise Miner within the healthcare industry setting. All clinical data sets are designed to follow the same data structure, data variable set, data characteristics, and methods of published research and industry applied experiential learning examples.

Audience Accessibility

Healthcare and analytics are among the fastest growing areas in industry and curriculum development. This textbook is intended for professionals, lifelong learners, upper-level undergraduates, graduate level students, and can be used for professional development courses, health informatics courses, health analytics courses, and specialized industry track courses. At the graduate level there are currently over 125 analytics programs for which this could be an applied elective or track course, along with over 100 informatics programs for which this could be a core course.

Sample University and Professional Education course titles and current coverage includes:

- Health Anamatics
- Health Informatics
- Health Information and Analytics Management
- Health Analytics
- Healthcare Analytics Management
- Evidence-Based Healthcare Management
- Healthcare Managerial Decision Making
- Applied Analytics in Healthcare

In previous courses, I have had the opportunity to enroll students from a wide variety of specialty areas with a strong interest in learning healthcare and analytics and have helped them be successful in the applied topics. This textbook follows my teaching approach in being accessible to a wide variety of backgrounds and specialty areas including industry professionals, administrators, clinicians, and executives. Examples of major specialty areas from prior enrollment include nursing, information technology, business, international studies, entrepreneurship, sports management, finance, biology, economics, marketing, accounting, and mathematics.

Learning Approach

You might be familiar with the 2015 Disney film, *Inside Out*, which follows the main character Riley, and her emotions of Joy, Sadness, Anger, Disgust, and Fear (Disney, 2017). Watch the following YouTube clip: "Long Term Memory Clip – Inside Out" https://www.youtube.com/watch?v=V9OWEEuviHE

During the film, Joy and Sadness find themselves stuck in endless banks of long-term memory and have trouble finding their way back to headquarters. That is, they do not know the pathway back. Similarly, suppose you are traveling through an endless forest. How do you find your way back? If you walk the path hundreds or thousands of times, you will find it easier each time to find your way back through a clear trail that you have made over time. After a while it will be easy to follow the trail back and find your way home. Human memory is like a nature trail: through frequent retrieval of information that you are creating a pathway, and if you retrieve the information enough, a clear trail forms. Many times along your journey, you might feel that remembering is impossible and you might be like Sadness – this will never happen! Instead, be positive like Joy – with repeated practice and determination that you will find the pathway! Learning takes tremendous effort. It is through this effort that the pathways and memory are built, increasing your intellectual capabilities. Synapses are connected in the brain, and by frequently retrieving memories that you are forming a path to that information. If you retrieve the memory enough times, a well-defined path forms.

Like Riley in *Inside Out*, mental models are psychological representations of real, hypothetical, or imaginary situations, and the individual representation that is used for reasoning. Mental models allow users to understand phenomena, make inferences, respond appropriately to a situation, and define strategies, environment, problems, technology, and tasks. Mental models influence behavior and create reasoning basis, which improve human decision making, by allowing pre-defined models which speed information processing. Mental-model maintenance occurs when new information is incorporated into existing mental models and reinforcement occurs. Mental-model building occurs when mental models are modified based on the new information. Achievement of both mental models is important to achieving quality and sustained performance. Similarly, health anamatics is intended to provide all stakeholders with high quality, easy to use, and relevant information for decision making. To measure the success, one might gauge whether health anamatics capabilities help users learn. Learning is defined as a purposeful remembering displayed through skillful performance, and is measured as potential change in performance behavior, as the change might occur at a point in time after the information is collected (Vandenbosch and Higgins, 1995; Woodside, 2010a). Health anamatics can be used to improve mental model development. In other words, it help users such as patients, clinicians, and administrators learn.

This textbook follows an experiential, integrative, and applied learning approach using techniques of practice and reflection to reinforce learning. Experiential learning has been included in classrooms as an improved way to educate and engage students as compared with traditional lecture-based learning (Chapman et al., 2016). Traditional education does not offer learners the opportunity to understand the importance of the learning content and real-world scenarios, thereby emphasizing the importance of having learners conduct real-world scenarios to learn and apply to future scenarios. Effective and quality higher education can be achieved only when the balance of academic and practical professional engagement is reached and integrated in a meaningful way (AACSB, 2017). Despite the value of integrative learning across all courses including general education, these student-centered techniques have had limited adoption throughout colleges and universities (Hora, 2017). Instructors and educators also have an important role in experiential learning, requiring individual engagement to facilitate the learning experience and to ensure knowledge generation. Advance planning of the experiential learning activity is critical more so than a traditional lecture, and a learning session might be customized on-the-go and provide opportunities for teachable moments during the session. Experiential learning can also assist with individualized instruction, as each individual has the flexibility to internalize the content to their own individual needs and reflect in a manner meaningful to them as individuals (Roberts, 2003). After following these best learning practices, what can you expect from the results? Research has found that the results include deeper learning and higher grades, which are both agreeable goals.

The empowered and engaged learning approach as shown in Figure 1.2 consists of three phases:

1. Capturing initially difficult concepts through the learning journal and rephrasing in your own words,
2. Communicating concepts through retrieval practice in varying scenarios, this phase is mental model maintenance, and
3. Connecting concepts to professional career areas, industries, and opportunities, this phase is mental-model building where the knowledge is connected to new domains and existing knowledge (Woodside, 2010a).

Figure 1.2: Three-Phase Learning Approach

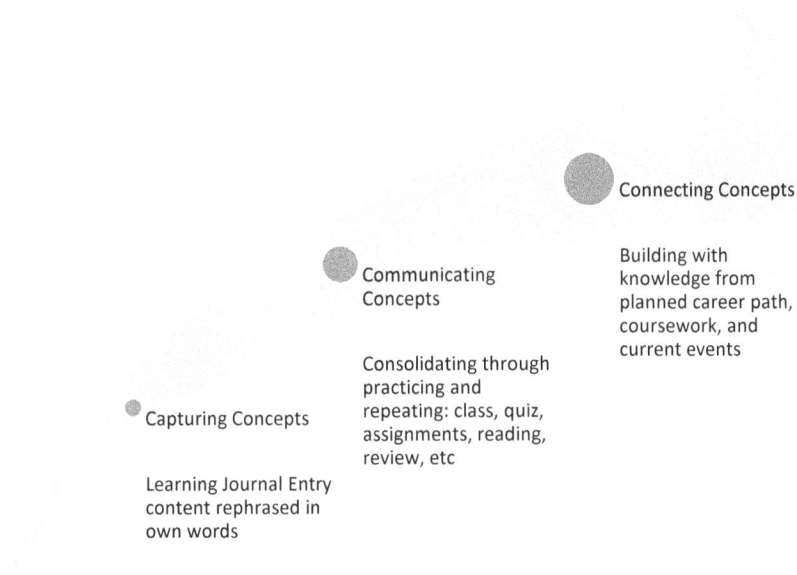

Connecting Concepts

Communicating
Concepts

Building with
knowledge from
planned career path,
coursework, and
current events

Consolidating through
practicing and
repeating: class, quiz,
assignments, reading,
review, etc

Capturing Concepts

Learning Journal Entry
content rephrased in
own words

The learning approach phases can also be thought of as three learning loops, or continuous learning, at each phase. Connections are a key component of experiential, integrative and applied learning that allows connections to be made between concepts and experiences throughout your other courses, professional knowledge, and events, to continually apply your learning to more complex issues and challenges. Over the course of one's career, you will likely change jobs and positions many times. To be successful, you must incorporate your prior knowledge and connect to your new environment to improve decision making and to adapt easily. Initially, the learning might not appear as evident as traditional learning methods would, such as assignments. However, over time the connections become strengthened through experientially based work. The Commission on Accreditation in Physical Therapy Education, the American Association for the Advancement of Science, and the Association of American Colleges and Universities all highlight the critical nature of integrative learning for students to be successful throughout their professional careers. Twenty-first century general education, liberal arts education, co-curricular and pedagogical innovations require effective instructional methods that are able to blend and cut across areas. These methods are the foundation of experiential, integrative and applied learning, and the overall health anamatics approach (Ithaca, 2017; AAC&U, 2017).Trying to solve a problem before being taught the solution leads to better learning, even when errors are made. Applied real-world simulations allow retrieval practices, and spaced and interleaved practice. Interleaved practice often feels slower than massed practice, and as a result, is unpopular and rarely used. Learners might see their grasp of each element coming more slowly and the compensating long-term advantage is not apparent to them. But research shows that mastery and long-term retention is better if you interleave practice, rather than if you mass repeat practice (Brown, et. al, 2014). In this textbook, a common methodology is used in which concepts are interleaved within each chapter. Variable practice is also better, and along with interleaved practice, helps lead to deep learning versus memorization. Reflection is another form of retrieval practice and individual reflection can lead to stronger learning: retrieval knowledge from memory, connecting to new experiences, and visually and mentally rehearsing what you might do differently. Reflection questions might include "What happened?", "What did I do"?, "How did it work out"?, and "What would I do differently next time?". In an effort to assist with learning as you read through this textbook, a summary of learning tips are included below based on best practices (Brown, et. al, 2014; Woodside, 2018a) and shows how this textbook will support those aims:

Table 1.1: Learning Best Practices

Learning Tip	Description	Textbook Alignment
Understand how learning works	Learning consists of three phases: capturing, communicating, and connecting.	At the end of each chapter, there will be a section for capturing your knowledge, practicing communicating your knowledge, and connecting your knowledge.
Spaced repetition and retrieval practice	After reading, ask what are the key ideas, how would I define them, and how do the ideas relate to what I already know. Set aside time every week to quiz yourself on the material. Compared to rereading, this is initially awkward and frustrating when the new learning is difficult to recall week to week. Nevertheless, when you work hard, you strengthen your learning pathways, and spaced practice is more productive than a single session of repeated practice. Spaced repetition and retrieval practice interrupts the process of forgetting and therefore improves learning.	Self-quizzing after reading each section of the textbook. Experiential Learning activities and applications.
Reflection	Reflection is a combination of retrieval and elaboration and generation to add layers to learning and skills. Elaboration is finding additional meaning by explaining concepts in your own words and how these concepts relate to industry events. Generation is trying to solve a problem before seeing the answer, or filling in a blank rather than choosing from a list, and experiential learning activities.	Learning journal entries - Reflect on what you learned during the previous week and how the class learning concepts connects to life outside of the class.
Effort	If learning is easy, then it is quickly forgotten. Some challenges during learning help to make the learning stronger and better remembered, as effortful learning changes the brain, making new connections and increasing intellectual ability. To achieve excellence in learning, you must strive to surpass your current level of ability, striving by its nature results in setbacks that provide the necessary information for learning mastery.	At first, experiential learning might feel difficult. However, through practice and effort, it leads to stronger long-term learning.

Learning Tip	Description	Textbook Alignment
Integrative Learning	According to the Association of American Colleges and Universities (2017), there is an increasing emphasis at a national level to improve student's experiential, integrative, and applied learning through engaged and empowered educational experiences. Integrative learning occurs across disciplines and is critical to deeper learning, instead of localized competencies for a specific field or role. In addition, learning through engagement and student construction of their own ideas is crucial to improve learning, retention, and ability to apply or integrate the knowledge in new areas (AAC&U, 2017). This type of education is one that empowers students to prepare for the wider world and develops transferable intellectual and practical skills, such as analytical and problem solving, with the ability to apply knowledge and skills in real-world settings (AAC&U, 2017).	Experiential and applied learning activities are included within each chapter, and across varying healthcare disciplines and real-world settings to further integrate healthcare, analytics, and informatics into health anamatics knowledge, skills, and abilities.
Rigor	Rigor is a common term that used in higher education for increasing industry preparation of graduates and for increasing academic quality. However, rigor is often defined differently across stakeholder groups and is implemented throughout curriculum in varying ways.	This textbook uses a real-world rigor approach. Real-world rigor is the attainable balance between the possible and impossible, resulting in the constructive conceptualization and realization of student empowerment, engagement and learning through capturing, communicating, and connecting real-world industry knowledge, skills, and abilities to successfully prepare students for their careers and future work environment. Real-world rigor is required to address the adaptability of higher education graduates in a fast-changing business environment and to address the industry competencies that are received in higher education in order to be successful in the future workforce. Changes, such as automation and new skills requirements, will also have major employment impacts in the future (Woodside, 2018a).

Experiential Learning Activity: Learning Journal

Following our learning approach, we will begin with our first learning journal entry. You might record the learning journal entries in an electronic document, a notebook, or a learning management system if available for your course. The learning journal entries will be completed during each class session or as you complete a portion of the textbook. Each learning journal entry should take approximately five minutes. Write as efficiently as possible and continuously for the full time period. You might go back later and edit or add to the learning journal entries as you continue to refresh the topics and build your learning pathways. The learning journal entries will be for your benefit as you proceed through the textbook, as each item is phrased in your own words. The learning journal initially falls into the first phase of capturing concepts. Throughout the textbook and practice, you will begin to consolidate your knowledge through communication, and lastly to connect the concepts through experiential, integrative, and applied learning in order to build your long-term knowledge.

For your first entry, provide your background and knowledge of healthcare, informatics, and analytics. Then rate this knowledge on a scale from 1-100. Lastly, list your goals upon completion of this course or text. For example, this might be your first health-related course and you are seeking to find your area of interest. Or you might have 20 years of experience within a healthcare clinical role and are seeking to expand your knowledge of analytics.

Learning Journal Topics

- Knowledge of Healthcare (1-100)
- Knowledge of Analytics (1-100)
- Knowledge of Informatics (1-100)
- Goals Upon Completion

Chapter 2: Health Anamatics

Chapter Summary

The purpose of this chapter is to describe the importance of the topics included throughout the textbook, and to introduce general healthcare analytics and informatics concepts. Prior healthcare, analytics, and data experience is not assumed, and this chapter is intended as an introduction for those who are new to the healthcare industry and analytics areas, along with a refresher for experienced readers in the field, through an introduction of a new anamatics concept.

Chapter Learning Goals

- Define health informatics, health analytics, and health anamatics
- Describe the primary systems and sources of data in the healthcare industry
- Explain the growth of data in the healthcare industry
- Understand the importance of anamatics in the healthcare industry

Health Anamatics

Health Anamatics

Health anamatics is the combination and use of health analytics and health informatics. From history, we know that health analytics or health informatics alone cannot fix healthcare. A comprehensive systems approach must be taken to improve healthcare, combining the transformative power of people plus health analytics and health informatics. Therefore, health anamatics is an interdisciplinary and integrative field involving the systems, technologies, and delivery to inform decision makers and to improve the value-based delivery of healthcare. Health anamatics as a result can be considered "The Art of Analyzing Health Information."

A summary of the Health Anamatics components is displayed in Figure 2.1.

Figure 2.1: Health Anamatics Overview and Components

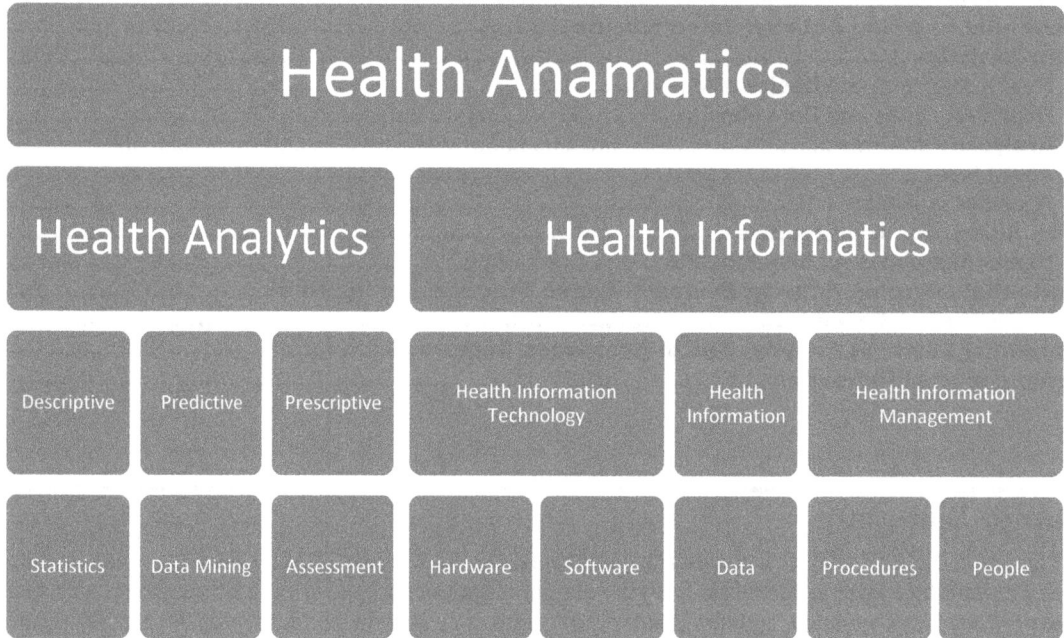

Health anamatics consists of health analytics and health informatics. With health informatics comprising primarily the technology systems, information, and management, health analytics is using the data and information captured within the technology systems to improve decision-making and the value-based delivery of healthcare.

In healthcare, value is often an ill-defined term, and can refer to the delivery, quality, availability, or cost of care (Okoye, 2015; Woodside, 2018b). Given the continued rising costs, countries around the world are seeking to redefine value in healthcare with economic and patient considerations. Despite significant technological changes over the last several decades, the business model and value proposition of healthcare have remained the same. Porter defines the value chain as a set of inputs, outputs, and processes that occur to produce a service. The value chain activities generate value within a resource, product, or service. The value chain can show the greatest value points, highest costs, or waste (Okoye, 2015). A traditional value chain consists of primary activities, which contribute directly to the product or service such as manufacturing and operations, and support activities that contribute indirectly such as human resources. The value chain includes interactions among the activities known as linkages (Kroenke and Boyle, 2017).

The healthcare value chain consists of payers, fiscal intermediaries, providers, purchasers, and producers. The 1990s saw investments in the healthcare value chain as a result of vertical and horizontal integrations, changes in federal healthcare laws, reimbursement pressures, and the rise of the internet. Many initial efforts were unsuccessful and instead led to consolidation among major members in the value chains, and there was an open question about whether the market consolidation resulted in improved or lessened levels of competitiveness. Healthcare regulations have been created to limit costs on support activities in the value chain. The 80/20 rule for insurance companies specifies that 80% of premiums must go to direct healthcare costs and quality such as patient care, manufacturing, and operations. No more than 20% of premiums, can be used on supportive administrative activities such as information technology, administration, and overhead (Healthcare.gov, 2018, Woodside and Amiri, 2018).

Health Anamatics and Broccoli

Before we discuss health anamatics further, let's first talk about broccoli. Broccoli is one of our world's super foods, and has been shown to prevent cancer, reduce cholesterol, detoxify the body, improve heart and eye health, maintain healthy digestion, and contributes to a decrease in joint damage (Szalay, 2014; The George Mateljan Foundation, 2016). Likewise, health anamatics is a super technology, and can improve healthcare, reduce costs, enhance outcomes of care, and help us lead healthier lives. Always remember, health anamatics, like broccoli, is good for you, so be sure to eat your broccoli and study health anamatics!

In the U.S., broccoli consumption per capita has approximately doubled from 1990 through 2012, and the rise is projected to continue through 2020. For adults, a recommended vegetable serving is 2.5 cups per day (Lamagna, 2016). This being a health-related course, we know that broccoli, like health anamatics, is good for us. As you read through the content of the textbook, feel free to have a second or third helping or 2.5 helpings of health anamatics per day, to reinforce and supplement the important concepts. Throughout the textbook, highlight important areas and identify those areas that you might wish to have a second or third helping of in order to improve your understanding or to repeat an important foundational concept. To help form this comparison between broccoli and health anamatics, think of terms ranges. There are probably folks that consume no broccoli, and those that consume above average amounts. President George H.W. Bush once said that his mother made him eat broccoli as a kid, and, now that he's president, he will no longer have to eat broccoli (Szalay, 2014; The George Mateljan Foundation, 2016)! Similarly, there are folks that have no knowledge of health anamatics, and those that have above average knowledge. By starting this book. you are heading in the right direction by eating your broccoli and feeding your mind with health anamatics. Through our continuous learning process, we'll also make the content easier to consume over time and, by the end, you'll wonder how you could ever do without health anamatics (and broccoli)!

Need for Health Anamatics

Health anamatics is required due to 1) increasing costs, 2) population changes, 3) government initiatives and incentives, and 4) increasing data capture and analytics capabilities. These ongoing cost pressures, quality focus, data captures, and disruptive innovations are causing every company to rethink their strategies for future growth. Consumer-driven demand and evidence-based medicine are changing the global healthcare markets. Even though costs and technology have been in discussion for over 40 years, changes in globalization are adding additional pressures to change, including aging population, emerging markets, and healthcare reform. The global healthcare profit pool is projected to grow from $520 billion in 2010 to $740 billion in 2020. Sector and regional factors will impact the profit pool. For example, the pharmaceutical and medical devices are projected to see slower growth and declining profit margins driven by patent expirations and pricing controls, and competition. By contrast, healthcare IT companies are projected to experience significant growth as organizations outsource functions and data demand increases (George et al., 2012).

Healthcare stakeholders promote the use of healthcare informatics, information systems, and analytics as a way to provide safe, affordable and consumer-oriented healthcare. This includes avoiding medical errors, the improved use of resources, accelerated diffusion of knowledge, reduction in access variability, consumer role advancement, privacy and data protection, and public health and preparedness. Health information systems have been shown to decrease billing issues, medical and drug errors, and improve patient health, use of medical evidence, cash flow and collections, paper cost, quality, safety, research, compliance, and preventative care. In other industries, information systems usage increases can be tied to improved quality and competitive advantage. Extracting, formatting, analyzing, and presenting this data can improve quality, safety, and efficiency of delivery within a healthcare system. Tools and applications for healthcare are required to analyze and coordinate information and intelligence between areas. With increasing costs and competition, healthcare organizations have increasingly turned to analytics to improve operating efficiencies. Analytics allows the organization to maximize the value of the information to reduce costs and improve quality (Woodside, 2013a). Health anamatics, or the combination of health analytics and informatics, provides the required value and addresses these stakeholder needs.

Increasing Costs

The first primary need for health anamatics is to address increasing costs because across nearly 60 countries, healthcare costs are projected to rise over 6% per year. In regions such as Asia and Middle East, development of healthcare systems and movement of universal health coverage will drive differences among counties in terms of spending. Continued pressure to reduce costs and improve quality is also expected. The U.S. has one of the highest per capita healthcare expenditures among developed nations, averaging over $9,000 per capita (Deloitte, 2016). Despite high expenditures, the U.S. had lower health outcomes, such as life expectancy, infant mortality, and chronic disease prevalence (Squires and Anderson, 2015). Historical healthcare costs have increased at a rate beyond economic growth over the past two decades, and healthcare costs are projected to become unsustainable by 2050, barring additional reforms. The global recession in the late 2000s helped slow healthcare costs even though costs are increasing once again, with the percentage of gross domestic product (GDP) spent on healthcare that is projected to increase from an average of 6% to 14% by 2060. Public funding contributes approximately 75% of overall healthcare funding, and is supported mainly through payroll taxes, which are projected to decrease with aging populations (Biernat, 2015).

Population Changes

The second primary need for health anamatics is to address ongoing population changes. The global population has surpassed 7 billion and is projected to continue to grow over the next several decades. In 2015, 60% of the global population was based in Asia, 16% in Africa, 10% in Europe, 9% in Latin America and Caribbean, and 5% in North America. China and India are the two largest countries with more than 1 billion people, and nearly 20% of the global population in each county. The gender breakdown globally includes 50.4% male and 49.6% female, with a median age of 29.6 years. Looking ahead over the next century, even though growth rates are slowing, the global population is still expected to reach 11.2 billion by 2100. Africa is projected to have the highest population growth rate followed by Asia, North America, and then Latin America. As an exception to future population growth, Europe is projected to have a decreasing population growth. By 2050, nine countries are projected to contribute to half the global population: India, Nigeria, Pakistan, Democratic Republic of Congo, Ethiopia, the United Republic of Tanzania, the U.S., Indonesia, and Uganda. By 2050, Asia and Africa would contribute just under 80% of the world's population or 5.3 and 2.5 billion, respectively. By 2100, Asia and Africa would contribute nearly 83% of the global population or 4.9 and 4.4 billion, respectively. Both Europe and Latin America and the Caribbean are projected to have decreasing population growth between 2050-2100. The median age is projected to be 46 by 2100. Again, China and India are projected to be the two largest countries, while reversing places with India by 2100 as the largest country by population (United Nations Department of Economic and Social Affairs, 2015).

In addition to growth, there are also increases in aging, and by 2050. nearly one quarter of the global population will be over the age of 60. This aging also varies by country. For example, Brazil, China, and India will have a longer period of several decades to reach 20% of the population over the age of 60. A concept known as "health aging" is gaining some traction within healthcare, allowing individuals to continue in a contributory capacity, and realigning government programs and health systems to meet the requirements of an aging population (World Health Organization, 2017). Life expectancy is also increasing globally, rising from 67 years for those born between 2000-2005, to 70 years in 2010-2015, to 77 years in 2045-2050, and 83 years in 2095-2100. As life expectancy increases, populations that are aged over 60 are among the fastest growing, increasing to 2.1 billion by 2050 and 3.2 billion in 2100, which is an increase of nearly 3 times. People over the age of 70 are projected to increase to 944 million by 2100, which is an increase of 7 times (United Nations Department of Economic and Social Affairs, 2015).

Government Initiatives and Incentives

The third primary need for health anamatics is to address government initiatives, which are increasingly intervening in order to reform healthcare. In the U.S., there have been several major government initiatives to reform healthcare over the last several decades. In China, a series of governmental health reforms were announced and aimed at improving safe, effective, convenient and low-cost healthcare to the greater than 1 billion population. These reforms were aimed at health insurance, primary care, hospital management, medications, and public health (Sussmuth-Dyckerhoff and Want, 2010). In the European Union, the trend has moved to decentralization and privatization in order to move resources and knowledge to local populations where they are better applied. Other tenants of reform include patient choice, public health, cost sharing, incentive systems, pharmacy cost reductions, restriction of hospitals and regional care, and improvement of quality and outcomes of care. However, many of these initiatives have been around since the 1980s (World Health Organization, 1996).

Although costs continue to rise, instead of medical science breakthroughs or additional government policy required, more timely and simpler solutions point to focusing on the complete cycle of care by aggregating and analyzing information at the patient level. Early incentive programs focused on financial rewards, which was commonly known as "pay for performance" or P4P. These programs relied on increased payments or penalties in an effort to move away from pay for service models, however these programs have had limited impact on the intended results of improving healthcare outcomes and reducing costs. Current incentive programs expand on the financial incentives to also include motivation, social influences, and public policy. One method is to accurately measure costs and compare the costs with the healthcare outcome, which, in other words, is an embodiment of value-based healthcare. Outcomes can include survival, ability to function, duration of the care, discomfort, complications, and recovery time. Provider incentives are also aligned with these outcomes. Instead of being paid for the number of services performed, providers are paid based on the health outcomes of the patients (Zezza et al., 2014; Kaplan and Porter, 2011).

Data Capture and Analytics Capabilities

The fourth primary need for health anamatics is to address increasing information and advanced analytics capabilities. Data collection is commonplace and often taken for granted by companies. In some cases, it is even seen as a waste product of the company operations. Other organizations relegate the responsibility to the technology department rather than to all areas of the organization to use as a valuable asset. In a PricewaterhouseCoopers (PwC) survey, 75% of organizations lacked the technology and skills to use data for a competitive advantage. Of the 1,650 organizations, only 4% were categorized as data elite, meaning that the organizations had strong data leadership and a culture of evidence-based decision-making, employed skilled analysts, and implemented advanced analytics tools (White, 2015).

Health Informatics

Health informatics is an interdisciplinary field comprising information technology, information systems, information management, computer science, decision science, information science, management science, cognitive science, organizational theory and healthcare services. Health informatics contains several subsets and related discipline names, including medical informatics, clinical informatics, business informatics, biomedical informatics, bioinformatics, nursing informatics, and dental informatics (HIMSS, 2014; AMIA, 2017). By 2020, the global healthcare information technology market alone is projected to reach nearly 229 billion with an annual growth rate of 13.4%. Health informatics software solutions are offered from McKesson, Allscripts, AthenaHealth, Epic Systems, GE Healthcare, Siemens Healthcare, Cerner Corporation, and Carestream Health (MarketWatch, 2016). Because health informatics is often a complex topic, there are numerous terms and definitions within the healthcare industry, almost a language unto its own. When hiring within the industry and working with the human resources area, the employer often finds it easier to locate candidates with healthcare knowledge and to teach them the analytics rather than to find someone with analytics knowledge and to try to teach them the healthcare knowledge. Throughout this textbook, I hope that you will gain knowledge about health anamatics and competitive career advantages.

Health informatics can be broken down into health information technology, health information, and health information management. Health information technology is the software, hardware, data storage, and user interface for management. The hardware includes the PCs, tablets, storage devices, and other medical devices. The software includes the applications such as the electronic medical records application. Health information includes the physical data generated through the software applications and stored on the hardware. Examples of data are patient data, diagnosis, procedures, lab results, and clinical notes. Health information management involves the analysis of information and combines business, science, and technology aimed at improving quality, business process, and accuracy and also protecting information. Health informatics is the combination of these areas, including the people, science, information technology, delivery, and outcomes (AHIMA, 2017a, AHIMA, 2017b).

Seen from another viewpoint, the healthcare environment can be portrayed as a universe with many different components, and may be viewed as a set of rotating planets. There are larger planets, asteroids, and so on, each being a part of the universe. Typically, these components in the universe are held together by gravitational forces, some forces weak and some forces strong, even though each of these forces act on one another. Similarly, in healthcare, we have components such as patients, providers, payers, pharmaceuticals. and governments. The forces holding these entities together can often be described through a political, economic, social, and technical means, influencing trajectories and a long-term route.

Figure 2.2: Healthcare Universe

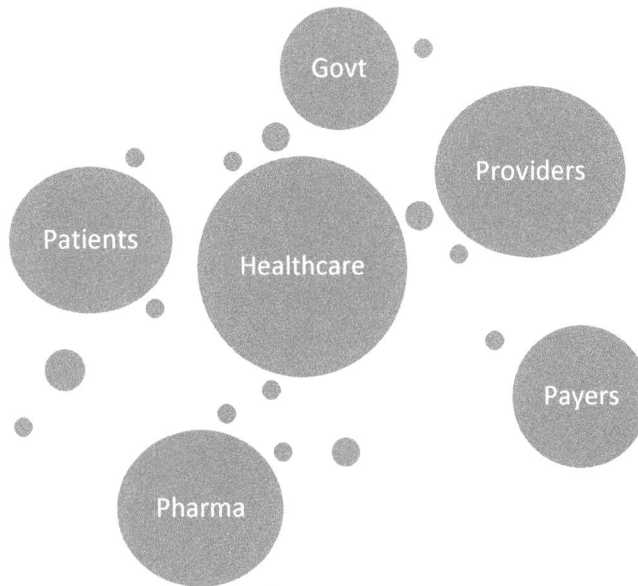

Much like the universe, technology usage within healthcare predates us all. Some early uses of healthcare technology dates back to 1250 when the first magnifying glass was developed, but was later used for eyesight correction and surgery (Bakalar et al., 2012). In the 1970s, the computerization of healthcare began to occur (mThink, 2003). Although healthcare information technology has been around for a long time, recent advances have accelerated the trend. This acceleration can be traced in part to Moore's Law, which states that the number of transistors per square inch on a computer chip doubles every 18 months (Kroenke and Boyle, 2016). Gordon Moore, who was a cofounder of Intel, developed the law in 1965 that still largely holds true today, despite calls for the doubling to end. As an impactful result of Moore's Law, the cost of processing power and storage is approaching $0, allowing advanced computerization capabilities that were unavailable in earlier decades. As evidence of processing power, the latest iPhone processor contains over 4 billion transistors (Apple, 2018). In comparing prices, the same processing power that costs less than $100 in today's iPhone would have cost over $500,000 in 1997 and a whopping $100+ million in 1983. New drug development, molecular modeling, cancer research, and other applications require massive processing power that is now readily available and cost-accessible today (Kroenke and Boyle, 2016).

However, despite these advances in computerization, when comparing the rise of healthcare costs to Moore's law costs, we see little impact to date from the increase in availability of processing power to significantly affect the rise of healthcare costs. The best angle that proponents have been able to claim is that the healthcare cost growth has been slowed. The National Health Expenditure Accounts (NHEA) has tracked total healthcare spending in the U.S. from 1960-2015 as a percentage of the gross domestic product (GDP) and per capita. U.S. total healthcare spending grew from 5% of the GDP and $146 per person in 1960 to 17.8% of GDP and $9,990 per person in 2015 (CMS, 2016a).

Figure 2.3: Healthcare Technology Cost Versus Healthcare Expenditures Chart

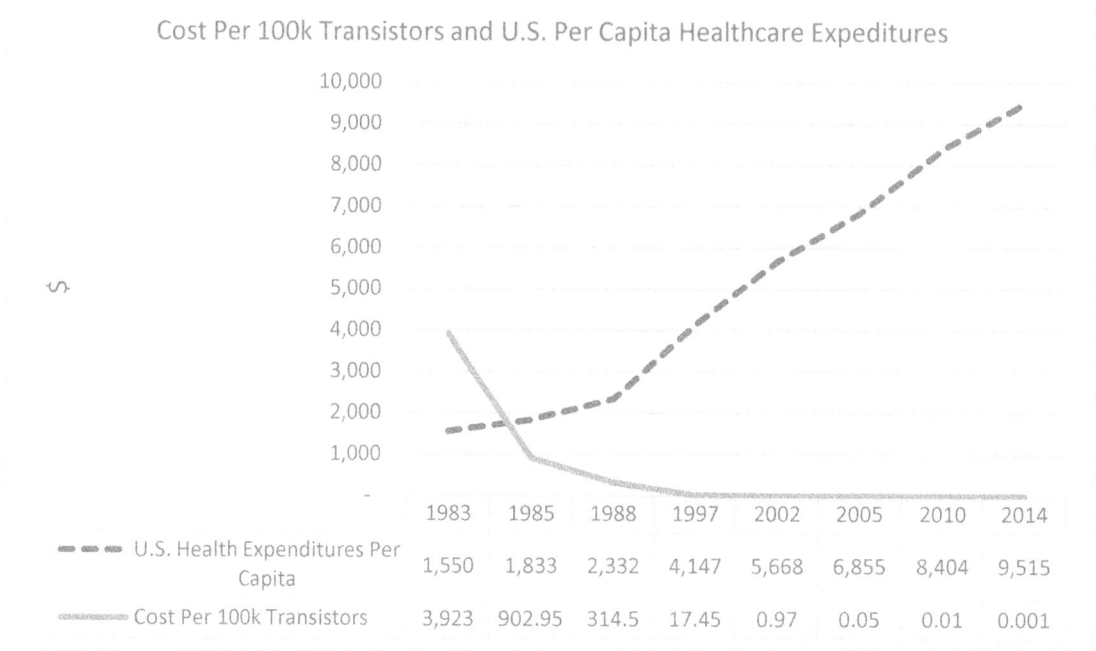

Cost Per 100k Transistors and U.S. Per Capita Healthcare Expeditures

	1983	1985	1988	1997	2002	2005	2010	2014
U.S. Health Expenditures Per Capita	1,550	1,833	2,332	4,147	5,668	6,855	8,404	9,515
Cost Per 100k Transistors	3,923	902.95	314.5	17.45	0.97	0.05	0.01	0.001

Typically, health informatics approaches are constrained to the technical level. Socio-technical theory describes the technical and social subsystems that exist in every organization. The technical system is responsible for taking inputs and converting them into system outputs and goals. The social system includes the people, jobs, and roles. The social and technical approaches aim for joint optimization of the subsystems and joint planning. With this combination, socio-technical configurations have been shown to lead to increased levels of business performance. Systems thinking is a conceptual framework to describe the relationships among social and technical parts, and it incorporates the entire system of elements. Elements of a system can include subsystems, activities, and so on, to form the whole entire system instead of its individual parts. All organizations and systems contain independent but interrelated components that function in complementary methods to complete tasks and goals. Often, in more complex systems, efficiency and quality are impeded. The achievement of this objective requires the close cooperation and collaboration between the parts of the organization and system. This concept applies to healthcare via providers and healthcare systems, which are highly complex due to different institutions, services, administrators, physicians, nurses, professionals, rules, social care, funding, cultures, and patients. Despite growing recognition of systems thinking in healthcare, implementation of these approaches remains challenging. Without full social and technical integration this often leads to service delivery issues, quality declines, patient satisfaction decreases, and cost increases (Woodside, 2013b).

To view and visualize health informatics another way, the health informatics areas can be described using a five-component information systems framework of hardware, software, data, procedures, and people (Kroenke and Boyle, 2017). Health information technology is at the hardware and software layer, and health information at the data layer. Health information management is at the procedures level and the people layer. Health informatics comprises all layers. Health information technology and health information are the technical components, and health information management is the social component. They are all integrated parts of a successful health informatics implementation.

Figure 2.4: Health Informatics Overview and Components

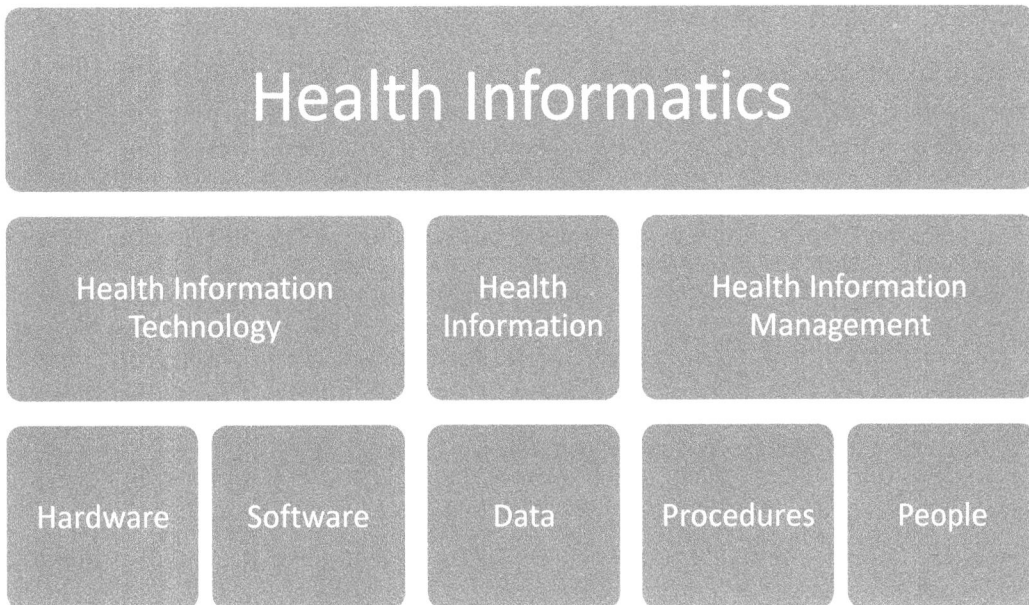

Health Informatics Systems and Hardware

Hardware within the healthcare industry is similar to those of other industries in that personal computers, tablets, laptops, and mobile devices are used to access software and data and to follow set procedures. Unique hardware within the healthcare industry includes medical hardware devices, commonly known as medical devices. In the U.S., the Federal Drug Administration (FDA, 2016a) oversees and approves medical devices before marketing and sale of the medical devices can occur. Medical devices can include pacemakers, surgical devices, ultrasounds, x-rays, and lasers (FDA, 2016b).

Health Informatics Systems and Sources of Data

Below is a brief summary of the health information technologies and health information generated from those technologies. A more detailed description is included within each of the chapters. The health information management aspect provides the bridge between the technology, information, and end users. These are organized by most common end users who can be providers, payers, patients, and government. Providers are the clinical delivery providers, typically physicians, hospitals, pharmacists, and nurses. Payers are the insurance companies, including government programs such as Medicare and Medicaid. Patients are the consumers or individual recipients of healthcare, and government constitutes the institutional entities that are responsible for managing the healthcare system from a legislative and societal standpoint.

Patient Systems and Sources of Data

Common patient systems include Personal Health Records (PHRs), Social Media, Mobile Devices and Wearable Technology, and Employee Records and Wellness Systems. One of the most commonly known and referenced patient sources of data is a PHR, in which patients are able to store, maintain, and manage their health information. Along with employer records, PHRs are typically updated on an annual basis or during an annual review. The most frequently captured and used forms of patient data include social media, mobile devices and wearable technology, which are always tracking and recording data, perhaps even unbeknownst to most end users.

Provider Systems and Sources of Data

Common provider systems include Electronic Health Record (EHR), Computerized Physician Order Entry (CPOE), telemedicine, and pharmacy information systems. EHRs are typically the starting point for information technology adoption and storage for providers. These systems store all the patient history and medical information such as diagnosis, procedures, and billing that occurs from a clinical delivery standpoint. The EHR systems replace the paper-based processes that many providers have used in the past. Many times, Electronic Medical Record (EMR) is used interchangeably with EHR. However, there is a slight distinction. EMR contains the medical and clinical information from a provider whereas the EHR contains both the provider data and patient data, which is similar to a PHR. In other words, an EHR can be seen as a combination of EMR + PHR = EHR. These EHRs are accessible across all providers, and EHR is becoming the more acceptable term (HealthIT.Gov, 2016a). Many CPOE systems provide a mechanism that helps reduce errors, provide alerts, and send information electronically. Pharmacy systems can use the CPOE system for prescribing medications and fulfilling requests.

Payer Systems and Sources of Data

Common payer systems include claims, eligibility, benefits, population health management, and financial and billing systems. The common payer systems are kept for purposes of claim management and payment. Payers must store and maintain large sets of data for claim review, including patient information, patient history, medical records, claim records, benefits and coverages, and eligibility. This comprehensive set of information is used to make decisions for claim payment. In addition, payers are increasingly using their patient and provider information to reduce the cost of care and to improve quality. Through population health management programs, payers outreach to at-risk or future-risk patients to avoid higher costs later on. Payers also reach out to providers through education or quality improvements to lower costs and improve patient quality.

Government Systems and Sources of Data

Common government systems include billing systems, funded healthcare institutions, health information exchanges, public health information, and legislative initiatives and protocols. Governments also collect a myriad of data about individuals. From government services, such as social security, driving licenses, organ donation information, health billing, and governmental insurance programs, governments store and maintain a great deal of information.

Health Information Management

Health information management involves the procedures and the people aspects of healthcare. For procedures, this can include improving efficiency and effectiveness through workflow enhancements, business process improvement, and Six Sigma review. As in all industries, people are the most critical component of any information system, even though we can use information and technology to make us more efficient and effective in the delivery of healthcare.

A McKinsey report recently reviewed the capability for automation in various industries and tasks within job areas. The easiest tasks to automate are well-defined activities or those with predictable physical work or data collection and processing, such as resource extraction and food services. In these areas, 80% of activities have automation potential. The most difficult tasks to automate are those involving the management of people, which provide only a 9% automation potential, and the application of expertise to decision-making, planning or work creation, which provides an 18% automation potential. In particular, the healthcare and education sectors are identified in the report with low automation potential. Although lower than other industries, healthcare as a whole still has an automation potential of 36%, including many administrative activities. Although automation potential decreases for healthcare professionals with patient interaction, such as nurses at 30% and dental hygienists at 13% automation potential. Learning and teaching deep expertise, including complex interactions, are more difficult, with 27% automation potential. To maintain a leadership position in their industry, top executives must first identify how automation can transform their organizations, identify tasks, and support a culture of future adaptability (Chai et al., 2016).

Following our review of health informatics, let's continue with an experiential learning activity on telemedicine. Review the description and answer the following questions.

Experiential Learning Activity: Telemedicine

Telemedicine

Description: Telemedicine is the ability to remotely provide care through online technologies, such as video, email, and teleconferencing, and has been available for several decades. Telemedicine is one of the health information technologies used by providers in caring for patients. Advantages of this approach include ability for a variety of providers and experts to reach patients in any location, 24 hours a day, and to provide and improve quality of care (Chen, 2000).

University of Pittsburgh Medical Center (UPMC) AnywhereCare allows online access 24/7 for a fee of $49 or less based on insurance coverage. The software promotes the benefits of no waiting, traveling, or scheduling to receive care. A variety of treatable conditions are identified such as colds, cough, infections, flu, and allergies (UPMC, 2017). Another telemedicine vendor ZocDoc has 6 million users a month. ZocDoc attributes its success to reforming the patient-provider relationship versus a purely transactional one. Many of the patients are recurring, and time is saved with appointments and cancellations both on the provider and patient side of the business (Court, 2017a).

Despite the promises, telemedicine has had less than expected adoption primarily due to the resistance of providers rather than any technological limitation, with concerns including patient relationships and trust when conducting care remotely (Chen, 2000). In the U.S., healthcare organizations with installed telemedicine technology have increased from 55% in 2014 to 61% in 2016. Hospital adoption has been slightly lower at 45% with an expected 2020 rate of 52%. Adoption of telemedicine technology has increased by approximately 3.5% per year. Adoption increases continue to be relatively small, but the upward trajectory indicates that organizations are coming to terms with the need to use telemedicine technologies to better serve patients (Fitzgerald, 2016). Challenges to telemedicine locally and internationally include technological factors, legal considerations, proven effectiveness, and user adoption (World Health Organization, 2010).

Describe 3 Benefits of Telemedicine

Describe 3 Barriers to Telemedicine

What treatments would work best for Telemedicine? Discuss your justification:

How can the anamatics framework be applied to Telemedicine?

- Health Information Technology

- Health Information

- Health Information Management

- Health Analytics

How could Telemedicine impact local health providers?

What are the global implications to Telemedicine?

Health Analytics

Health Analytics is an interdisciplinary field involving business intelligence, statistics, decision science, information science, information technology, and healthcare services. Whereas health informatics is concerned with the storage and use of data, health analytics is concerned with the use of data for decision-making. Healthcare organizations are under both industry and government pressure to reduce costs and improve quality of care delivery. With ever-increasing amounts of data, healthcare organizations are identifying the importance of analytics for decision-making as organizations seek to compete and differentiate themselves through data-based decisions. Analytics includes the mathematical, algorithmic data processing through techniques including text mining, natural language processing, and visual analytics. Health analytics incorporates these technologies and skills to deliver business and clinical insights in order to improve health outcomes, quality, and costs. Through business intelligence and analytics, organizations can improve healthcare in a variety of industry areas including life sciences, insurance, providers, and public health. The growth of data and analytics in healthcare has created strong demand for professionals who have knowledge of both clinical informatics and information science (Simpao et al., 2014; SAS, 2016a).

Healthcare generates 30% of the world's data according to estimates, with 65% of survey respondents indicating their data storage will grow at a rate between 25-50% per year, driven by imaging files, electronic health records, personal health records, and scanned documents. Big data in healthcare reflects the volume, growth, and types of data, as well as the tremendous opportunities available to unlock the potential value. In one report, big data in healthcare was expected to be valued at $300 billion in the next decade, with annual growth between 1.2-2.4 exabytes per year. Despite this growth, full integration and coordination has yet to occur, and few healthcare providers have developed a formal strategy for handling the increasing amounts of data (DeGaspari, 2010; Hughes, 2011; Milliard, 2012; Woodside, 2013a).

By 2021, the global health analytics market is projected to reach nearly $25 billion with an annual growth rate of 27%. The increase is underlined by governmental programs, electronic health records, healthcare costs, population health management, personalized health, social media, and technology. The use of analytics in precision and personalized medicine, increasing focus on value-based medicine and cloud-based analytics, increasing number of patient registries, and emergence of social media and its impact on the healthcare industry provide significant growth opportunities in the market. Potential obstacles to this growth include the lack of qualified professionals, analytics solutions, and gaps in the care continuum between providers and payers. The lack of skilled professionals, high cost of analytics solutions, and operational gaps between payers and providers might hinder the growth of this market during the forecast period. In addition, healthcare analytics solutions are currently fragmented with solutions from IBM, Optum, Cerner, SAS, Allscripts, McKesson, MedAnalytics, Inovalon, Oracle, Verisk Analytics, and Health Catalyst (Bresnick, 2015; MarketWatch, 2016). Healthcare systems have used analytics to drive quality and performance improvements through quantitative and qualitative decision-making. Clinical services supported by analytics include home health monitoring, population health management, value-based accountable care, disease management, outcomes improvement, length-of-stay reduction, and readmissions reduction. In addition, analytics can support administrative services including financial management, billing fraud detection, marketing, geo-targeting, patient education, resources allocation, and risk assessment (Simpao et al., 2014).

Health Analytics and Decision-Making

Decision-making is a broad topic, which has gained considerable adoption within organizations through decision support and business intelligence systems. Decision theory is used to model human decision-making, and identify how people make or should make decisions. Decision theory mainly uses normative or prescriptive approaches, which assume that the person is logical, rational, and fully informed. Because people do not always act in logical ways, there are positive or descriptive methods aimed at determining what people will actually do. The normative decision creates hypotheses for testing against the descriptive result. These concepts are applied to decision support systems, which for these purposes includes any computerized systems that assist human decision-making. Decision-making consists of three phases: finding the need for a decision, finding possible alternatives, and choosing an alternative. These are also referred to as intelligence, design, and choice. Today's organizations spend a majority of their time engaged in intelligence and alterative design activity, and a small amount on choice (Woodside and Amiri, 2015).

There are generally three theories behind a decision. Positive theory considers each decision as a single item and attempts to describe the decision as precisely as possible. Positive theory is commonly associated with technical operations, where problems must be solved for the short-term. Normative theory lends itself to individuals that use alternatives to create long-term organizational success. Normative theory is associated with institutional managers who have strategic views of the firm. Behavioral theory incorporates components of both positive and normative theories and is based on individual perception and bounded rationality. Behavioral theory is commonly associated with organizational management who acts as an intermediary between the technical and institutional managers to improve the organization. (Woodside and Amiri, 2015).

Figure 2.5: Health Analytics Overview and Components

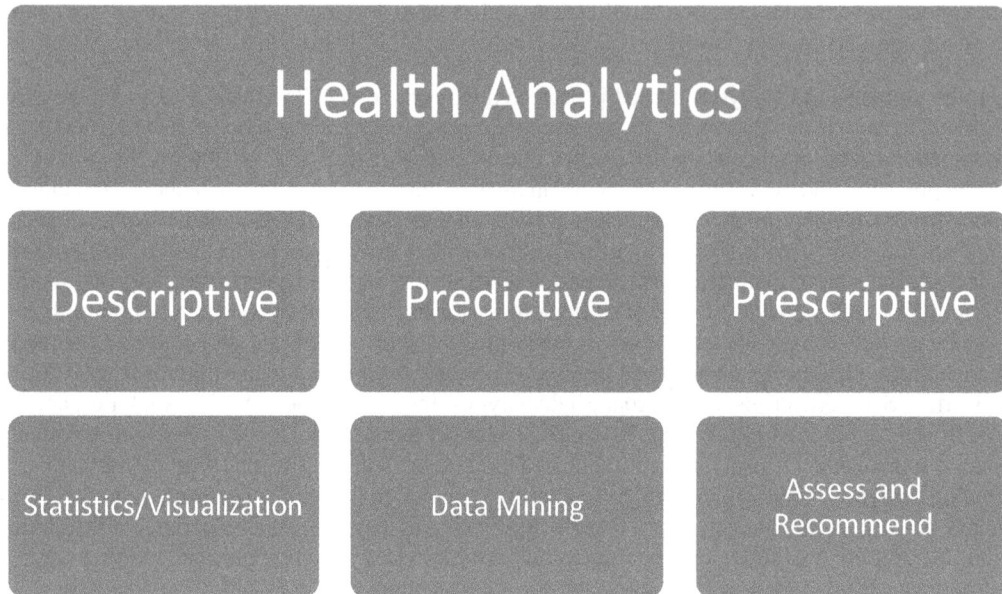

Health Analytics		
Descriptive	Predictive	Prescriptive
Statistics/Visualization	Data Mining	Assess and Recommend

Similar to the three theories of decision-making, there are generally three types of analytics: 1) descriptive, 2) predictive, and 3) prescriptive. Descriptive analytics describes past information or answers the question of what has happened. Predictive analytics describes future information or what can happen. Prescriptive analytics describes future decision-making or what should happen. Prescriptive analytics integrates both descriptive and predictive analytics to recommend the best solution. After running the analytical models, you can assess the overall results and provide a recommended approach. Decision- making occurs similarly at three different levels of the organization, 1) operational, 2) managerial, 3) strategic. Operational decision-making is day-to-day or hour-by-hour and often focuses on historical data. Managerial decision-making is longer term over weeks or months and involves a combination of historical data for future decision-making. Strategic decision-making occurs over quarters and years and is prescriptive in nature (Bertolucci, 2013).

Figure 2.6: Three Types of Analytics and Related Decision-Making, Theory and Approach

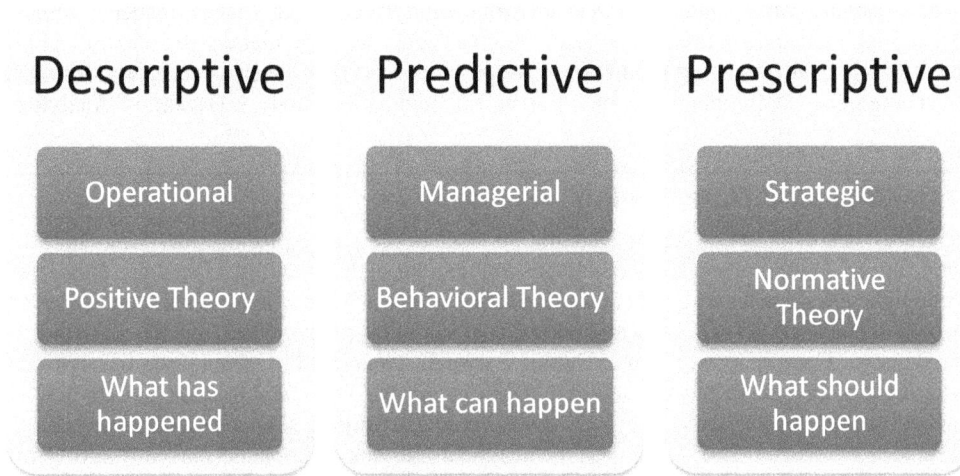

Descriptive

- Operational
- Positive Theory
- What has happened

Predictive

- Managerial
- Behavioral Theory
- What can happen

Prescriptive

- Strategic
- Normative Theory
- What should happen

To categorize analytics another way, analytics maturity can be thought of in three phases. The first phase involves using information for intelligence; these forms of information for intelligence are typically related to going beyond instinct-driven decision-making and moving toward data-driven decision-making along with visualization of information. Analysts in this phase typically still spend the majority of time preparing data. The second phase currently underway for most organizations involves big data and use of sophisticated analytics tools, such as SAS, which can also include predictive forms of analytics. Companies can spend more time analyzing data during this phase. The third phase, as companies move forward, is prescriptive analytics. Companies can spend more time taking actions on the data analytics during this phase. Key components to enabling this final phase include combing various disparate forms of data, using embedded analytics to improve decision-making, using cross-disciplinary data teams to involve both the business and technical users, and chief analytics officers, using all forms of analytics (descriptive, predictive, and prescriptive), and using a culture of smart data-driven decision-making (Davenport, 2013). Using a healthcare scenario, consider an example where a patient schedules a physician office visit. The patient describes the symptoms, such as fever, cough, and so on. The physician then uses this information to predict the sequence of events such as escalation to a disease state. The physician then prescribes the expected outcome to return to wellness such as through medication, rest, or activity avoidance.

Given the wide adoption of health information technology, such as EHRs, over the last several years, organizations have access to increasing amounts of data for analytics. An algorithm can use prior historical data to identify factors for readmission rate, and the same algorithm can then be used to quantify the percent at risk for readmission. However, the final level of prescriptive analytics would help decide whether discharge is possible, is home care required, or is a post-discharge intervention necessary to prevent readmission. Parkland Health and Hospital System in Texas used EHR data to find heart-failure patients with a higher risk of being readmitted. These higher risk patients were then prescribed interventions including education, follow-up telephone interventions to ensure medication compliance, and follow-up appointments. The patient interventions successfully reduced readmissions by 26% (Parikh et al., 2016). Researchers at University of California, Davis are using EHR data for early detection of sepsis. This is important due to the high mortality rate and common late detection. Physicians at Massachusetts General Hospital are using an analytics system to measure surgical risk to improve patient safety. The system searches national databases and provides a visualization to the surgeon in the form of red, yellow, and green risk dashboard indicators. Analytics allows a description of past information, such as diabetic patients or asthma patients, and predicting future episodes, such as a diabetic patient forgetting to take medication or an asthma patient encountering an environmental trigger. At Kaiser Permanente, researchers use analytics to predict diabetic patients who have the greatest chance of dementia. Beyond only clinical improvements, analytics is also being used to improve provider workflows such as reducing wait times and bottlenecks in emergency departments, by means of staffing and patient analysis (Health IT Analytics, 2015).

Health Analytics and Data Mining

As a component of health analytics, health data mining is the process of exploring, analyzing, and uncovering meaningful patterns and trends by reviewing data through various mechanisms. This data is captured through health informatics and health information systems. Data mining is still considered a relatively new and evolving field of study and draws from statistics, mathematics, machine learning, and artificial intelligence. The importance of data mining has been accelerated in recent years through the exponential growth of data as well as the decreasing cost to capture, store, and process data (Shmueli et al., 2010). Healthcare organizations are increasingly investing in data mining services to improve quality, service, and cost (Fickenscher, 2005). Many of the healthcare service components currently suffer from lengthy delays and additional stakeholder requirements, which limit real-time information accessibility for decision-making and improvements. Table 2.1 describes a set of possible healthcare service and data mining application components and improvement objectives.

Table 2.1: Healthcare Data Mining Applications

Data Mining Application	Description
Hospital readmission	Identification of the factors leading to hospital readmission, predictive modeling of readmission, and successful methods to reduce readmission rates.
Machine learning outcomes	Development of machine learning algorithms to learn and predict outcomes automatically, significantly reducing costs and increasing quality.
Geographic information systems	Identification of the spatial and temporal trends through cluster analysis to predict healthcare use and costs. This research can be applied to new population management and database applications.
Pay for performance	Formulation of incentive programs that can be used for providers and healthcare entities as a cloud service.
Fraud detection	Pattern detection of patients, providers, and healthcare entities for fraud as well as training or errors improvement.
Risk management	Through classification, validation of existing and identification of potential future care management needs for patient conditions based on imaging, notes, and historical medical records.
Wellness	Application architecture for a patient-centric framework to improve virtual health management and monitoring.
Social media	Development of social media tools and evaluation of current tools for improving member's general health and condition-specific areas.

Data Mining Application	Description
Workflow	Monitoring of clinical and non-clinical activities for improving process intelligence through best practices and efficiency of care.
Communities of Care	Analysis of social media, web logs and outcomes information for measuring self-management of conditions by the patient and community of patients.

Analytics Platforms

Gartner publishes an annual Magic Quadrant for Advanced Analytics Platforms. Gartner named SAS as a leader on the quadrant and SAS scored highest among vendors on the ability to execute. Leaders in the quadrant have a strong market record and impact market growth and direction. From the report, by 2020 predictive and prescriptive analytics will consume 40% of new investments by companies, and over 50% will compete using advanced analytics, which will cause industry disruptions. Advanced analytics uses quantitative methods, statistics, descriptive analytics, predictive analytics, data mining, machine learning, simulation, and optimization to analyze data and provide insights. The software platforms for advanced analytics provide model development, data source access, data preparation, data exploration, data visualization, model deployment, model management, and high-performance capabilities. SAS is the most common choice when choosing an advanced analytics platform and has the widest range of advanced analytics capabilities and use cases. SAS is reported as the "gold standard" for advanced analytics platforms. Customers choose SAS for the intuitive user interface, gained productivities, quality of products, degree of flexibility, ability to incorporate a variety of data, and availability of skill sets (Kart et al., 2016).

About SAS

SAS is the global leader in analytics solutions with installations in over 83,000 sites in 149 countries around the world. SAS is used by 94 of the top 100 companies in the 2016 Fortune Global 500 and provides customers with THE POWER TO KNOW. SAS has been considered among FORTUNE magazine's best companies to work for in the United States since 1998. SAS had its origins at North Carolina State University for use in agricultural research and stood For Statistical Analysis System or "SAS". In 1976 SAS was formally founded as a way to use statistical analysis across all industries, applications, and vendor platforms. In 2000, SAS focused on web-based and hosted solutions, now known as SAS OnDemand for Academics (SAS, 2016b).

SAS is used across a variety of industries and applications areas including healthcare. Taipei Medical University and Highmark are two examples of how SAS has been used in healthcare settings across the globe. Taipei Medical University in Taiwan uses SAS analytics to analyze and monitor performance across their hospital system by hospital, department, and physical areas. The analytics capabilities allowed the hospital to detect and correct errors leading to process improvements. Highmark uses SAS analytics to develop decision tree models that identify patient outcomes based on symptom, history, and demographic factors. The analytics system reviews claims for potential miscoded or missing diagnosis codes, which cause reimbursement to be missed or underreported. An example would be a diabetes patient who has treatments and prescriptions, such as insulin, that are included in their charts, even though the patient is missing the diabetes diagnosis code (SAS, 2017a).

Analytics in Action

Dynamic capabilities describe the changing environment and the role of strategic management in adapting the organization to the changing environment (Teece, et al., 2003). Dynamic capabilities of a firm describe their ability to adapt over time to take advantage of disruptive technologies. The competitive global nature and increasing technology require an enhanced paradigm in order to realize competitive advantage. In the past, many companies followed a resource-based strategy, which, alone, was insufficient for supporting competitive advantages. Successful global organizations have been able to assertively and rapidly match the external environment requirements over time, which leads to a source of competitive advantage. Organizations that are able to develop and use internal and external capabilities, along with the development of new capabilities in response to the environment, are able to realize competitive advantages. Organizations can gain a competitive advantage through their managerial and organizational processes aligned to their position and potential avenues of value creation, capitalizing on best practices and learning. These dynamic capabilities and processes are inherently complex and difficult to replicate by competitors, leading to strategic advantages (Teece, 1997).

Analytical capabilities can be applied as an enabler to allow an organization to generate value. SAS enables and empowers the activities and the intersection of data, discovery, and deployments, which lead to analytics in action and organizational value. Other competitors might have gaps between data, discovery, and deployment, and are unable to maximize their value (SAS, 2016c; Tenconi, 2016). The following is a summary of the data, discovery, and deployment activities.

1. Data includes the sources and combination for insights such as financial data and social media data.
2. Discovery includes the ability to prototype and test analytical models such as prediction and machine learning models.
3. Deployment includes the implementation and operationalization of validated analytical models to end users.

Health Anamatics Architecture

When implementing health anamatics, an often overlooked part is the architecture. Generally, a systems architecture includes a data repository (such as a data file, database, or data warehouse), an analytics platform, and a user interface. These components can be installed locally or hosted remotely using a cloud-based architecture such as SAS OnDemand for Academics.

SAS allows access to a variety of data sources such as relational database management systems. Supported vendors and formats include ODBC, Oracle, Teradata, DB2, and SQL Server. The support of third-party vendors allows SAS to connect directly to data sources for production analytics. For ad hoc analytics, we are able to use direct files, eliminating the need for a formal database or data warehouse to be set up first, saving time and allowing decision-making while long-term components are implemented (SAS, 2016d).

Figure 2.7: Health Anamatics Architecture

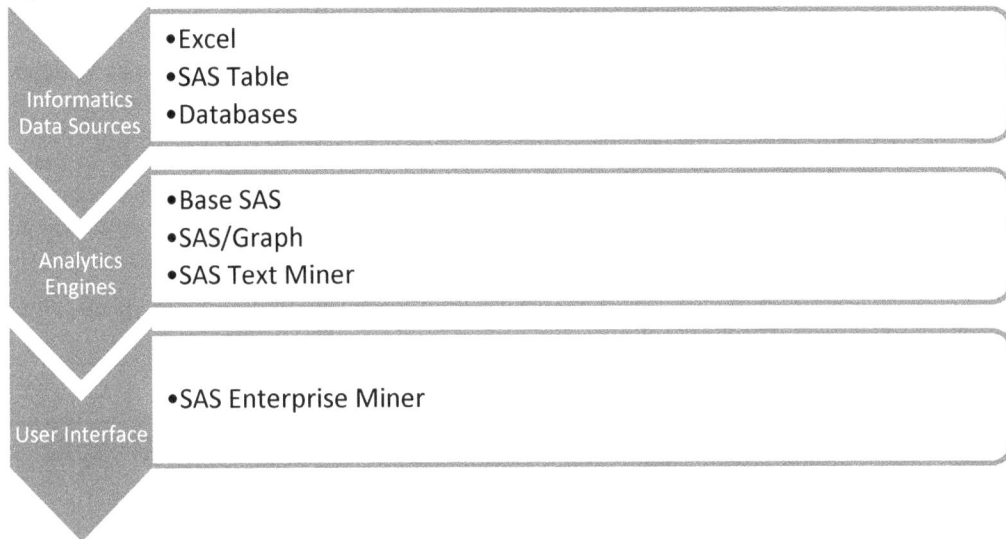

Informatics Data Sources
- Excel
- SAS Table
- Databases

Analytics Engines
- Base SAS
- SAS/Graph
- SAS Text Miner

User Interface
- SAS Enterprise Miner

Health Anamatics Architecture and Cloud Computing

LinkedIn provided and annual list of skills that employers are looking for in candidates. The top 2 global skills were 1) Cloud and Distributed Computing, and 2) Statistical Analysis and Data Mining (Fisher, 2016). Cloud computing enables convenient on-demand access to an elastic set of shared computing resources. Cloud computing can be broken down into three categories: Software as a Service (SaaS), Infrastructure as a Service (IaaS), and Platform as a Service (PaaS). SaaS is anticipated to grow the fastest and is typically what users see and interact with directly. The cloud computing market is anticipated to be nearly $200 billion in 2020 according to Forrester, up from $58 billion in 2013 (Seeking Alpha, 2015).

Software as a Service (SaaS) is software that can be deployed over the internet and is licensed to customers typically on a pay-for-use model. In some cases, a service might be offered at no charge if supported from other sources such as advertisements. SaaS is quickly growing and double-digit growth of 21% is anticipated, with a forecast of $106 billion in 2016 (Seeking Alpha, 2015). Cloud computing support SaaS by providing scalable and virtualized services to the end user via a simple web browser. A third party manages the computing infrastructure and provides the SaaS. Salesforce.com, Google Apps, Amazon, and Facebook all provide cloud computing offerings. Cloud computing allows organizations to reduce IT capital costs and buy computing on an as-needed basis. There are economies of scale through shared use of systems and resources by multiple customers. Cloud computing reduces the entry barriers by eliminating software distribution and site installation requirements. Cloud computing capabilities permit organizations to develop new business models and sources of revenue through on-demand services (Woodside, 2010b).

Infrastructure as a Service (IaaS) is a method for deploying infrastructure such as servers, storage, network, and operating systems. These infrastructure services are offered on-demand in a secure environment. This IaaS capability allows customers to use resources on-demand instead of purchasing the resources up-front. There are a few technologies that IaaS uses, including virtualization, enterprise information integration (EII), and service-oriented architecture (SOA). SOA is used as the access point for all systems through web services, and XML is used for the data representation. SOA promises improved agility and flexibility for organizations to deliver value-based services to their customers. Virtualization creates a virtual version of a computing platform, storage, or network. Unlike SaaS, users are responsible for managing the operating system components along with data and middleware (Seeking Alpha, 2015).

Platform as a Service (PaaS) is generally considered the most complex of the three categories of cloud computing. PaaS is a computing platform that also allows the instantiation of web-based software applications without the added complexity of purchasing and maintaining software and infrastructure. Based on market studies, PaaS is anticipated to reach $44 billion in revenue by 2020, with 16% of overall cloud services by 2018 (Seeking Alpha, 2015). Some of the advantages of PaaS include efficiency of development and deployment cycles, capacity on demand, portability between platforms, simplified mobile application creation, and increased business value (Mehta, 2015).

Following our review of health analytics, let's continue with an experiential learning activity on evidence-based practice and research. Review the description and answer the following questions.

Experiential Learning Activity: Evidence-Based Practice and Research

Evidence-Based Practice and Research

Description: Well-known gaps exist today between theory, discovery, and implementation, which can lead to reduced patient care from a quality and efficiency standpoint. Within health anamatics, the health informatics capabilities provide the capture of evidence, and the health analytics capabilities allow the ability to analyze large amounts of captured data to make decisions. Therefore, evidence-based practice (EBP) combines one's experience with best practices that are based on evidence to improve patient care. There is a five-step process for EBP: 1) Asking a searchable question, 2) Searching for evidence, 3) Critically examining the evidence, 4) Changing practice as needed based on the evidence, and 5) Evaluating the effectives of change (Hebda and Czar, 2013). This five-step process is continuous over time in an effort to incorporate the latest and best ongoing practices.

Your clinical team in a regional physician office. You and your colleagues have been discussing a patient question of whether stretching is or is not needed before exercising to prevent injury. On a positive note, according to a U.S. Centers for Disease Control (CDC) annual health survey, over 50% of respondents reported that they achieved the recommended physical activity of 150 minutes of moderate activity (such as walking or yoga) or 75 minutes of intensive exercise (such as running or weightlifting per week) (Ellen Foley, 2017). Describe how you would use evidence-based practices for your study idea. Following the five-step process, research the question and search the evidence to support your decision-making. For example, one question to consider would be: Should you stretch before exercising?

Provide your initial thoughts based on your existing knowledge and gut feeling or instinct:

Describe the benefits of using EBP:

Evidence-Based Practice and Research

Discuss a qualitative method of research that you evaluated for your decision:

Discuss a quantitative method of research that you evaluated for your decision:

Discuss three methods for using information technology in support of your study:
1.
2.
3.

Provide your recommendations to your clinical team:

Find and discuss one outside example of evidence-based medicine or evidence-based practice. Also describe the application and supporting technologies:

Health Anamatics Careers

Now that you have completed an overview of health anamatics, our next step will be to review potential healthcare career paths. Over an individual's working career, most will change positions, organizations, and even careers completely. Based on a national U.S. Bureau of Labor Statistics (BLS) (2017) survey, individuals held an average of 11.9 jobs between ages 18 and 50 (BLS, 2017). However, the number of actual jobs held often conflicts with personal estimates for the number of jobs that will be held in a lifetime to be between only 2 and 5 (Kurtz, 2013). The number of jobs held is attributed in part to new technology such as online career sites including LinkedIn, which increase access to new positions and the ability to recruit new talent directly from the career sites, along with the exponential pace of change in many industries as part of the fourth industrial revolution (Kurtz, 2013; Schwab, 2016). To be successful in a varied and changing set of environments, individuals must develop adaptability in their decision-making capabilities. Integrative, experiential, and applied learning facilitates making connections to concepts and experiences in order to apply the knowledge to novel and complex challenges. It also provides students with a better understanding of the world and improves flexibility and adaptability (Ithaca College, 2017; Woodside, 2018a).

To provide additional information about my background, my undergraduate degree and coursework was in computer information science, and my graduate and doctoral course work was in business and information systems. My first full-time job accepted while finishing up my MBA was in healthcare and analytics. I was initially hired as an Information Analyst primarily responsible for supporting an insurance claims and recovery area. When graduating, I had no initial plans to be in the healthcare field and analytics fields, as I was currently working in the financial field supporting technical areas including networks, databases, and software applications. However, upon further reflection, joining the healthcare analytics field was one the best decisions and best areas to choose given its growth, and probably should have been expected.

Hopefully, throughout your education, you might be given the opportunity to complete coursework designed to specifically prepare you for a healthcare career. Despite your plans, most of you might find yourself in the career area of health anamatics, given the job potential over the next several decades, and you might have already anticipated this result. Learning and positioning yourself relatively early will provide an advantage for your career path. To give additional perspective on the possibility of finding yourself in a health anamatics career, healthcare-related employment growth projections are 19% growth through 2024 in the U.S., which is faster than the average for all occupations, and with a median annual wage of $62,610 in 2015 compared with median for all occupations at $36,200. The growth is attributed to aging populations and governmental health reform providing additional healthcare access (BLS, 2016a). Even during the last recession, healthcare growth continued, whereas all other industries combined declined (Wright, 2013).

The healthcare and social assistance sector will be the largest employment sector during the next decade through 2024 and representing nearly 1 out of every 7 jobs. The healthcare sector will also surpass government and business sectors as the largest sector. Healthcare support occupations such as healthcare practitioners and healthcare technical occupations are projected to be the two fastest growing groups through 2024 and are expected to contribute the most new jobs (1 out of every 4 new jobs) (BLS, 2016b). Based on a Glassdoor study of 5.1 million online jobs within 25 industries, healthcare had the most open jobs with nearly 800,000 and $45.1 billion in annual salaries (LaMagna, 2017). Healthcare demand is also expanding internationally in areas such as Southeast Asia, with driving forces of population growth rates, along with epidemiological shifts of infectious disease prevalence to that of chronic diseases more closely following Western countries (Yap, et al, 2016).

There are many clinical and non-clinical career opportunities in healthcare where knowledge of anamatics can provide value. These career opportunities consist of the following:

- Direct care providers such as physicians, physician assistants, and nurses
- Allied health professionals such as therapists, pathologists, pharmacists, laboratory technicians, dietitians, social workers, home health aides, and medical assistants
- Health information management professionals such as health information administrator compliance specialist, data specialist, coding specialist, coding coordinator, and transcriptionist
- Health information technology professionals such as project manager, analyst, applications coordinator, data quality manager, architect, and information security analyst
- Other administrative positions such as privacy, security, and compliance officers, research, insurance specialist, finance, accounting, marketing, and human resources specialists (Gartee, 2011).

Experiential Learning Activity: Health Anamatics Careers

Health Anamatics Careers

Description: For our experiential learning activity on health anamatics careers, choose one healthcare system of interest, and list three benefits of the system. Find an example of how an organization is using this healthcare information system and a career path that would use this system of interest.

Health Anamatics Careers

This system might be clinical (pharmacy, and so on) or administrative (financial, and so on) systems. For example, Humana, a health insurance company, is using mobile systems including the iPhone and Apple Healthkit software to track patient fitness activity and to more accurately determine healthcare insurance costs, provide employer-based incentives, and improve overall population health. Career paths that use this information can be an insurance actuary, human resource director, or healthcare case manager.

- Indeed.com
- Monster.com
- Careerbuilder.com
- Professional Job Titles
- Professional Organizations
- Organization Name
- Skills

Healthcare System Summary:

Healthcare System Benefits:

Organization Example:

Career Path Example:

From the earlier discussion about automation and from your selected area of interest, which aspects of the career path and healthcare system have the potential to be automated, and which aspects would be difficult to automate.

You have successfully completed the health anamatics introduction chapter and experiential learning activities! For our final chapter component, you will complete the learning journal reflection from the key chapter ideas and topics.

Learning Journal Reflection

Review, reflect, and retrieve the following key chapter topics from memory and add them to your learning journal. This can take effort and seem difficult at first. However, effortful reflection and retrieval helps build learning pathways to more easily find the way to and from your existing knowledge in long-term memory. Some difficulties encountered during retrieval help make the learning stronger and better remembered. Effort changes the brain, making new connections and pathways and increasing intellectual ability.

For each topic, list a single sentence description/definition. Connect these ideas to something you might already know from your experience, other coursework, or a current event. This follows our three-phase learning approach of 1) Capture, 2) Communicate, and 3) Connect. After completing, verify your results against your learning journal and ensure that all topics are included in your learning journal and update as needed.

Key Ideas – Capture	Key Terms – Communicate	Key Areas - Connect
Health Anamatics		
Health Informatics		
Health Information Technology Hardware, Software		

Health Information
    Patient, Provider, Payer,
    Government Data		

Health Information Management Procedures, People		
Health Analytics		
Descriptive		
Predictive		
Prescriptive		
Health Anamatics Architecture		

Chapter 3: Sampling Health Data

Chapter Summary

The purpose of this chapter is to introduce you to and get started with SAS Enterprise Miner software. Previous SAS experience is not assumed, and this chapter is intended as an introduction for those who are new to SAS Enterprise Miner. This chapter includes the formal process SEMMA and develops sampling skills within the first step of the SEMMA process. This chapter also includes experiential learning application exercises about health and nutrition sampling and claim errors rare-event oversampling. We begin this chapter with the "Sample" phase of SEMMA as shown in Figure 3.1.

Figure 3.1: Chapter Focus - Sample

Sample   Explore   Modify   Model   Assess

Chapter Learning Goals

- Describe the standard SEMMA process
- Understand analytics platform architecture
- Describe the SAS Enterprise Miner interface
- Apply common SAS Enterprise Miner start-up functions
- Develop data sampling skills

Health Anamatics Process

In the front matter, we discussed the importance of spaced repetition and practice to improve learning. The use of memory devices can help with the creation of mental structures for knowledge retrieval. Healthcare and analytics are full of acronyms, which can be useful for the creation of mental structures for knowledge retrieval, and serve as memory devices. Standardized data mining processes have been developed and followed to systematically conduct health anamatics projects. Examples of these processes include SEMMA for data mining and BEMO for big data mining (SAS, 2014; Woodside, 2016a). In this chapter, we introduce the memory device of SEMMA, which will be used, repeated, and practiced throughout subsequent chapters.

SEMMA is an acronym for Sample, Explore, Modify, Model, and Assess as shown in Figure 3.2. The SEMMA data mining process was developed by SAS Institute and is included in SAS Enterprise Miner. In the Sample step, data tables are created that contain enough information to be significant but small enough to process efficiently. In the Explore step, the analyst looks for relationships, trends, and anomalies to improve overall understanding. In the Modify step, data is transformed and chosen to prepare for the next step. In the Model step, analytical tools and methods are used to predict outcomes. In the Assess step, findings are reviewed among different predictive models (SAS, 2014; Woodside, 2016a).

Figure 3.2: SEMMA Health Anamatics Process

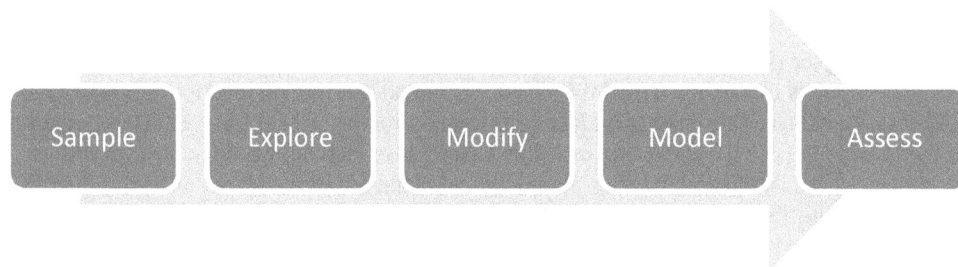

Health Anamatics Tools

SEMMA can be used with a set of tools from SAS for a repeatable, standardized data mining process. As a health anamatics professional, consider yourself a miner. In this case, instead of mining the earth to find valuable minerals, you are mining data to find valuable information for decision making. Across industries, finding this valuable information has a measurable value and can be used to save costs, improve efficiencies, and remain competitive. Within healthcare, finding this information can be the difference in saving lives and therefore has potentially unmeasurable value.

The mining capability has evolved over time with increasingly sophisticated health informatics and analytics systems. Similarly, consider many years ago, as a minerals miner, you might have used pick axes and shovels to mine for minerals. The goal was to move though mountains of dirt to find valuable minerals. Over time, these tools have evolved to make miners more efficient. Now, instead of pick axes and shovels, you can use bulldozers and equipment that makes the performance of tasks much more efficient.

Likewise, in health anamatics, you might consider yourself a miner, but, instead of minerals, you are mining for valuable information. You might have to move through mountains of data to find that valuable information. Over time, the informatics tools to capture data and the analytics tools to mine the information have also improved to make you more efficient as an analyst. Best practices, including process steps and tools, have also emerged to further contribute to the output capabilities of health anamatics professionals. Our primary health anamatics toolset that will be used throughout this textbook is SAS Enterprise Miner.

SAS Enterprise Miner

SAS Enterprise Miner is a graphical user interface-based software for data mining and analytics. SAS Enterprise Miner allows efficient development of descriptive and predictive analytics models. Its interface closely follows the SEMMA process steps, and the model is developed using a visual process flow diagram. The interface is intuitive and user-friendly for end users with limited programming or statistical knowledge. SAS Enterprise Miner also has capabilities for advanced statistical programming for end users with advanced knowledge. It allows various users to collaborate to build the model, increase accuracy, and allow results to be shared easily among team members. Advantages of SAS Enterprise Miner include the ability to build data mining models more quickly with a GUI, improve the analytics lifecycle with an end-to-end toolset, develop reusable process flow diagrams, and communicate results easily with users from a wide variety of statistical and business backgrounds (SAS, 2015a). A summary of the primary SAS Enterprise Miner interface components is included below and shown in Figure 3.3.

- Menu –This contains shortcuts such as File, Edit, View, Actions, Options, Window, and Help.
- Project –This contains the data sources, diagrams, and model packages.
- Properties –This contains the details of the nodes or selected items.
- Process –This contains the SEMMA process tabs and nodes.
1. Diagram –This contains the project workspace and process flows with nodes.

Figure 3.3: SAS Enterprise Miner Overview

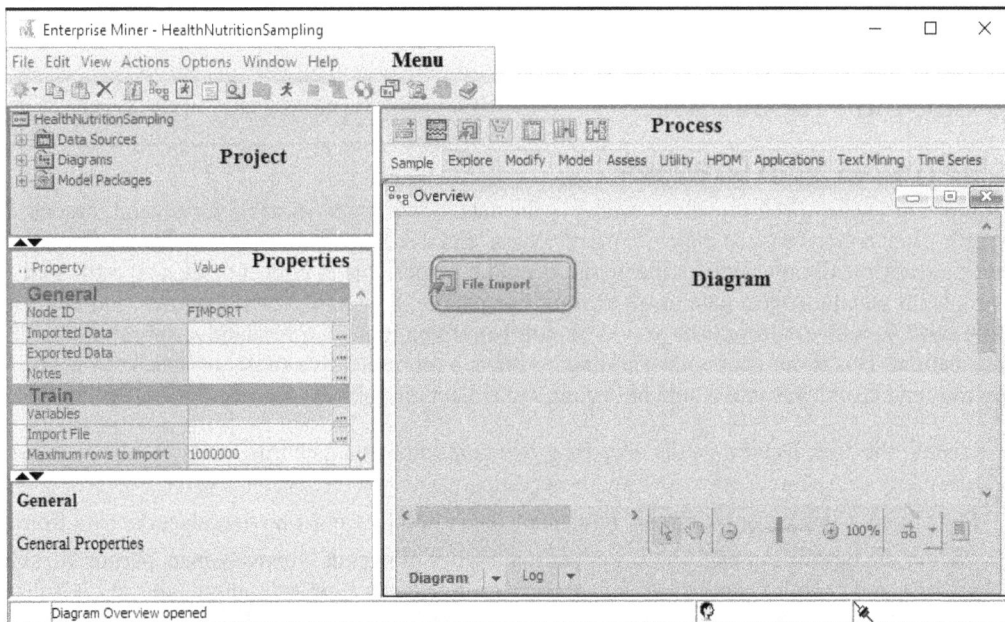

The SAS Enterprise Miner software is organized via a set of tabs according to the SEMMA process. Each tab contains a set of nodes that can be used in the diagram to generate the model. The set of SEMMA tabs and nodes will be covered in additional detail throughout the textbook. Table 3.1 shows the full set of

SEMMA tabs and nodes used. The items in bold-face type are covered in this textbook. Advanced tabs are also available beyond the standard SEMMA process. The Text Mining and HPDM tabs will also be covered in this textbook. The nodes that were selected for this text represent a frequently used set of components to support many healthcare applications and maximize learning efficiency.

Table 3.1: SAS Enterprise Miner Tabs and Nodes

Sample	Explore	Modify	Model	Assess
Append	Association	Drop	AutoNeural	Cutoff
Data	Cluster	**Impute**	**Decision Tree**	Decisions
Partition	DMDB	**Interactive**	Dmine	**Model**
File Import	**Graph Explore**	**Binning**	**Regression**	**Comparison**
Filter	Link Analysis	Principal	DM Neural	**Score**
Input Data	Market Basket	Components	**Ensemble**	Segment
Merge	**MultiPlot**	**Replacement**	Gradient	Profile
Sample	Path Analysis	Rules Builder	Boosting	
Time Series	SOM	**Transform**	LARS	
	Stat Explore	**Variables**	MBR	
	Variable		Model Import	
	Clustering		**Neural Network**	
	Variable		Partial Least	
	Selection		Squares	
			Rule Induction	
			SVM	
			Two Stage	

SEMMA: Sample Process Step

Sample Process Step Overview

The first phase of the SEMMA process is sampling the data. Sampling is valuable for several reasons: 1) reduce costs, 2) increase speed, 3) improve effectiveness, and 4) reduce bias. Suppose that you were performing a study on all the people in the world. Would it be possible to collect data on an estimated eight billion people? If you did collect data on all eight billion people, how long would that take, and how much would that cost? Would your results be very different than if you took a representative subset of the same world population? This is our goal with sampling: to select a representative subset of data at a fraction of the time, cost, and challenges that would be required to collect the full set. A process for sampling includes:

1. Identify the data: Determine the objectives and goals and data gathering method such as survey, secondary data source, interview, and observation.

2. Identify the population: Determine how much data to collect and who to collect the data from.

3. Choose the type of sample: There are four main types of samples: convenience, purposive, simple random, and complex random. A convenience sample is the simplest sample type, but it is also non-probabilistic and non-reliable. A purposive sample is also non-probabilistic and moderately reliable. For a simple random sample, the nth item would be randomly selected, and would require your carefulness to ensure that lists are not preordered or sorted, which can cause some bias. A complex random stratified sample is typically the best form and allows effective and efficient sampling across groups and areas.

4. Identify the sample size: Although many selected samples are based on sound statistical significance and principles, many samples are selected based on business constraints. Money is often the greatest determinant, followed by time and resources. Often these constraints take precedence over the desire to obtain a statistically significant level and low error rate. For certain clinical trials, statistical significance is important. However, for process improvement activities, for example, an efficient sample can be used to improve time to results (Kendall and Kendall, 2014). Online sample calculators are also available that are useful in determining statistical sample error rate and size requirements.

Sample Tab Enterprise Miner Node Descriptions

File Import Node

Within each tab in SAS Enterprise Miner, there are a set of nodes that we will use to conduct actions on the data set and include in our data mining process flow. The first node that we'll discuss under the **Sample** tab is the **File Import** node. After data is identified and collected, the file can be imported into SAS using the file import node. The file import is used to select external files (such as text, Excel, comma-separated values (CSV), and database sources) to bring into the diagram.

Figure 3.4: SAS Enterprise Miner File Import Node

The first step in importing the file is identification of the variables and data types, including the roles and levels. When developing analytical models, each variable has a role within SAS (SAS, 2003). A brief summary of each of the roles and levels is included in Table 3.2.

Table 3.2: SAS Enterprise Miner Data Types and Roles

Data Type	Name	Description	Example
Role	ID	Unique identifier for each record	1,2,3
	Input	Variable that explains the target	Age
	Target	Output variable	Nutrition level
	Rejected	Not used variable	Temporary ID
Level	Binary	Variable with two potential values	0/1, True/False, Yes/No
	Unary	Variable with one potential value	Negative
	Interval	Variable that is numeric and continuous	Claim costs
	Nominal	Variable that is categorical and unordered	Coverage Type
	Ordinal	Variable that is categorical and ordered	Age Group

Sample Node

Following file import, the Sample node can be added to allow a subset sample to be selected from the data set.

Figure 3.5: SAS Enterprise Miner Sample Node

Data Partition Node

Partitioning is the term used in data mining that indicates the splitting of the data source. The data is typically split into two sets: a training set and a validation set. In some cases, a third test set is also generated. The training data set is first run through the model for fitting. The validation set verifies that the model is not overfit, and the test set is for a final assessment.

Figure 3.6: SAS Enterprise Miner Partition Node

Now that we have covered a general overview of SAS and the nodes, let's get started with the SAS software setup.

SAS OnDemand for Academics Setup

SAS software might be installed directly on a computer or in a computer lab. However, as another option, SAS software can be used without any required installation through SAS OnDemand for Academics. It allows an interactive delivery platform for accessing SAS software statistical analysis, data mining, and forecasting. It is available online and at no charge for independent learners, students, or instructors. Advantages to using the online clients include no software installation or maintenance requirements, immediate access, and use of SAS from anywhere with an internet connection. The SAS OnDemand for Academics SAS Enterprise Miner software includes equivalent products that are hosted on the SAS server instead of in a local installation. The SAS OnDemand for Academics setup connects with your cloud-based architecture model using information from our first chapter (SAS, 2015b).

To begin the SAS OnDemand for Academics setup, we will navigate to the website, and follow the steps:

Step 1: Navigate to https://odamid.oda.sas.com.

Step 2: Click **Sign-in** if you have an existing SAS profile and account. Most users will need to register for an account in step 3.

Step 3: Click **Register** for an account if you are a new user, and complete the registration form.

Step 4: Sign in to SAS On Demand for Academics.

After login, you will be presented with a dashboard page. The dashboard page will provide a list of applications and any planned events such as scheduled maintenance for the OnDemand system. If you are enrolled in a course, your instructor can create a new course, and a link will be provided for your student registration. As a student, you will use your instructor-assigned course link to register for the course.

System Requirements

SAS OnDemand for Academics for Enterprise Miner requires an up-to-date web browser and Java Runtime Environment (JRE). Most computers that use Windows already have an up-to-date JRE installed. If needed, the Java software can be installed from http://www.oracle.com/downloads and visiting the Java SE download page. For additional requirements, support, and frequently asked questions, visit: http://support.sas.com/software/products/ondemand-academics/ and http://support.sas.com/ondemand/caq_new.html.

Experiential Learning Application: Health and Nutrition Sampling

After SAS has been set up, we'll now run our first SAS experiential learning activity. For our first sampling activity, you want to conduct a study similar to the Health and Nutrition survey, where a representative sample of 5,000 individuals are selected for additional interviews and stratified by age group, gender, and race-ethnicity. The Health and Nutrition survey program was designed to measure the health and nutrition status of both adults and children in the U.S. and has been conducted since the 1960s (NHANES, 2017). Since the survey contains both interviews and examinations (including medical, dental, physiological, and laboratory tests of patients), it would be difficult to conduct the annual study on the entire population, and therefore a sample of 5,000 individuals are selected.

The data for the Health and Nutrition survey contains age group, gender, and race/ethnicity with possible values listed below.

Data Set File: 3_EL1_Health_and_Nutrition_Sampling.xlsx

Variables:

- IndividualRecordID - unique identifier
- AgeGroup - (18-24, 20-24, 21-24, 25-44, 45-64, 65+)
- Gender - (Female, Male)
- Race - (Hispanic, Non-Hispanic Asian, Non-Hispanic Black, Non-Hispanic White, Other)

Step 1: Sign in to SAS Solutions OnDemand for Academics, or Open SAS Enterprise Miner Local Installation.

Step 2. Open the SAS Enterprise Miner Application (click the SAS Enterprise Miner link).

Step 3. Create a New Enterprise Miner Project (click New Project…).

Figure 3.7: Create New Project Step

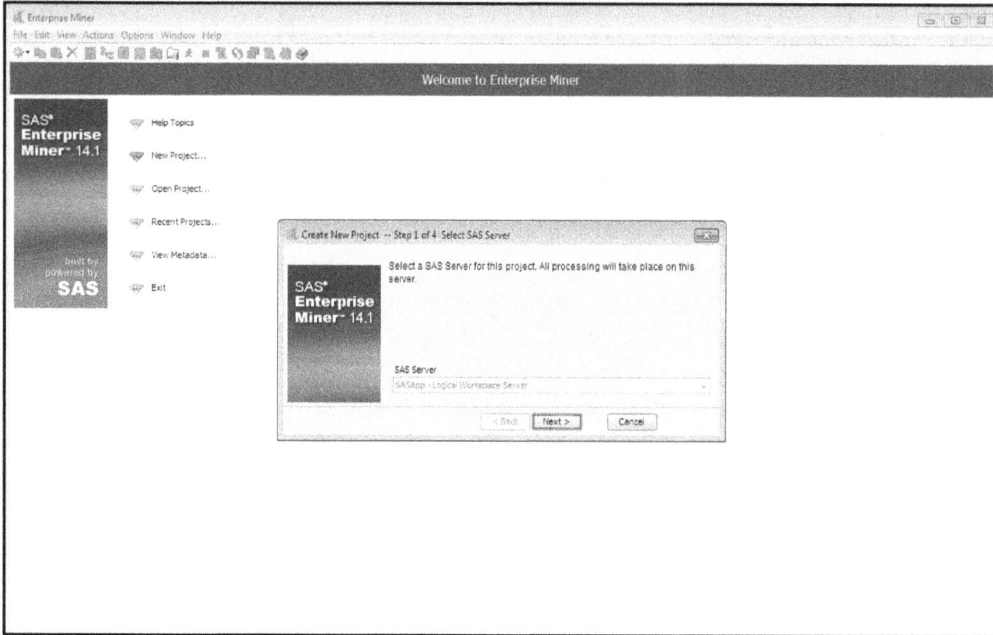

Step 4: Use the default SAS Server, and click Next.

Step 5: Add a project name, such as **HealthNutritionSampling**, and click **Next**. Generally, as best practices for naming projects, use mixed-case typing, or in other words capitalize the first letter of each word, and avoid spaces or other special characters in the project names and file names.

Step 6: SAS will automatically select your user folder directory is using SAS Solutions On-Demand (If you are using a local installation version, choose your folder directory), and click **Next**.

Figure 3.8: Create New Project Step

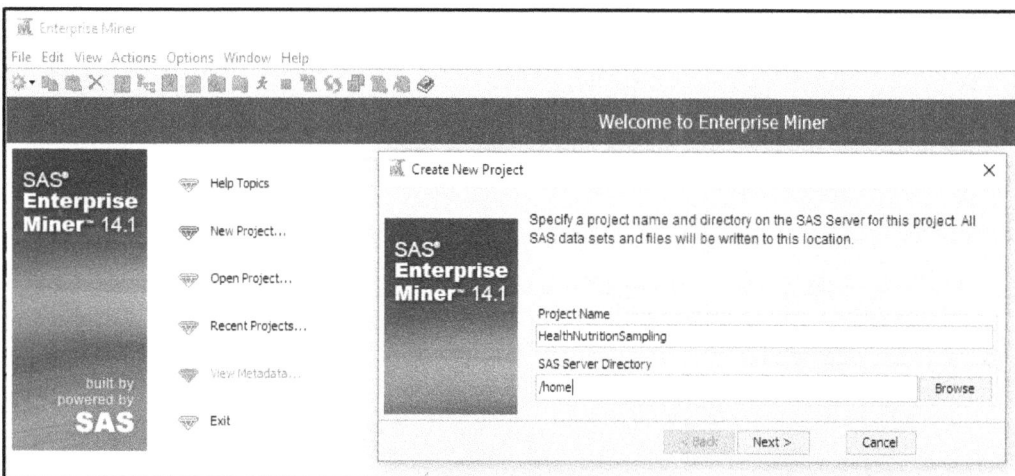

Step 6: SAS will automatically select your user folder directory if you are using SAS On-Demand for Academics (If you are using a local installation version, choose your folder directory), and click **Next**.

Figure 3.9: Create New Project Step

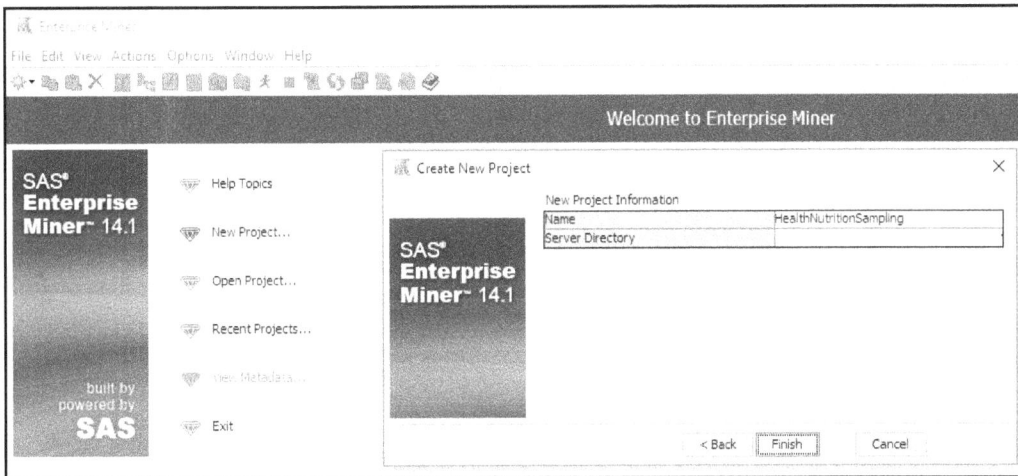

Step 7: Create a new diagram (Right-click **Diagrams**).

Figure 3.10: Create New Diagram

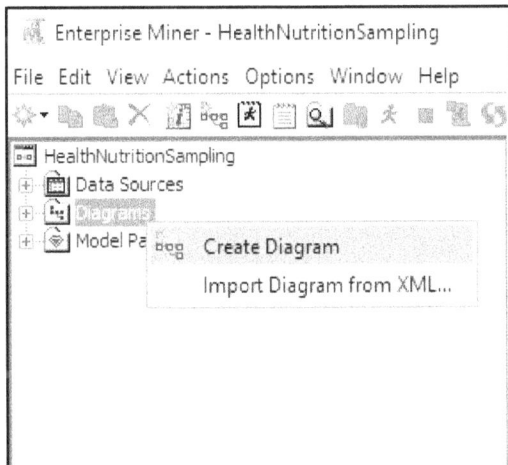

Step 8: Add a new diagram name and, again, as a best practice, use mixed case. The diagram can contain the same name as the project name. Click **OK**.

Figure 3.11: Add Diagram Name

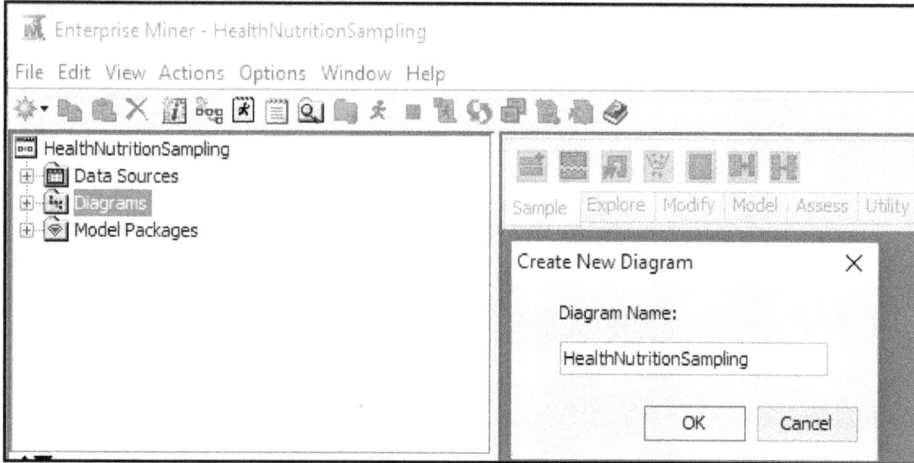

Step 9: Add a File Import node (click the **Sample** tab, and drag node into the diagram workspace). If you position your pointer over an icon, tooltip text will appear, such as when positioning your pointer over the **File Import** icon, the tooltip text for the File Import node will appear.

Figure 3.12: Add a File Import Node

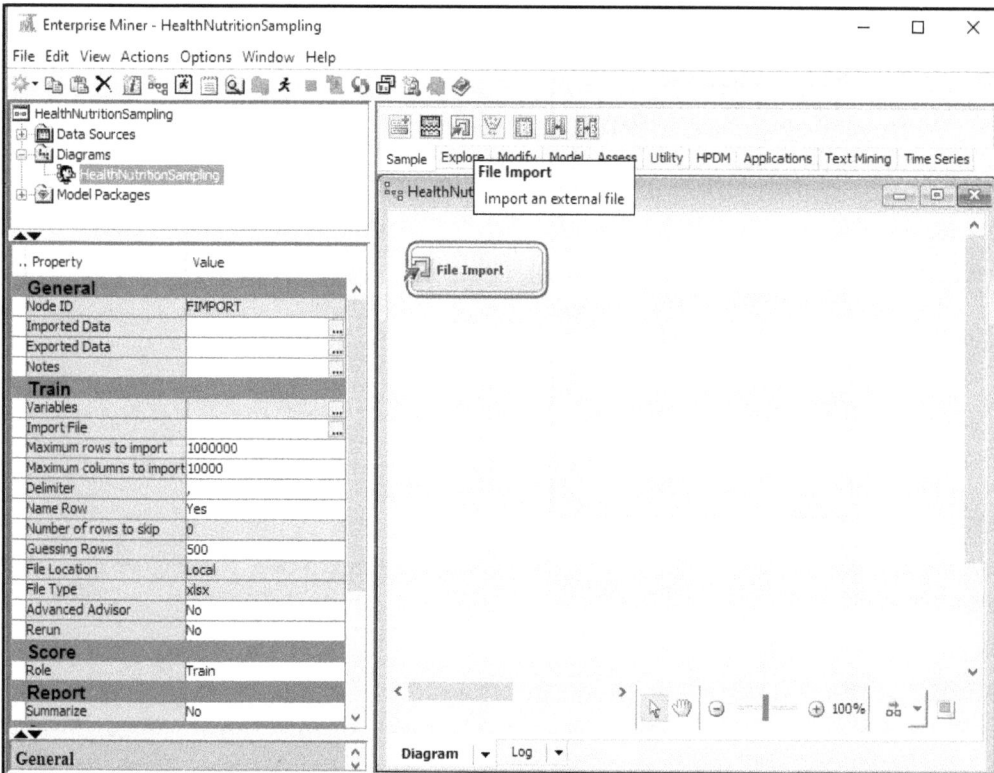

Step 10: Click the File Import node, and review the property panel on the bottom left of the screen.

Step 11: Click **Import File** under the Train properties, click the ellipses and navigate to the Chapter 3 Excel File *3_EL1_Health_and_Nutrition_Sampling.xlsx*.

Figure 3.13: Browse for Import File

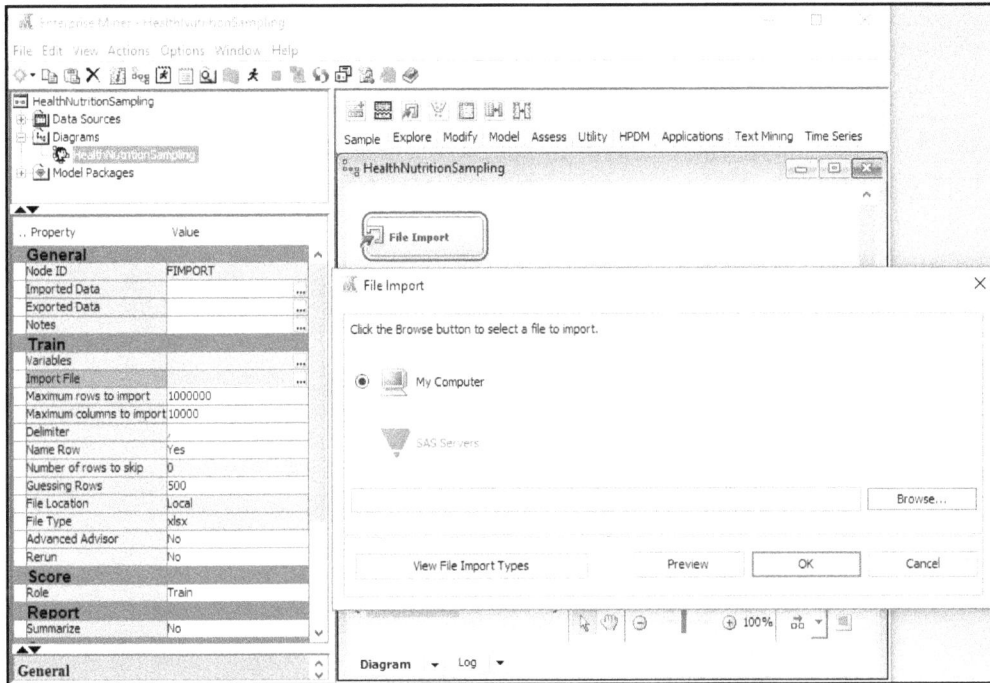

Step 12: Click **Preview** to ensure that the data set was selected successfully, and click **OK**.

Figure 3.14: Preview Import File

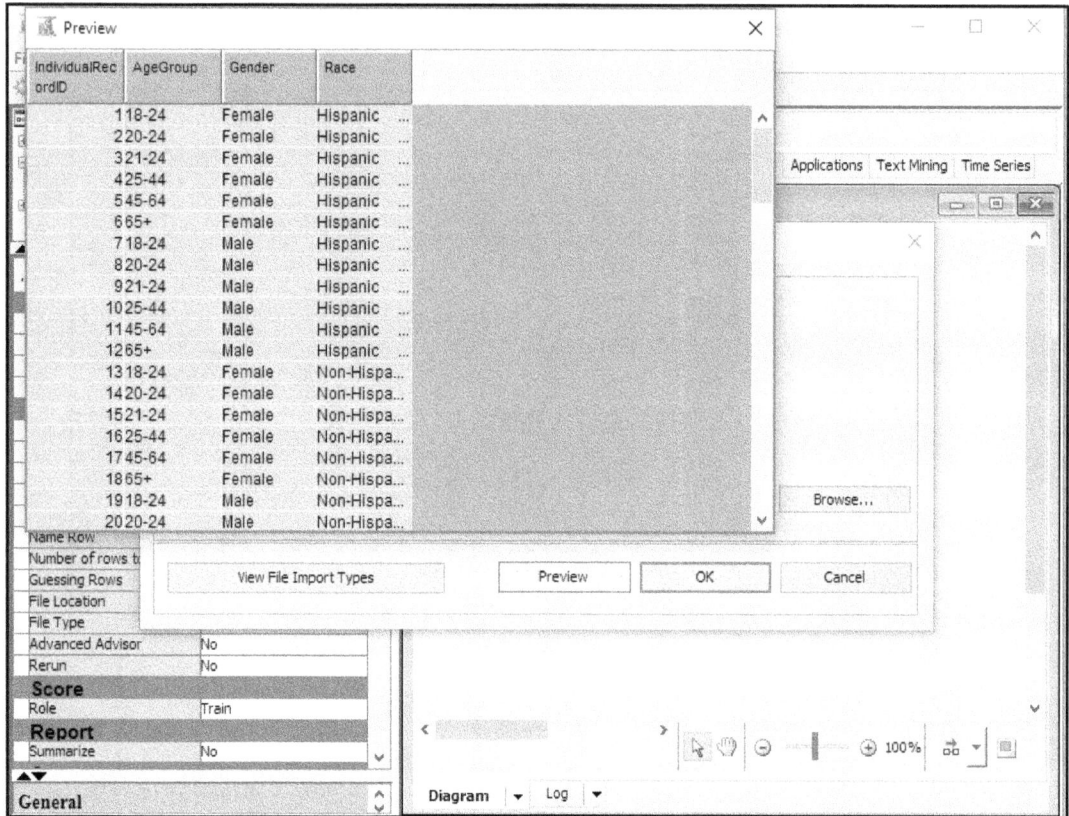

Step 13: Right-click the File Import node and click **Edit Variables**.

Figure 3.15: Edit Variables

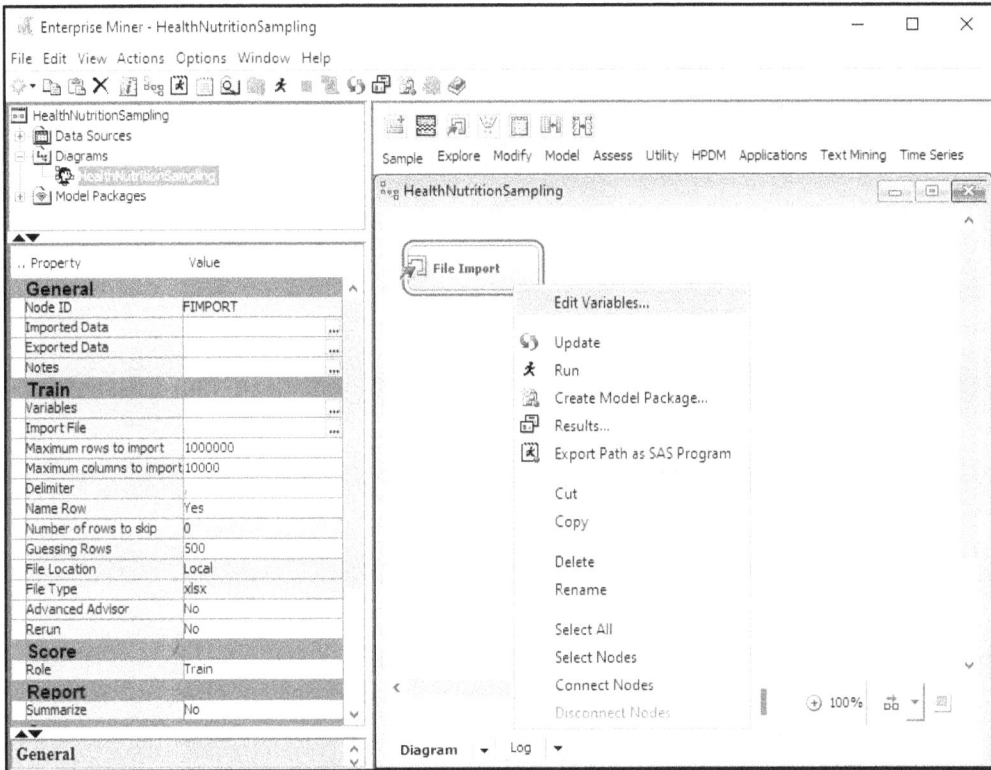

Step 14: For this step, we will use our knowledge of variables that was covered previously in this chapter. The variables are set to ensure that the subsequent steps in our process run as anticipated. Often, an error can occur if the variables are not set although the variables can be updated at any time and rerun. Set IndividualRecordID to the **ID** variable role. Set the AgeGroup, Gender, and Race variables to an input role and a nominal level. Click **OK**.

Figure 3.16: Edit Variable Roles and Levels

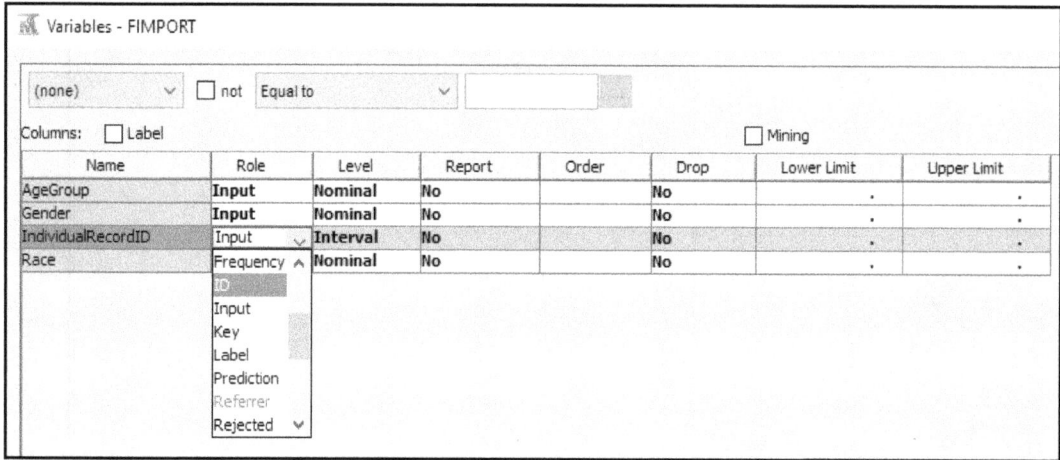

Step 15: If we are unsure of a variable or want to investigate further to verify the role and level assignments, we can also explore each variable individually. As one variable example, we might want to review the gender variable to ensure that our data set contains the correct values. Select and highlight the Gender variable name. Click **Explore**.

Figure 3.17: Select Gender Variable

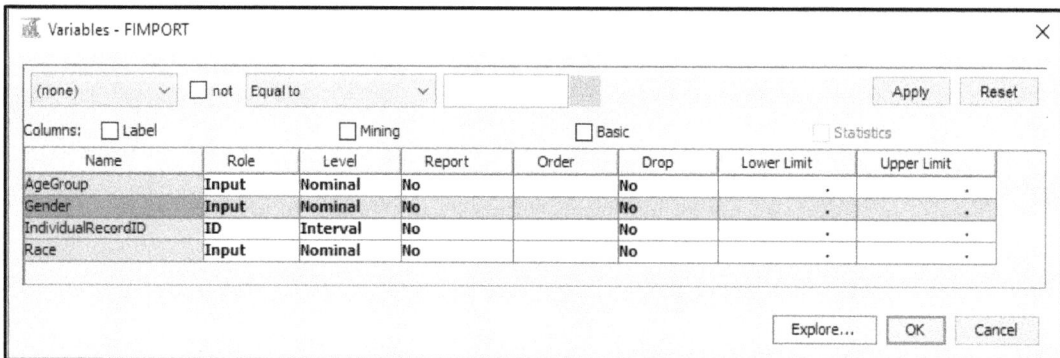

Step 16: Explore the Gender variable records' frequency. From the exploration, we find that our data set has a relatively even set of male and female gender records. Then close the **Sample Properties** window.

Figure 3.18: Explore Gender Variable

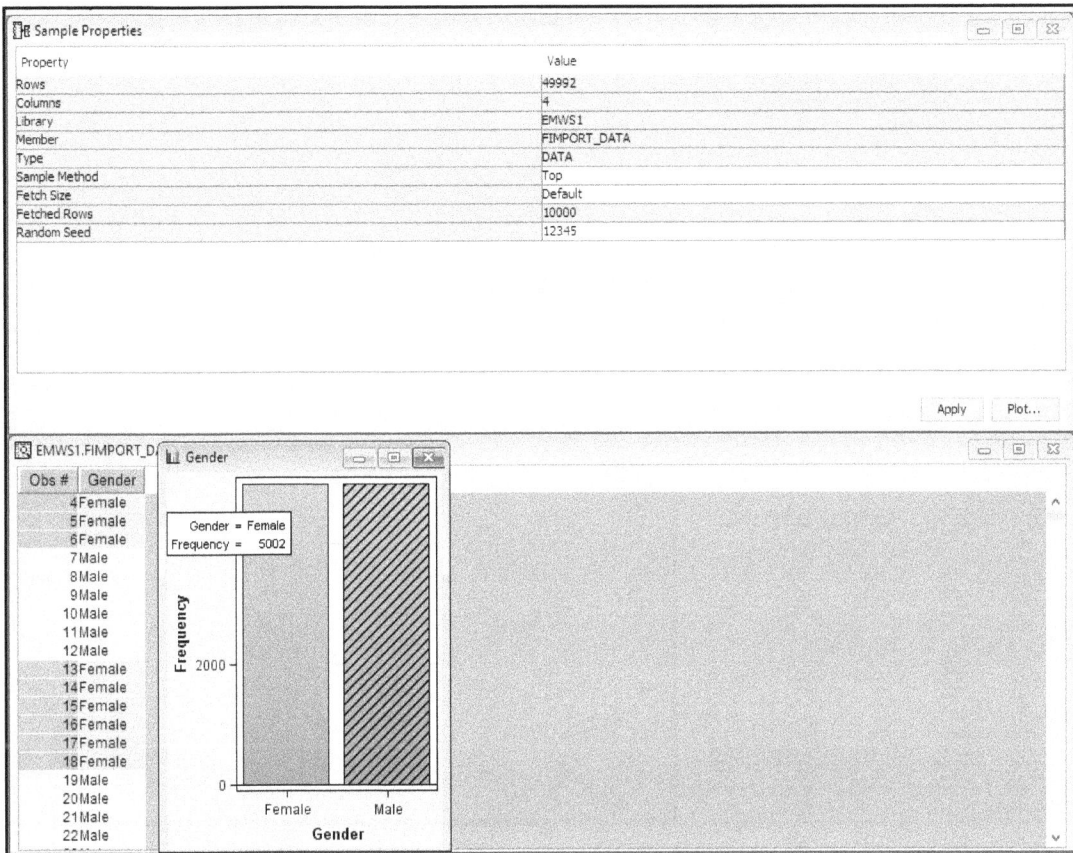

Step 17: Our next step will be to add the Sample node from the **Sample** tab. After adding the node, we want to connect the File Import node to the Sample node. The connection can be made by dragging the arrow from the File Import node to the Sample node. The arrow indicates the process flow, and generally we flow from left to right, like when reading a book. The process flow means that the File Import node would run first, and, once completed, the Sample node would run second. The process flow makes logical sense from a process standpoint since we would need to first import our data before we could sample the data.

Figure 3.19: Add Sample Node

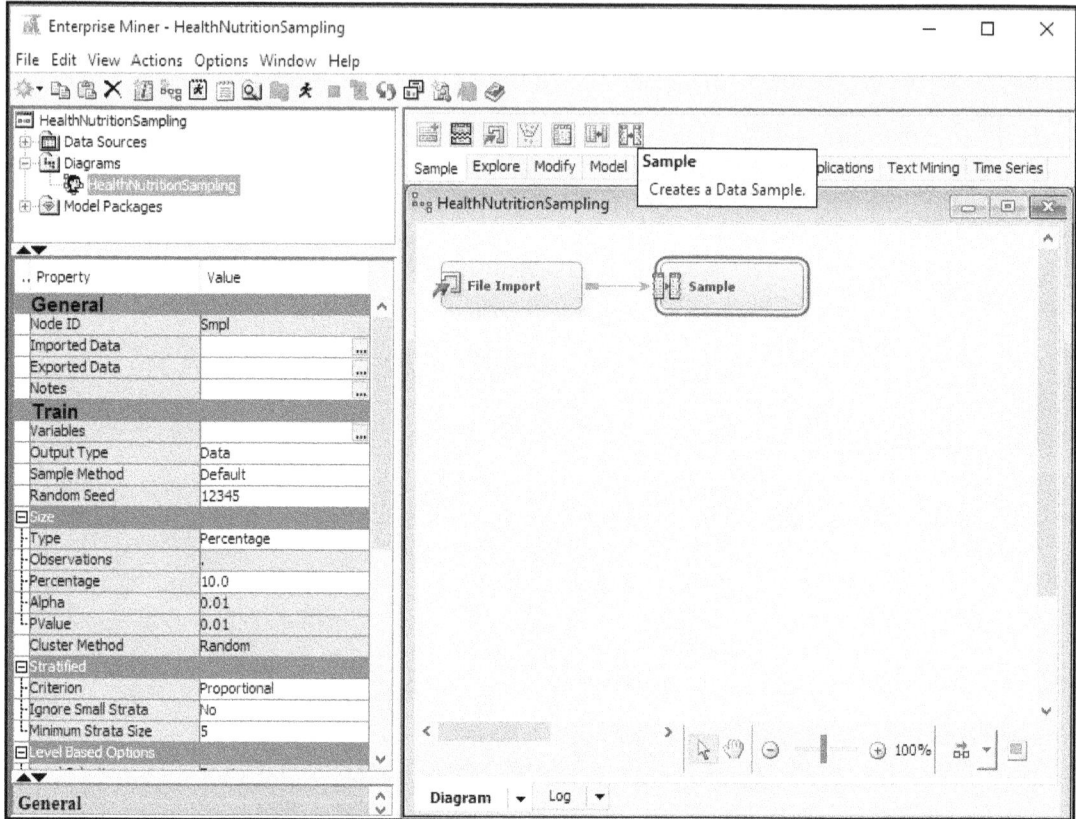

Step 18: Right-click **Sample Node**, and click **Run**. Most of the actions can be found by right-clicking. However, as alternatives, you can also use the toolbars.

Figure 3.20: Run Sample Node

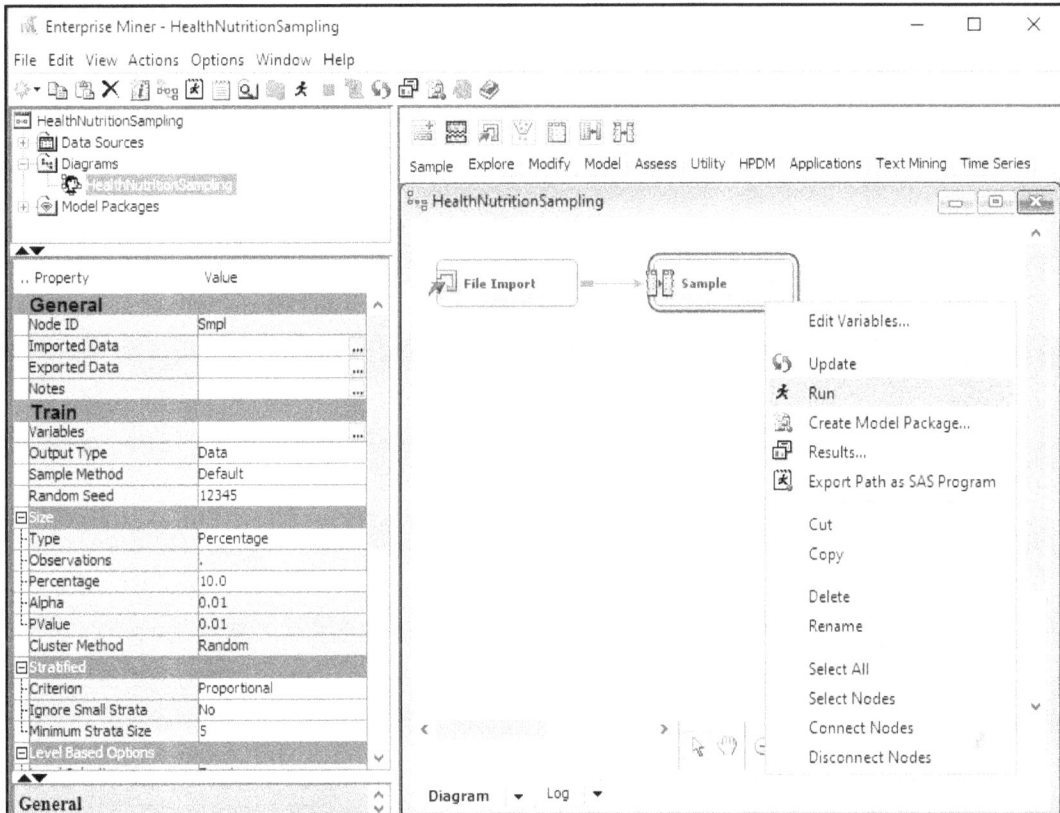

Step 19: Click **Results**. Notice the green checkmarks on the bottom right corner of the nodes. This indicates that the process ran successfully. If you receive a red x in the bottom right corner, this indicates a problem with the process, and you can simply go back to review the node setup, make any adjustments, then run again.

Figure 3.21: Sample Results

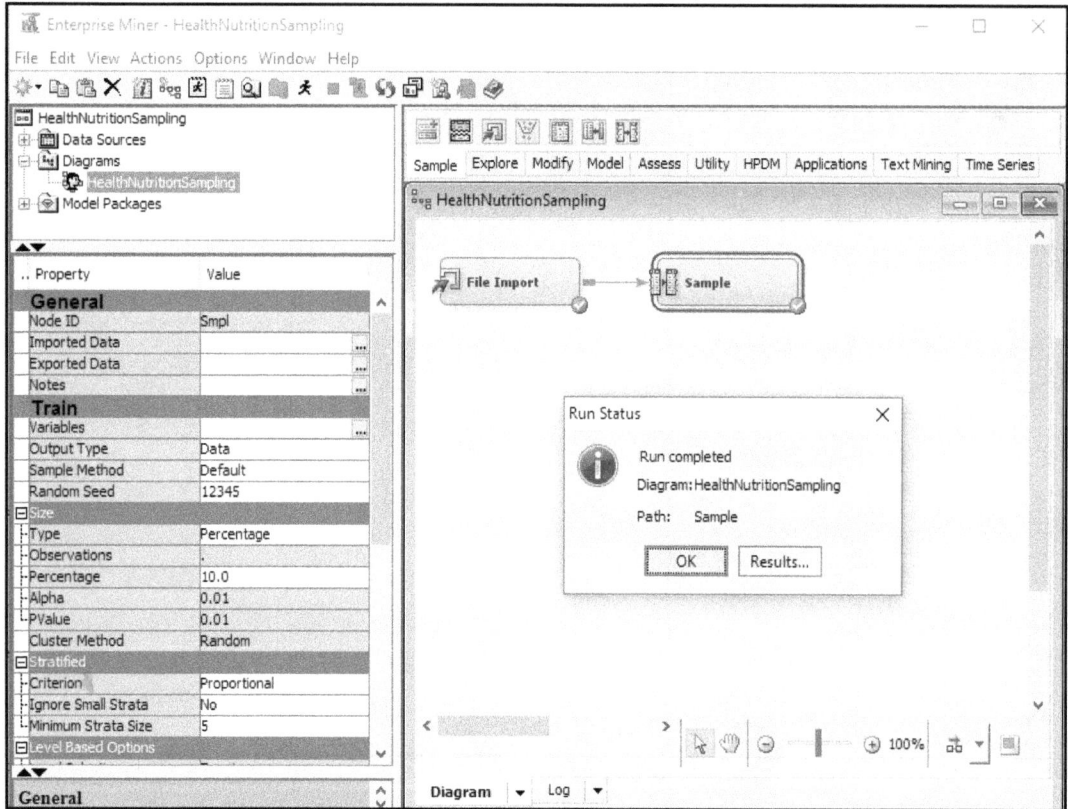

Step 20: View the results. Note that the total number of observations is 49,992, and the sample number of observations is 4,999. The number is a 10% sample, or 4,999 out of 49,992.

Figure 3.22: View Sample Results

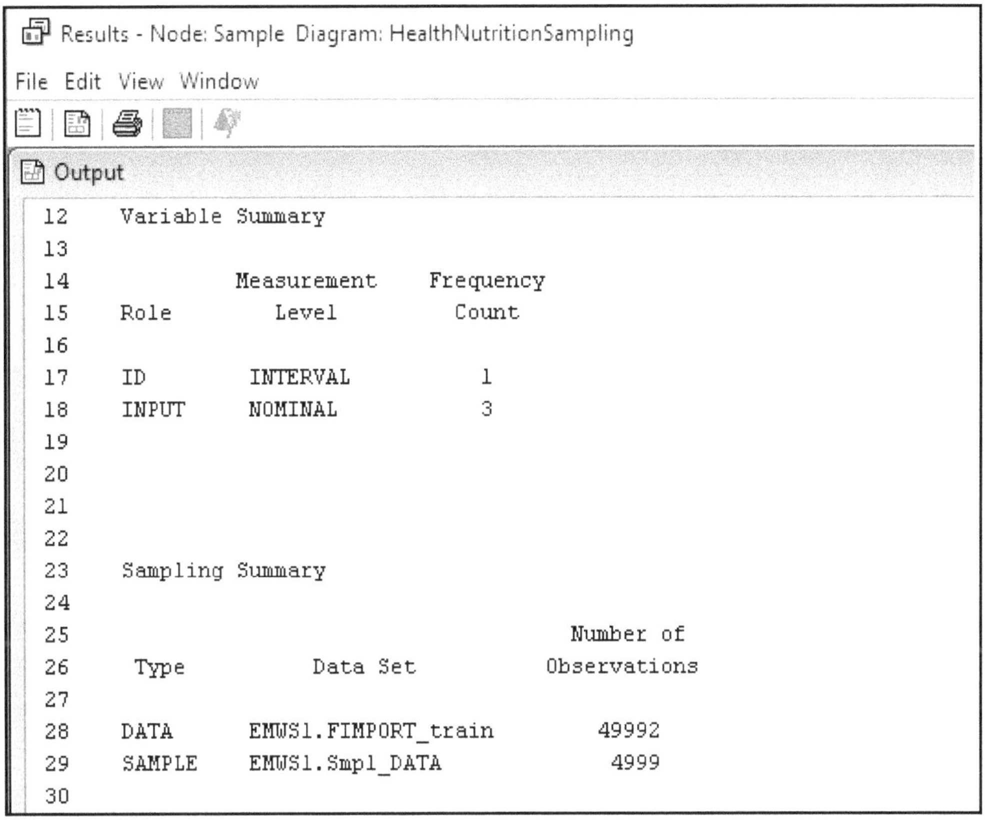

```
 Results - Node: Sample  Diagram: HealthNutritionSampling

File  Edit  View  Window

 Output
  12      Variable Summary
  13
  14              Measurement     Frequency
  15      Role       Level          Count
  16
  17      ID         INTERVAL          1
  18      INPUT      NOMINAL           3
  19
  20
  21
  22
  23      Sampling Summary
  24
  25                                      Number of
  26       Type           Data Set       Observations
  27
  28      DATA       EMWS1.FIMPORT_train     49992
  29      SAMPLE     EMWS1.Smpl_DATA          4999
  30
```

Step 21: To find where the 10% sample was generated from, click the Sample node. Notice the Size property in the Property window on the bottom left. Under the Size property, the Type is set to Percentage and the Percentage is set to 10.0%. These properties can be adjusted and the Sample node can be rerun to achieve different results. Also, in the Properties section under General, there is an Exported Data property. Click the ellipses.

Figure 3.23: View Exported Data

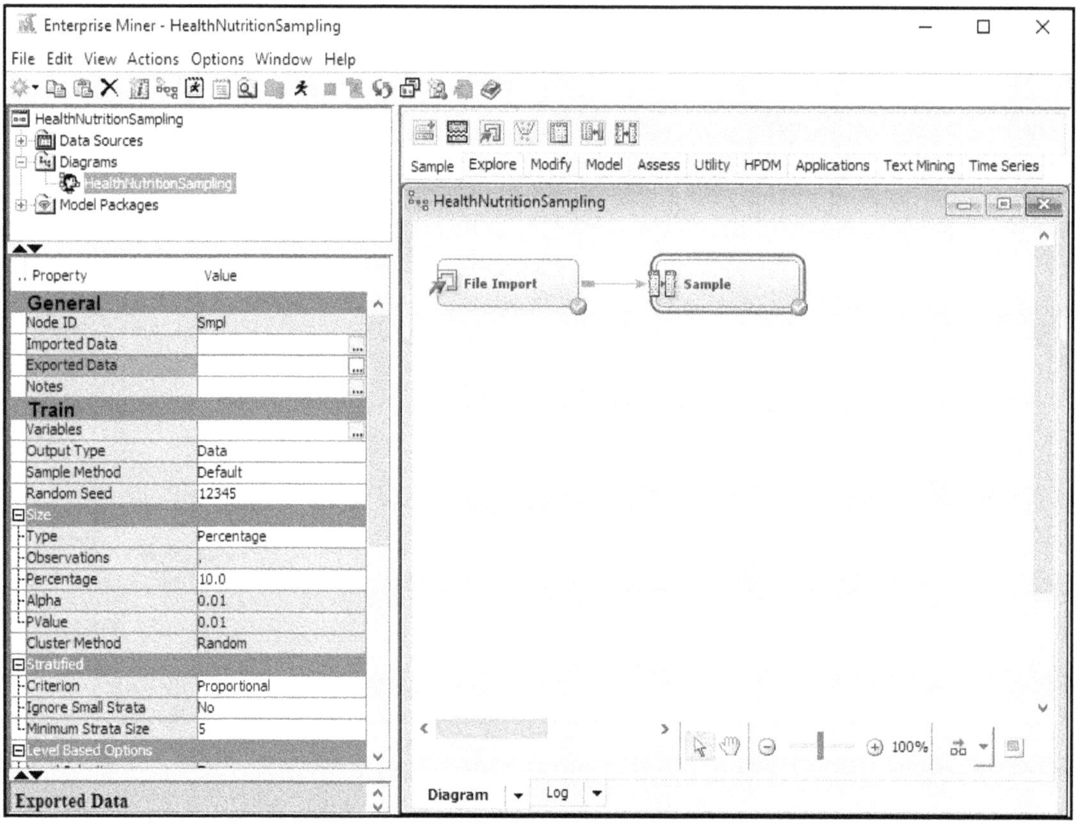

Step 22: After clicking the Exported Data ellipses, you will see the data set for the sample. Click **Explore**.

Figure 3.24: Explore Exported Data

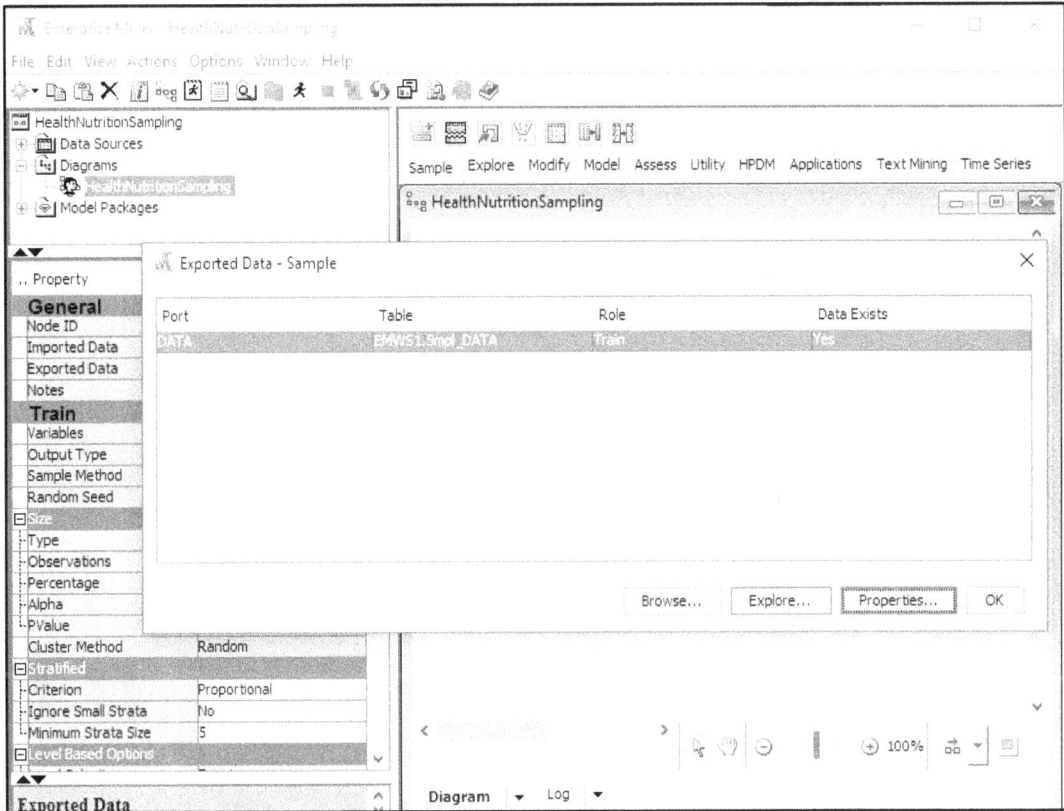

Step 23: After clicking **Explore**, you will see the individual records that have been sampled. The total is 4999, which matches the 10% sampling.

Figure 3.25: Exported Data

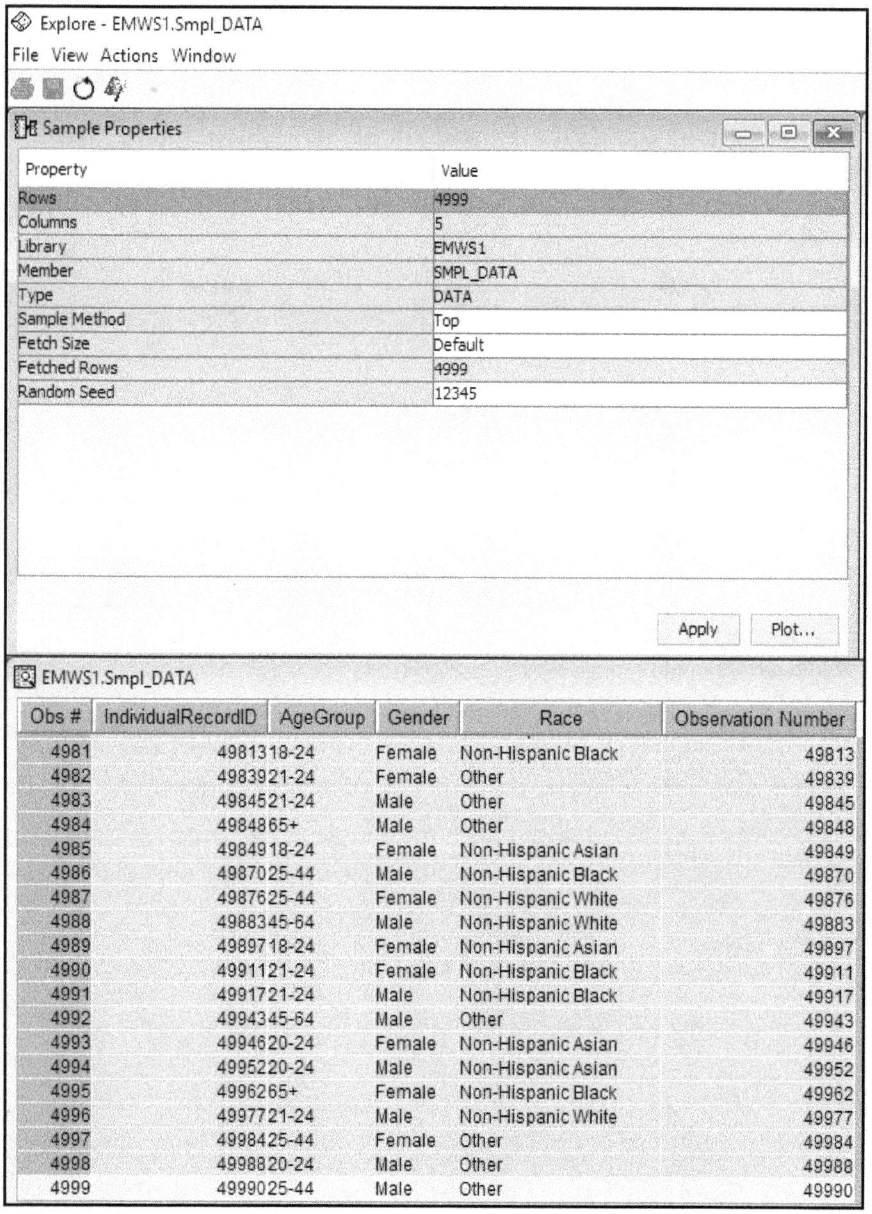

Property	Value
Rows	4999
Columns	5
Library	EMWS1
Member	SMPL_DATA
Type	DATA
Sample Method	Top
Fetch Size	Default
Fetched Rows	4999
Random Seed	12345

EMWS1.Smpl_DATA

Obs #	IndividualRecordID	AgeGroup	Gender	Race	Observation Number
4981	49813	18-24	Female	Non-Hispanic Black	49813
4982	49839	21-24	Female	Other	49839
4983	49845	21-24	Male	Other	49845
4984	49848	65+	Male	Other	49848
4985	49849	18-24	Female	Non-Hispanic Asian	49849
4986	49870	25-44	Male	Non-Hispanic Black	49870
4987	49876	25-44	Female	Non-Hispanic White	49876
4988	49883	45-64	Male	Non-Hispanic White	49883
4989	49897	18-24	Female	Non-Hispanic Asian	49897
4990	49911	21-24	Female	Non-Hispanic Black	49911
4991	49917	21-24	Male	Non-Hispanic Black	49917
4992	49943	45-64	Male	Other	49943
4993	49946	20-24	Female	Non-Hispanic Asian	49946
4994	49952	20-24	Male	Non-Hispanic Asian	49952
4995	49962	65+	Female	Non-Hispanic Black	49962
4996	49977	21-24	Male	Non-Hispanic White	49977
4997	49984	25-44	Female	Other	49984
4998	49988	20-24	Male	Other	49988
4999	49990	25-44	Male	Other	49990

Experiential Learning Application: Health and Nutrition Data Partitioning

As a next step in our sampling step of the SEMMA model, we will partition the data. In this example, a training set of 50%, a validation set of 30%, and a test set of 20% will be used. The partition node will be useful in subsequent chapters when we want to train a model with part of the data and then validate and test the model with another part of the data. If we used all the data to train the model, the model would know all cases. By using separate data sets, we can validate and test how the model would perform with new and otherwise unknown set of data. Using a validation and test set would allow the model to be more generalizable for new problems and data. The goal is to build a model that would perform well given new sets of data and scenarios in the future. If we built a model that worked extremely well for our analysis, but then worked poorly on new data, then we can say that we overfit the model. This means that we tried to perfectly fit the model for only the available data, but then the model was not generalizable or applicable to new sets of data.

Step 1. Under the **Sample** tab, Select the Data Partition node, and add it to the workspace. Connect the Sample node to the Data Partition node.

Figure 3.26: Add Data Partition Node

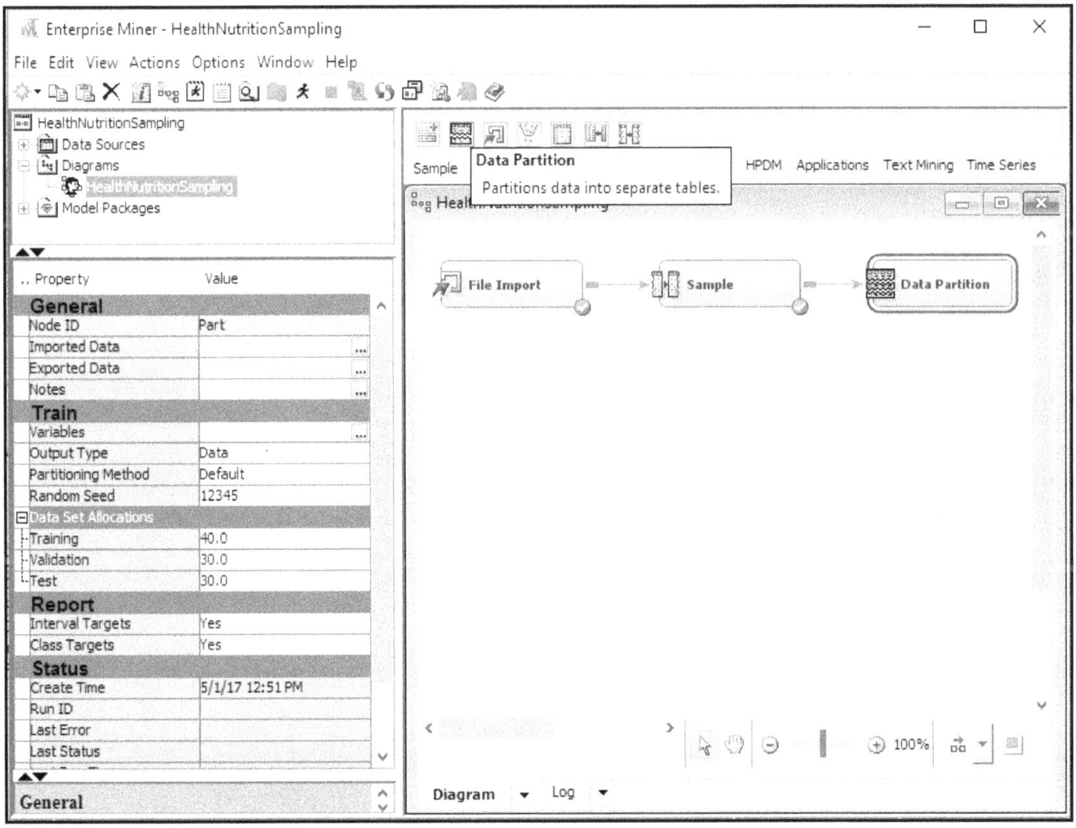

Step 2. Select the Data Partition node and review the Property panel. For this application, we want to create three data sets: Training, Validation, and Test. We will partition or split our data with 50% for Training, 30% for Validation, and 20% for Test. Verify that the values are set for 50, 30, and 20 in the properties.

Figure 3.27: Set Data Partition Node Data set Allocations

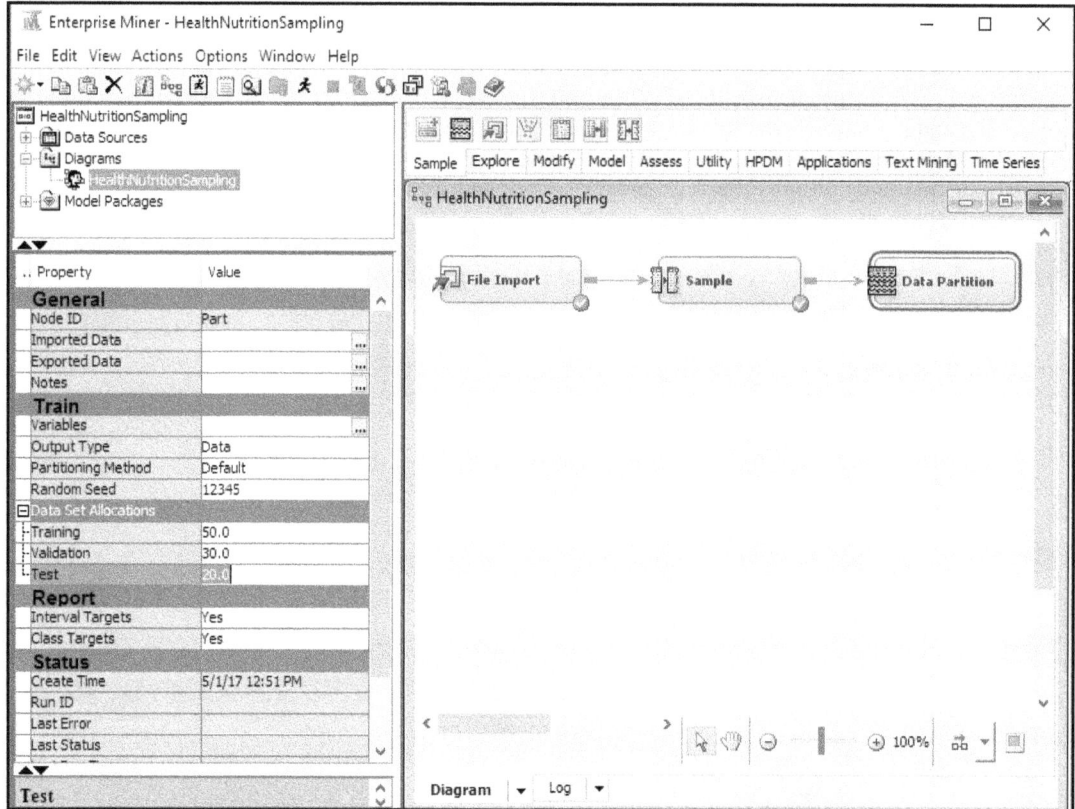

Step 3. Right-click and run the Data Partition node.

Figure 3.28: Run Data Partition Node

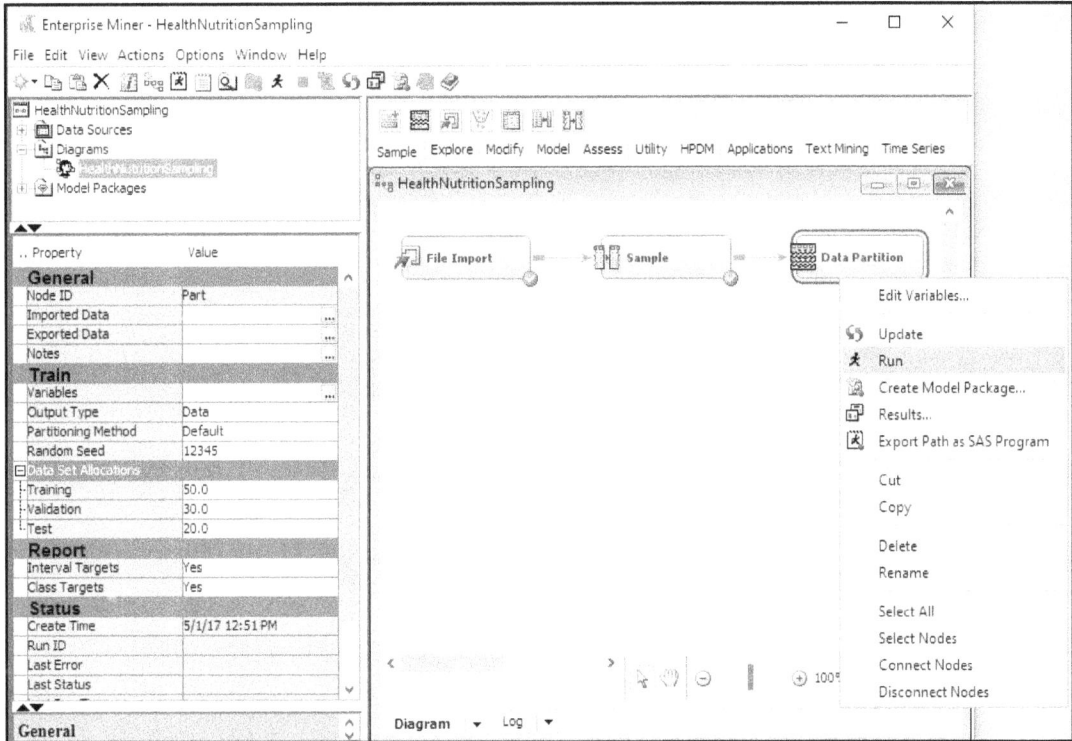

Step 4. Click **Results** after running the Data Partition node. If you accidentally close the Run window, no worries, you can simply right-click the Data Partition node and click **Results**.

Figure 3.29: View Data Partition Node Results

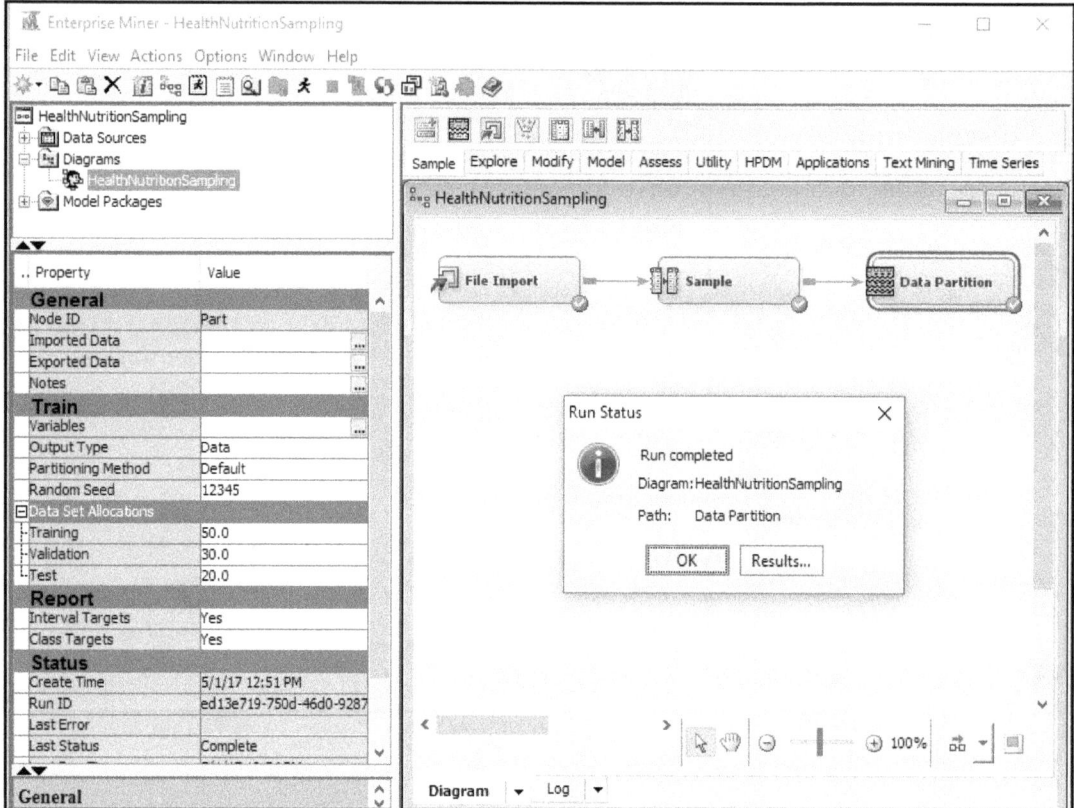

Step 5. From the results, we see that three data sets were created: Train with 2500 observations, Validate with 1500 observations, and Test with 999 observations. A quick calculation will verify the expected number. From our original data set sample of 4999 records, we set the Train data set at 50% or 2500/4999, Validate was set at 30% or 1500/4999, and Test was set at 20% or 999/4999.

Figure 3.30: Data Partition Results

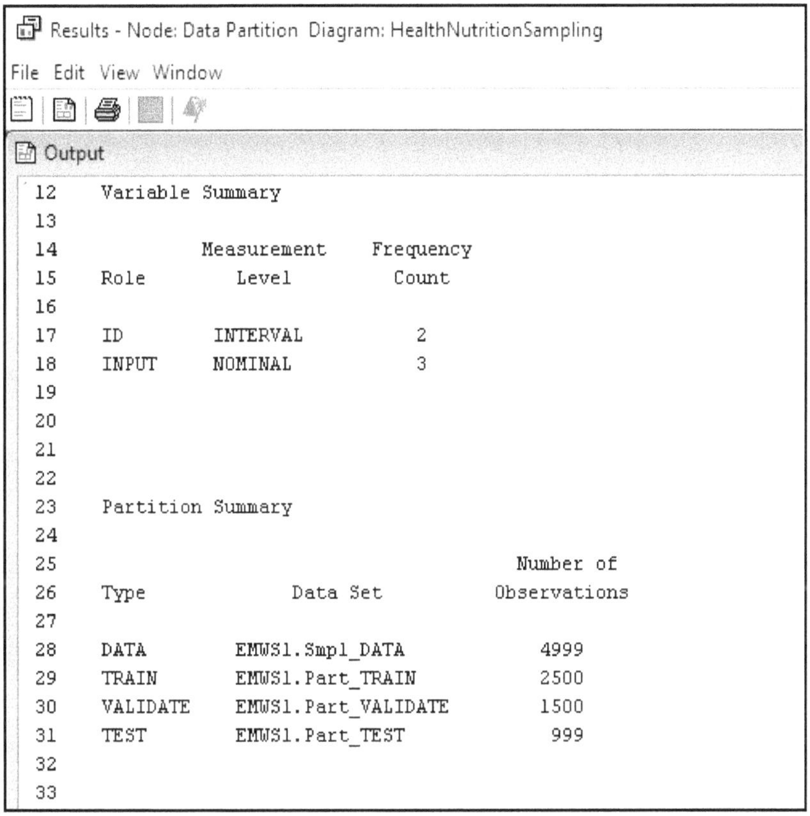

```
 Results - Node: Data Partition  Diagram: HealthNutritionSampling

File  Edit  View  Window

 Output

 12     Variable Summary
 13
 14              Measurement      Frequency
 15     Role       Level           Count
 16
 17     ID         INTERVAL          2
 18     INPUT      NOMINAL           3
 19
 20
 21
 22
 23     Partition Summary
 24
 25                                  Number of
 26     Type             Data Set    Observations
 27
 28     DATA        EMWS1.Smpl_DATA      4999
 29     TRAIN       EMWS1.Part_TRAIN     2500
 30     VALIDATE    EMWS1.Part_VALIDATE  1500
 31     TEST        EMWS1.Part_TEST       999
 32
 33
```

Step 6: To find the records in each of the data sets, click the Data Partition node. In the properties section under General, click the ellipses next to Exported Data.

Figure 3.31: View Data Partition Exported Data

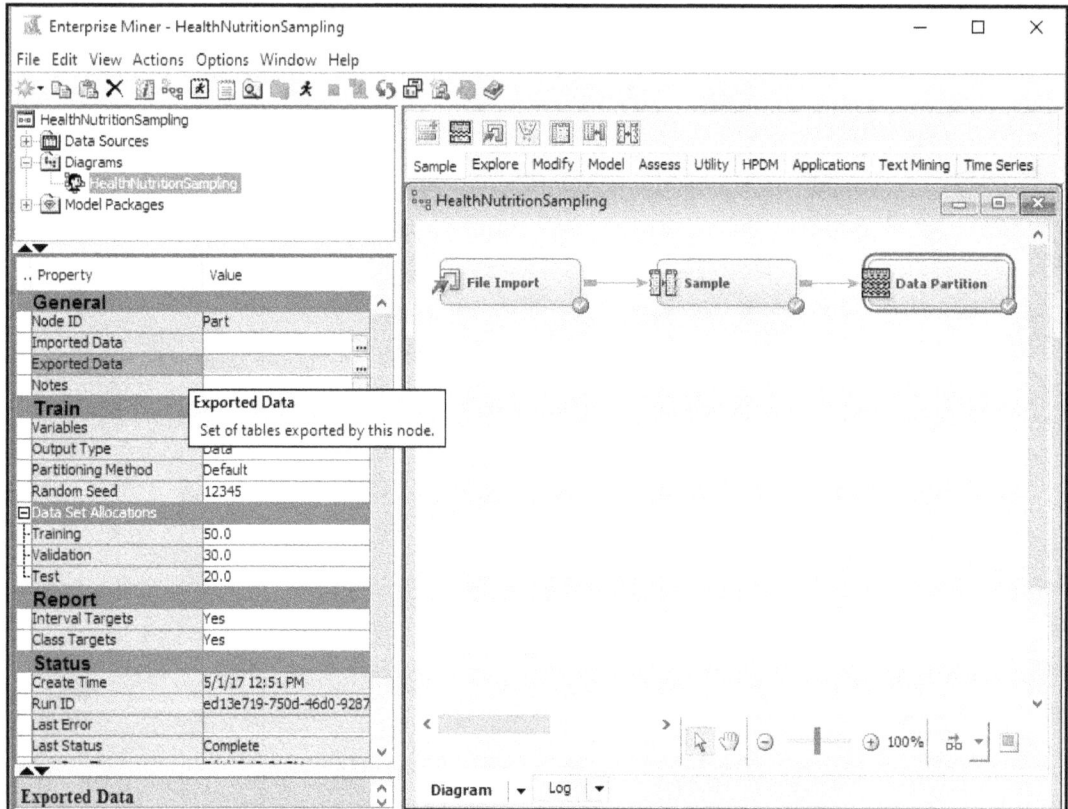

Step 7: After clicking the ellipses next to the Exported Data property, you will see the listing for each of the data sets. To view each data set, click Train, Validate, and Test, and then click **Explore**.

Figure 3.32: Data Partition Exported Data

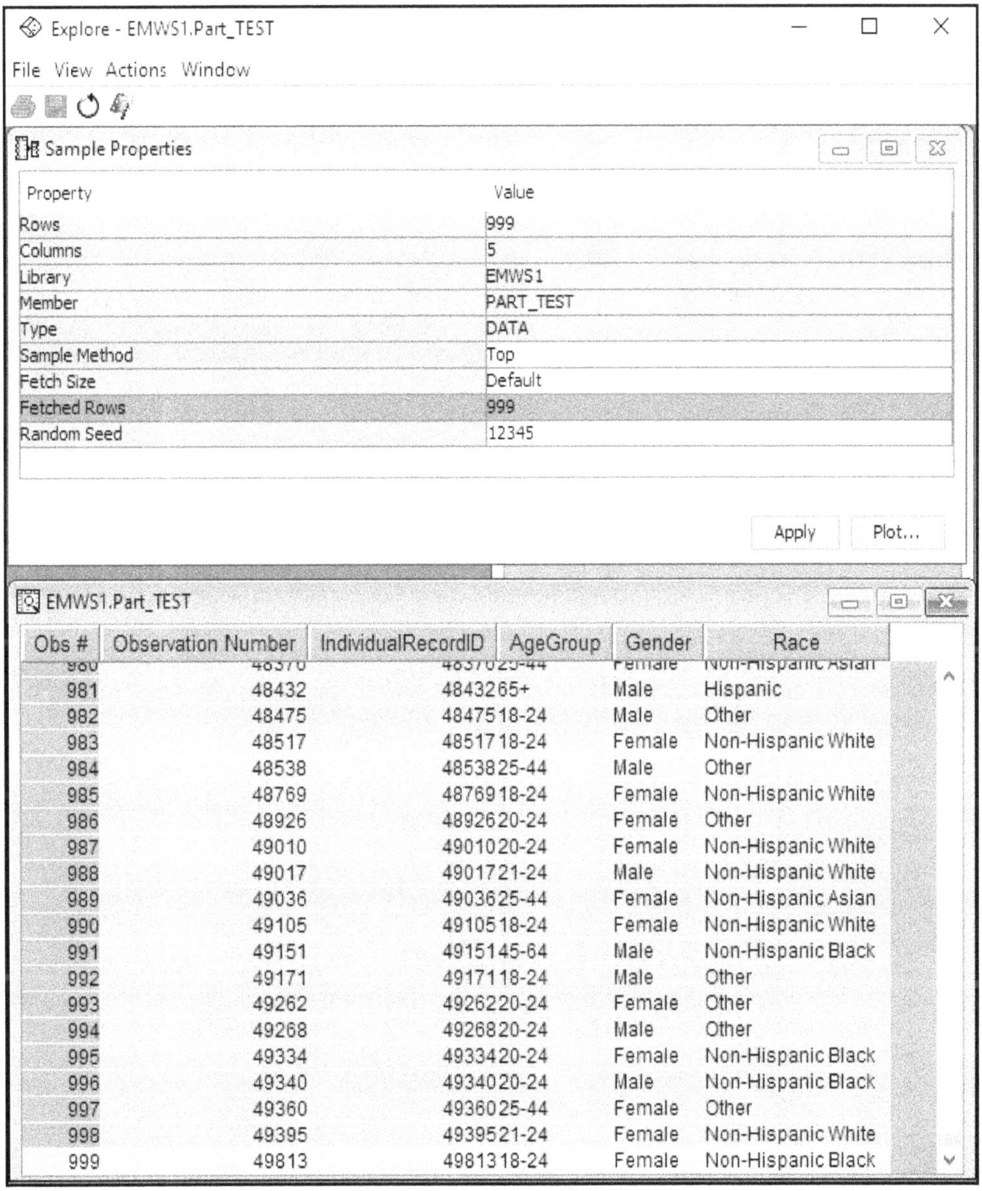

Obs #	Observation Number	IndividualRecordID	AgeGroup	Gender	Race
980	48370	4837025-44	Female	Non-Hispanic Asian
981	48432	4843265+	Male	Hispanic
982	48475	4847518-24	Male	Other
983	48517	4851718-24	Female	Non-Hispanic White
984	48538	4853825-44	Male	Other
985	48769	4876918-24	Female	Non-Hispanic White
986	48926	4892620-24	Female	Other
987	49010	4901020-24	Female	Non-Hispanic White
988	49017	4901721-24	Male	Non-Hispanic White
989	49036	4903625-44	Female	Non-Hispanic Asian
990	49105	4910518-24	Female	Non-Hispanic White
991	49151	4915145-64	Male	Non-Hispanic Black
992	49171	4917118-24	Male	Other
993	49262	4926220-24	Female	Other
994	49268	4926820-24	Male	Other
995	49334	4933420-24	Female	Non-Hispanic Black
996	49340	4934020-24	Male	Non-Hispanic Black
997	49360	4936025-44	Female	Other
998	49395	4939521-24	Female	Non-Hispanic White
999	49813	4981318-24	Female	Non-Hispanic Black

Experiential Learning Application: Claim Errors Rare-Event Oversampling

For our experiential learning application, you will use your knowledge of sampling to run the following scenario. A research study reviewed the differences between two methods for verifying self-funded medical claims. The two methods reviewed were 1) verifying 100% of claim exceptions or errors that were identified from all 54,000 claims and 2) a random sampling of 300-400 claims. The claims data was collected from two Fortune 100 corporations. The researchers found that the second random sample method missed many of the claim errors that were identified through the 100% of claims review with the first method. Some of the missed errors contained $200,000 - $750,000 in claim dollar values. To catch all the claim errors is an important goal for self-funded plans for ensuring lower expenditures. The process goal can follow a Six Sigma methodology, which is a quality improvement process, that ultimately seeks to obtain 0 defects or errors. The researchers applied a five-step protocol for the claims review. The steps are: 1) All paid claims were loaded into a data warehouse. 2) Eligibility data and benefits (such as covered services, co-pays, administrative agreements, and rules) are reviewed. 3) Industry standards from Centers for Medicare and Medicaid Services (CMS) and others are used to detect exceptions. 4) The data is compared against best practices and administrative processes. 5) An on-site review is conducted for any claims identified for validation (Sillup and Klimberg, 2011).

Data Set File: 3_EL2_Claim_Errors.xlsx

Variables:

- ClaimID, Unique Claim Identifier
- PatientID, Unique Patient Identifier
- BillingID, Unique Claim Billing Identifier
- AmountPaid, Total Claim Amount Paid
- Exception, Error Identification No/Yes, False/True, 0/1

Below is a summary of the data set identified:

- Total amount claims paid: $13,106,411
- Total claims records: 54,192
- Total amount of exception claims: $2,243,030
- Total exception claim records 8,299

The 300-400 claim sample did provide a statistically significant sample of exceptions and errors. However, all exception claims were not identified. In addition, a secondary step of root cause analysis should be conducted to resolve the errors for continuous improvement over time. The continuous improvement allows the movement toward error-free payments and meets the optimal balance of cost-quality healthcare delivery and best use of limited healthcare resources. Follow the SEMMA process (see figure) for your experiential learning application. A template has been provided below that can be reused across future projects. For this application, you will also need to set property settings within the Sample node for rare-event oversampling (SAS, 2017b). Additional information has been included in the template.

Data Set:	Role	Level
ClaimID		
PatientID		
BillingID		
AmountPaid		
Exception		

Title	Claim Errors - Rare Event Oversampling
Introduction	Provide a summary of the business problem or opportunity and one or more key objectives or goals. Create a new SAS Enterprise Miner project. Create a new Diagram.
Sample	Data (sources for exploration and model insights) Identify the variables data types, the ID, input and target variable along with the levels during exploration.

Data set:	Role	Level
ClaimID		
PatientID		
BillingID		
AmountPaid		
Exception		

Add a FILE IMPORT node
Provide a results overview following the file import:
Input / Target Variables
Level of Variables

Add a SAMPLE node
Note: to achieve rare-event oversampling, click the Sample node, set the Sample Method property to Stratify, click the ellipses next to the Variables property above Sample Method, and set the target variable Claim_Exception as the stratification variable. If the stratification option is grayed out, double-check your variable roles and levels under the File Import node. On the main diagram, click the Sample node on the Size property, and set the Type equal to

Title	Claim Errors - Rare Event Oversampling
	Percentage, and set the Percentage = 100. This will select 100% of the exception records. We also want to select an equal number of non-exception records to review. On the Stratified property, set the Criterion property to Equal. This option selects all the claim exceptions plus a random sample of non-identified Claim-Exceptions. Set the Oversampling Adjust Frequency property to No. Provide a results overview following the sampling: Generate a DATA PARTITION. Provide a results overview following the data partition:

Learning Journal Reflection

Review, reflect, and retrieve the following key chapter topics only from memory and add them to your learning journal. For each topic, list a one sentence description/definition. Connect these ideas to something you might already know from your experience, other coursework, or a current event. This follows our three-phase learning approach of 1) Capture, 2) Communicate, and 3) Connect. After completing, verify your results against your learning journal and update as needed..

Key Ideas – Capture	Key Terms – Communicate	Key Areas - Connect
SEMMA		
SAS Enterprise Miner		
SAS Solutions OnDemand		
Sample Process Step		
File Import Node		
Sample Node		
Data Partition Node		
Data Roles		
Data Levels		
Survey Sampling		
Rare-Event Oversampling		

Chapter 4: Discovering Health Data Quality

Chapter Summary

The purpose of this chapter is to review data quality and develop data discovery skills using SAS Enterprise Miner with regard to the Explore and Modify capabilities within the SEMMA process. This chapter also includes experiential learning application exercises on heart attack payment data and data quality exploration. We continue this chapter with the Explore and Modify phase as shown in Figure 4.1.

Figure 4.1: Chapter Focus - Explore and Modify

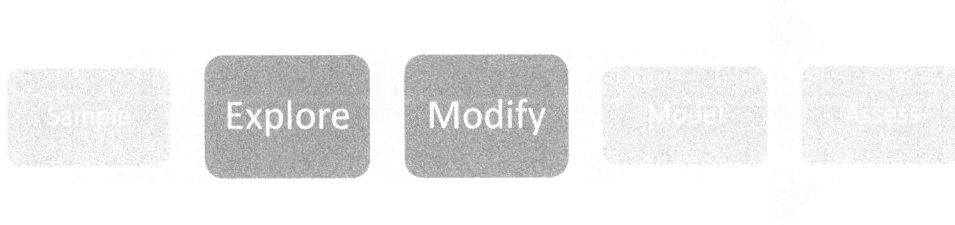

Chapter Learning Goals

- Understand healthcare quality
- Describe data quality
- Define the Explore and Modify process steps

- Develop data exploration skills
- Apply SAS Enterprise Miner Explore and Modify data functions

Healthcare Quality

Medical errors are the third leading cause of death among patients, and adverse events occur in up to one-third of all admissions, with only 1 of 2 patients receiving recommended treatment (Makary and Daniel, 2016). The Centers for Disease Control and Prevention (CDC) generate an annual list of the causes of death in the U.S. However, these causes are associated with an International Classification Disease (ICD) code, which does not include human-related or system-related causes such as communication errors, diagnostic errors, judgment errors, and skill errors. John Hopkins researchers have estimated over 250,000 deaths occur annually in the U.S. due to medical errors (Makary and Daniel, 2016).

Due to the concerns over medical errors, various entities (business groups, insurance organizations, state legislatures, health policy groups, and the U.S. Congress) have focused on improving healthcare quality within the U.S. for several decades. Congress passed the Healthcare Research and Quality Act of 1999, requiring a national quality report to be established. The Agency for Healthcare Research and Quality (AHRQ) identifies two key public challenges: improving quality of healthcare, and ensuring that no one is left behind in quality improvement. As part of AHRQ, required annual congressional reports are used to examine healthcare system quality and disparities through key measures. The reports identify several improvement areas such as recommended cancer screenings, nutrition counseling, environment changes to improve asthma-related conditions, and recommended diabetes screenings. Overall, access and quality of care is often determined based on racial, ethnic, or economic grouping. AHRQ has turned to more advanced information systems capabilities through web-based tools and improved data, graphing, and reporting. These capabilities increase access to and insights into quality trends. The state snapshots functionality allows graphical dashboard comparison of measures and helps identify quality improvement areas (Brady et al., 2007; Naidu, 2009; Weaver and Hongsermeier, 2004).

The Institute of Medicine (IOM) released the "Crossing the Quality Chasm" report in which six aims of a high-quality healthcare system are identified. The report states that quality healthcare systems should be: 1) safe, 2) effective, 3) patient-centered, 4) timely, 5) efficient, and 6) equitable. The National Health Services (NHS) in the U.K. similarly has focused on quality since 1998, with the one distinction from the U.S. being that the government is the purchaser of healthcare services. In the U.K., those with economic means can go outside the public healthcare system to receive care, defying the equitable aspect of quality care. Similarly, the six key areas for U.K.'s NHS include health improvement, fair access, effective delivery, efficiency, patient and caregiver experience, and health outcomes (Weaver and Hongsermeier, 2004).

Despite the charge and activities toward quality, leaders still struggle to determine which data to measure for quality purposes, which quality framework to use, and the amount of resources and technology to implement quality initiatives. Resistance to mandatory reporting and adoption of standardized safety practices have occurred due to medical and legal concerns. In the U.S., healthcare services are often not purchased based on quality (Weaver and Hongsermeier, 2004). Variability in healthcare throughout the world is well-known, resulting in differences in health outcomes. A further challenge is translating and developing worldwide evidence for local applications. Assessment and improvement of healthcare quality is a key example of evidence-based methodologies. These methods and metrics are used by the National Committee for Quality Assurance (NCQA), the Joint Commission on Accreditation of Healthcare Organizations (JCAHO), and similar accrediting organizations. The information (or evidence) along with experts are used. However, in many clinical areas, such as mental health, the evidence is underdeveloped.

Quality of healthcare delivery and rapid, measurable, and sustainable improvements are a high priority for health systems. Like quality in most services, healthcare quality is difficult to measure because of inherent intangibility, heterogeneity, and inseparability features. Quality in healthcare is often more challenging to determine than in financial or aviation industries because the output is individually perceived quality of life. Successful quality improvement has been demonstrated at a state and national level. However, most measures of quality are improving at only a modest pace, with a 3.1% improvement rate over the last three years (Clancy et al, 2005; Woodside, 2014).

Beyond direct patient care, quality of services can be linked to financial and competitive improvements. In the healthcare industry, quality perceptions from medical patients have been found to account for up to 27% of financial variances in earnings, revenue, and asset returns. In addition, negative word of mouth, advertising can cost upward of $400,000 in revenue loss throughout a patient's lifetime (Naidu, 2009). Third-party groups (such as the Joint Commission on the Accreditation of Health Care Organizations (JCAHO) for hospitals, the Agency for Healthcare Research and Quality (AHRQ), National Committee for Quality Assurance (NCQA), Health Employer Data Information Set (HEDIS), and the National Quality Forum) develop outcomes against a set of benchmarks. The use of benchmarks is a Six Sigma method to easily compare performance and improve continuously (Weaver and Hongsermeier, 2004). Outcomes measurement is the assessment of population health through process and outcome measures. The Institute of Medicine (IOM) has identified healthcare quality as the degree to which desired health outcomes (such as improvement in health and patient experiences) are achieved (Woodside, 2013c).

Quality improvement approaches have been adopted within the healthcare industry in an attempt to improve quality of care. Most of the activities to date have focused on manual activities without a direct link to the data within the healthcare information system. Support systems can provide patient outcome information and clinical pathways to assist patient care, and identify factors that influence quality and treatment. Data mining, which allows for knowledge discovery from large sets of data, can be used to identify patterns or rules to improve healthcare quality. Patient characteristics (including age, gender, department, disease class, and quality indicators) were used as part of decision tree analysis to determine in-patient quality factors. An index score can be developed to identify how quality rates compare to overall proportions, and which segments to focus on (Chae 2003).

Quality improvement proponents have recommended a healthcare quality information system (HQIS) to accurately measure and manage healthcare quality information collection, analysis, and reporting. Outcomes and treatment-cost improvements exist, and the public is seeking more information about healthcare quality in order to make informed decisions. To ensure successful HQIS, hardware, software, and a complete information framework of standard data fields and processes must be developed. A patient-centered data structure is often implemented, linked by a patient ID, and used to manage quality of care. Quality typically includes both a technical dimension and a functional dimension. Patients might be unable to accurately determine technical quality, and, typically, functional quality is used as the primary method. Despite much literature on healthcare quality, few tools exist for assessing and managing quality. Within the healthcare services review sector, direct patient care is affected, and key components of quality are supported through anamatics capabilities. Information quality is vital as for organizations to reduce uncertainty and enhance their decision-making capabilities. System quality is addressed through information processing needs or the supported communication requirements that are based on individual and unit interactions as well as system support for information processing capability through use of technologies. Service quality can also be described through information processing capability through the information technology support that is given to end users to assist with performance outcomes (Woodside, 2014).

Value-Based Healthcare

The Affordable Care Act (ACA) has several components that are targeted toward improving quality in healthcare. High-quality healthcare requires technology tools and systems to realize the ACA goals and objectives (Woodside, 2014). As a result of legislation and healthcare quality focus, the Health and Human Services (HHS) Secretary announced a move from traditional fee-for-service payments to quality or value-based payments. HHS set a goal of 50% of alternative payment models, such as Accountable Care Organizations (ACOs), by the end of 2018 and 90% of Medicare payments to quality models. In a fee-for-service model, each service provided is reimbursed a fixed fee. An ACO is a group of providers, such as physicians or hospitals, that provide coordinated quality care to a group of patients. The ACO model for providers aims to deliver higher quality care while reducing costs through eliminating duplicated or unnecessary services (HHS, 2015).

As part of the American Recovery and Reinvestment Act of 2009 (Recovery Act), the Centers for Medicare and Medicaid Services (CMS) have awarded incentive payments for "Meaningful Use" through an electronic health record (EHR) to improve quality, safety, efficiency, outcomes, research, and efficiency. Meaningful use sets objectives, comprehensive guidelines, and core objectives that eligible providers and hospitals must follow to be qualified for the CMS incentive programs. In stage 1, all medications should be entered by a computerized physician order entry (CPOE) system, and a drug-to-drug and drug-to-allergy interaction check should be implemented within the system. In addition to quality checks, the guidelines also include minimum data capture for completeness. For example, the following vital signs are required to be recorded and charted: height, weight, blood pressure, BMI, and a plot of children growth charts for ages 2-20, including BMI. A summary of the three stages of meaningful use, timeline, and key components are included in Figure 4.2. It should be noted that stage 2 (which was originally scheduled for completion in 2014) was extended to 2016. Stage 3 (which was originally scheduled for completion in 2016) was extended to begin with an option for 2017 for providers with two years completed under stage 2, based on recommendations from industry (Reider and Tagalicod, 2013; HealthIT.gov, 2017a; HealthIT.gov, 2017b; CMS, 2017a).

Figure 4.2: Meaningful Use

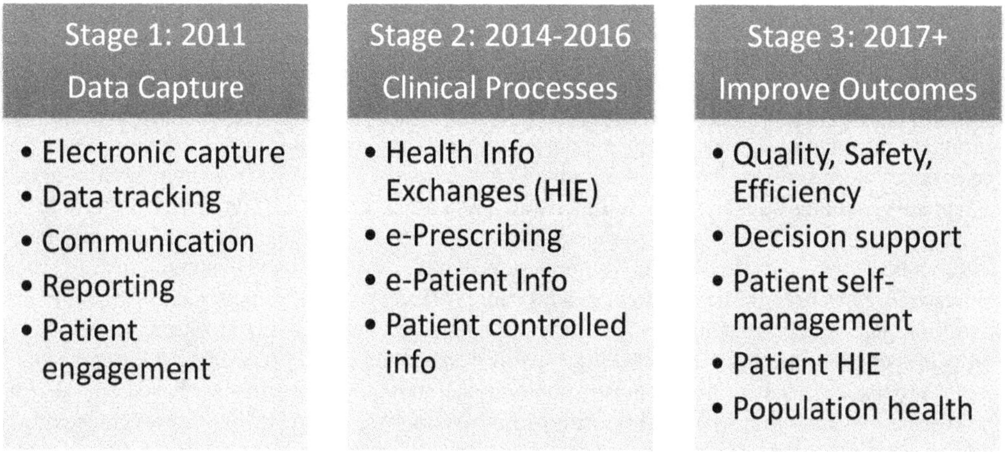

Stage 1: 2011 — Data Capture	Stage 2: 2014-2016 — Clinical Processes	Stage 3: 2017+ — Improve Outcomes
• Electronic capture • Data tracking • Communication • Reporting • Patient engagement	• Health Info Exchanges (HIE) • e-Prescribing • e-Patient Info • Patient controlled info	• Quality, Safety, Efficiency • Decision support • Patient self-management • Patient HIE • Population health

The impact of healthcare efficiency and effectiveness is evident from an estimated $300 billion in annual value created through improved use of analytics by McKinsey Global Institute. IBM estimates a mid-size health plan can improve annual economic benefit by over $600 million if using all available anamatics (information technology and analytics available). The overall goals of value across all industries including healthcare is to simultaneously increase revenue and reduce costs. In healthcare, cost reduction can be generated through operational efficiencies, and revenue can be increased through improving clinical outcomes and engaging patients in healthcare (McNeil, 2015). As one example, HealthGrid was awarded the Microsoft Innovation award for their solutions. Their intelligent technology allows providers and patients to improve information sharing for decision-making, which is a key component of the Meaningful Use goals across all stages. HealthGrid enabled Family Physician Group (FPG) to shift to value-based care and meet Meaningful Use guidelines, resulting in additional appointments per provider, which is a significant reduction in clinical care gaps, and improved care plan compliance. HealthGrid and FPG estimate a 9:1 return on investment (HealthGrid, 2017).

Healthcare Data Quality

Today, healthcare organizations are becoming increasingly computerized, thereby capturing increasing amounts of data in various places. Extracting, formatting, analyzing, and presenting this data can improve quality, safety, and efficiency of delivery within a healthcare system. With the growth of healthcare data, one of the commonly overlooked traits of data in healthcare deals with the veracity or the quality of the data, leading to significant numbers of preventable medical errors and limited application of recommended care guidelines. In order to ensure the healthcare quality dimension and appropriate evidence-based care, the underlying data sources must be of high quality to realize the full value. Healthcare information systems and technology allows the capture of data, analysis, and use of decision support capabilities (Weaver and Hongsermeier, 2004). Analytics applied to poor data can lead to faulty decisions. This is the concept of garbage in garbage out (GIGO). Analytics can also be used to detect and correct data errors (Simpao et al, 2014), which we will use throughout this chapter to Explore and improve overall data quality in order to lead to quality decision-making. In an IBM report, one in three business leaders do not trust the information to make decisions. Poor data quality costs the U.S. an estimated $3.1 trillion a year (Woodside 2014; IBM, 2014). Data quality can be described as accurate, complete, consistent, legible, relevant, timely, and unique. These attributes ensure high quality data and information (HIQA, 2012; Baltzan, 2015). The following table lists the healthcare data quality attributes, descriptions, and examples of common data quality errors and issues.

Table 4.1: Healthcare Data Quality Attributes

Healthcare Data Quality Attribute	Description	Data Error/Issue Example
Accurate	All data values are correct.	A patient's year of birth is set as 1957 instead of 1975.
Complete	All data values are fully populated.	A secondary diagnosis code is missing.
Consistent	All data values follow standards.	Some diagnosis codes use the ICD-9 rather than the ICD-10 standard.
Legible	All data values are readable	Notes are written in erasable ink and smeared.
Relevant	All data pertains to the decision.	A report is developed by providers outside of your service area.
Timely	All data values are up-to-date.	The latest patient allergies are included in their record.
Unique	All data values are not duplicated.	A claim is billed twice for the same service.

Experiential Learning Activity: Healthcare Data Quality Check

Healthcare Data Quality Check

Description: Identify the healthcare data quality attributes in the claims data below and provide a recommendation where a computerized method can be used to improve the quality:

#	PatientID	Patient Date of Birth	Diagnosis Code	Date of Service	Procedure Code	Provider ID	Amount Billed
1	7549824	9/9/2099	*****	2/1/2017	99213	1045265490	$175
2	7549824	5/1/2000	7549824	1/1/2001	99213	1045265490	$450
3	95487156	2/31/1955	S92.4	3/1/2017	325	1045265490	$325

#	Healthcare Data Quality Attributes	Recommendation to Prevent
1		
2		
3		

Healthcare Data Quality Case Study

Health informatics initiatives increasingly require quality impact as the basis for justification and value. As healthcare organizations seek to continually improve overall quality outcomes and adopt new capabilities, health informatics plays a critical role in providing such capabilities as well as streamlined access to summarized information for decision-making support. Clinical Decision Support Systems (CDSS) are implemented in order to improve patient safety and quality of care. CDSS often suffer from the quality of the underlying data. Sweden is viewed as an early adopter of many healthcare information technology systems and has an EHR hospital adoption rate of 100%, with one computer per employee on average. Despite this, EHR systems might not be fully used within the emergency care setting due to user resistance and usability factors. A study was performed on vital-sign data quality within a set of Swedish emergency departments to determine the factors influencing data quality. Summary factors included people and process components, and information technology components. Only half of the emergency departments had a fully electronic vital-sign documentation process, and documentation was found to lack currency and completeness. To improve data quality, a five-step approach was followed: 1) Standardize the care process -- the aim was to improve completeness of records and establish a process to re-check and document vital signs every 15-30 minutes on identified patients to improve currency. 2) Improve digital documentation -- EHRs were identified to make better use of documentation and integrate the systems to improve complete and current vital signs. 3) Provide workflow support -- Mobile solutions were developed for the EHR to improve usefulness and portability of paper-based methods, which also improved completeness and currency. 4) Ensure interoperability -- EHR systems improved the exchange of vital signs. Previous versions had different keywords and templates without standardized data reference models. 5) Perform quality control -- ongoing management feedback on quality was important to improve completeness and training through staff review of data quality and error sources (Skyttberg et al., 2016).

Six Sigma Health Data Quality

Six Sigma is used across various industries and thousands of organizations as a method to improve and set the highest standards for quality. Six Sigma uses a data-driven methodology to identify and eliminate defects or errors and improve to six standard deviations of the mean using a statistical basis. From statistics, two standard deviations from the mean would incorporate approximately 67% of all cases. Three standard deviations would incorporate 95% of all cases, and six standard deviations would incorporate 99.9997% of all cases. To successfully achieve Six Sigma quality standards, organizations must achieve six standard deviations of the mean, incurring no more than 3.4 defects or errors per million opportunities. A defect or error is any instance that does not meet a customer's expectations. To begin a Six Sigma project, one can follow a DMAIC process (Define, Measure, Analyze, Improve, and Control). Organizations that are committed to Six Sigma use teams to work on problems, have management support for Six Sigma quality across the organization, and have individuals trained in statistical data thinking (ASQ, 2016; iSixSigma, 2018). In one case study, the data quality within the Dutch National Intensive Care Evaluation (NICE) registry was reviewed. Results showed that 4.0% of data was incomplete and 1.7% was inaccurate for automatic data collection, and 3.3% was incomplete and 4.8% was inaccurate for manual data collection. As with other studies, this would put the error rate and quality between 3 and 4 sigma, again with an end goal of 6 sigma, or 3.4 defects per million opportunities (Arts et al., 2016).

Previous studies have focused primarily on patient outcomes as a measurement for healthcare quality. Although patient outcomes represent a core quality dimension, healthcare organizations must also contend with other, often conflicting, quality dimensions. The desire to meet shareholders return on investment expectations must be balanced with the desire to improve quality of care and outcomes for patients. In addition, neither quality of care nor quality of return can be fully realized without a deliberate effort to achieve quality in employee satisfaction. Health anamatics initiatives are increasingly requiring quality impact as the basis for justification and value. Healthcare data capabilities and processes enable individual organizations to focus specifically on those areas that have the greatest impact on the multi-faceted organizational quality. As healthcare organizations seek to continually improve overall quality outcomes and adopt new capabilities, health anamatics plays a critical role in providing such capabilities as well as streamlining access to summarized information for decision-making support. To achieve organizational quality, healthcare entities must balance the needs of various stakeholders. Financial returns must be weighed against quality of care provided to customers and training provided to employees. Within service sectors (such as healthcare), employee requirements, resulting interaction with customers, and impact on shareholder returns all must be considered with a greater degree. Other industries have experienced similar challenges in balancing stakeholder requirements (Woodside, 2013c).

Due to competitive markets and globalization, organizations must continually seek to add value to their services, deliver additional profits, and exceed customer expectations. Satisfied employees tend to be more involved and dedicated to quality. Presenting positive employees leads to a positive customer attitude toward a given product. Dissatisfied or hostile employees create hostile customers regardless of the organization's tasks and service performance. Customer satisfaction has also been shown to have a long-term profit impact. Satisfied customers are more likely to purchase in greater quantity with greater frequency and add additional services. Satisfied customers are also more willing to pay premiums and are less-price sensitive than non-satisfied customers (Hurley 2007; Yee 2008). Within a given organization, there is often significant variability at a business unit level. In past studies, units that score above the median on employee and customer satisfaction measures were found to be 3.4 times more effective financially as measured by total revenue, performance targets, and year-over-year gain in sales and revenue. If the focus is on employees only, the business unit can be too inwardly focused. If the focus is on customers only, employee satisfaction will erode over time. Unchecked quality of the customer and employee experience can create issues. Often the variability goes unnoticed or unmanaged, and revenues and profits are bled off and growth is stagnant. In many organizations, the objective of achieving a unified corporate culture and brand goes unrealized. Performance must be continuously improved, and feedback given at the lowest level of variability and specificity (Fleming 2005; Woodside, 2013c).

While recognizing the need for quality, Total Quality Management (TQM) and other quality initiatives such as Six Sigma have often initially led to high levels of turnover within an organization. Organizations must find ways to continue to motivate and retain employees, with employee participation being a critical component to quality, and the associated personal enrichment, which goes along with direct involvement and decision-making. In an employee improvement process, individual transformation and understanding of knowledge will occur. The individuals will add new meaning to events and interactions. They will have a judgment and decision basis for organizational transformations, and they will assist others to move into similar practices and capabilities, leading to improved organizational quality (Connor, 1997, Woodside, 2013c).

Following our review on healthcare data quality, we will review data sources and exploration further for our experiential learning activity.

Experiential Learning Activity: Public Data Exploration

Public Data Exploration

Description: There are many different types of public data sources available. These data sources include primary data sources and secondary data sources. Primary data sources are those that are collected directly from patients or providers. The primary data collection can occur through monitoring devices, tests, physician notes, forms, and surveys. Secondary data sources are those that are collected after patient or provider care, and are developed through collecting information from medical records, claims, government agencies, quality improvement reports, and study data.

Visit a public data site and find a public data set, examples:
- General Health Data: https://healthdata.gov/
- Florida Health Data: http://www.flhealthcharts.com?aspxerrorpath=/charts/default.aspx
- Vital Statistics Data: https://www.cdc.gov/nchs/data_access/ftp_data.htm
- Public Health Data: https://wonder.cdc.gov/
- Core Data Elements: https://www.cdc.gov/nchs/data/ncvhs/nchvs94.pdf
- National Cancer Data: https://www.facs.org/quality-programs/cancer/ncdb/qualitytools
- Outcomes Reporting Data: https://www.cms.gov/Medicare/Quality-Initiatives-Patient-Assessment-Instruments/OASIS/DataSpecifications.html

Select a data set and describe the data set summary or specifications:

Identify key data elements in the data set, including the types of variables:

Identify any potential quality issues that can occur in the data set:

Now that we have covered healthcare data quality components, we can now begin our formal process to Explore and Modify the data using SAS Enterprise Miner.

SEMMA: Exploration

Exploration Process Step Overview

During the exploration process step, the data is reviewed and described, and quality issues are identified.

Figure 4.3: Exploration Step

Data Exploration – Data Discovery and Descriptive Statistics

The first phase of the SEMMA process is sampling the data covered in the previous chapter. In the second phase of the SEMMA process, data exploration occurs. When exploring data, we typically start with describing the data. That is, we can run a set of descriptive statistics on the data (Anderson et al., 2002). Even if you are familiar with many of these measures already, a brief description is included as follows:

- Count: This is the total number of records or occurrences within a data set or population sample. For example, there can be a set of 100 patient records.
- Average: Also known as the mean, this measure provide a central point for the data set. To calculate the mean, you add all the occurrences and divide by the count of records. If three claims are billed at $100, $150, and $200, the average is ($100+$150+$200) / 3 = $150.
- Median: The median is similar to the average, being a measure of a central point for the data set. However, the median uses the middle of the data if it were arranged in order. The median might be

a preferred measure over average if there are data values outside the typical values. If three claims are billed at $100, $150, and $14,700, the average would result in ($100+$150+$14,750) / 3 = $5,000. A median or midpoint would be the center or $150. This might be a more reasonable value despite the one outlier value. In cases where there are an even count of records, the average of the middle two records would be used for the median.

- Mode: The mode is the number that occurs most often in terms of count. If the claim billed amount of $150 appears 75 times out of 100 records, the mode would be $150.

- Minimum: The minimum represents the smallest data value for the variable in the data set. If three claims were billed at $100, $150, and $200, the minimum value would be $100.

- Maximum: The maximum represents the largest data value for the variable in the data set. If three claims were billed at $100, $150, and $200, the maximum value would be $200.

- Range: The range represents the smallest and largest data values for the variable in the data set. If three claims are billed at $100, $150, and $200, the range of values would be $100-$200.

- Variance: The variance represents a measure of the distance between the actual value and the average value for the data.

- Standard Deviation: The standard deviation represents a measure of the distance between the value and the average value for the data, using the variance, and is the square root of the variance. If three claims are billed at $100, $150, and $200, the standard deviation would be $50.

- Outliers: The outliers represent data values outside the norm. As a rule of thumb, any data values outside two standard deviations from the mean can be considered as outliers. In some cases, these might be valid and require secondary inspection. In other cases, these might be the result of data entry errors or miscoded data values. An outlier is a value that is typically several standard deviations outside the average or mean. In a population distribution, 67% of all records are within two standard deviations of the mean, 95% are within three standard deviations of the mean, and 99% are within four standard deviations.

- Skewness: Measures the symmetry of the distribution of the data set. If a data set is normally distributed, the skewness is zero. A typical range for skewness is -2 to 2. If we say that the data set is skewed left, there will be a negative value for skewness, and the left side tail is longer than the right side tail. If we say that the data set is skewed right, there will be a positive value for skewness, and the right side tail is longer than the left side tail (NIST/SEMATECH, 2017).

- Kurtosis: Measures the tail of the distribution, or the number of outliers as compared with a normal distribution. In other words, a high level of kurtosis indicates a higher level of outliers. Low kurtosis indicates an absence of outliers (NIST/SEMATECH, 2017).

Explore Tab Enterprise Miner Node Descriptions

Stat Explore Node

The Stat Explore node is used to display the descriptive statistics of a data set, including univariate, bivariate, and correlation.

Figure 4.4: Stat Explore Node

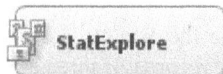
StatExplore

Graph Explore Node

The Graph Explore node is used for graphical or visualization of the data. The node can be used to create additional graphs such as pie charts, line charts, or box plots to detect outliers.

Figure 4.5: Graph Explore Node

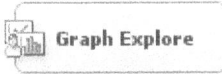

Graph Explore

Multiplot Node

The Multiplot node is used for graphical representation (or visualization) of the data to identify information, such as histogram distributions, to detect skewness and kurtosis. Histograms are an effective charting method for determining skewness and kurtosis or for identifying outliers.

Figure 4.6: MultiPlot Node

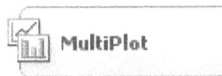

MultiPlot

Experiential Learning Activity: Health Data Surveillance

Health Data Surveillance

Description: The CDC (Centers for Disease Control and Prevention) analyzes BRFSS (Behavioral Risk Factor Surveillance System) data for metropolitan and micropolitan statistical areas (MMSAs) to provide localized health information that can help public health practitioners identify local emerging health problems, plan and evaluate local responses, and efficiently allocate resources to specific needs. The BRFSS is a system established in1984 that collects information about U.S. residents through telephone surveys. Completing more than 400,000 surveys per year, the BRFSS is the largest continuously conducted health survey system (CDC, 2016a).

https://www.cdc.gov/brfss/

From the BRFSS page, click Prevalence Data and Data Analysis Tools, and then click Prevalence and Trends Data.

Review the page section on the right labeled Explore BRFSS Data by Topic.

Explore the Data by Class and Topic -> Select a Class Category (for example, Cholesterol Awareness) and Topic Sub-Category (for example, Cholesterol High), Click GO.

The chronic disease indicators (CDI) allows public health professionals and policy makers to access standard data for chronic diseases and risk factors in order to evaluate public health interventions. The CDI is a joint effort by the Centers for Disease Control (CDC) and state health departments to develop a standard set of health indicators. The CDI includes 124 indicators within 18 category groups: alcohol, arthritis, asthma, cancer, cardiovascular disease, chronic kidney disease, chronic obstructive pulmonary disease, diabetes, disability, immunization, mental health, nutrition, older adults, oral health, overarching conditions, reproductive health, school health, and tobacco. Explore the state data, describe the data, Explore the data in geographic, table, and graph format. Describe the data, and recommend an action plan for public health practitioners.

List the category and factor that you are reviewing.

List the data variables included in the output.

Which states have the highest or lowest rates.

Provide a summary action plan recommendation for public health practitioners based on your findings above (priorities, locations, best practices, methods, and so on).

SEMMA: Modify

Modify Process Step Overview

During the third process step, data is modified after detecting any anomalies during the exploration process step. Often this step corresponds with correcting poor data quality. The data is updated to make the results more accurate, or the data is standardized to make the input more consistent. Although these functions and capabilities are available to correct data during the process, best practices would seek to have high levels of data quality in the original source data, thereby eliminating the requirement to further correct data during analysis. The stage in which data modification occurs can be important since model results may change based on the data updates during the modification stage. Because the source system data can also be used for a variety of purposes, such as patient-physician interaction, you would not want to wait until analysis to clean the data and ensure accuracy.

Figure 4.7: Modify Process Step

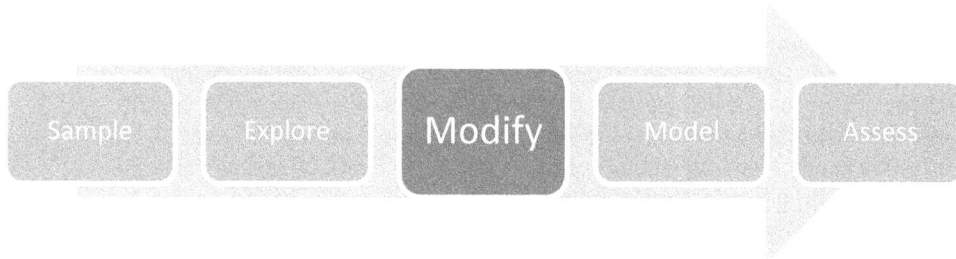

Sample   Explore   Modify   Model   Assess

Modify Tab Enterprise Miner Node Descriptions

Impute Node

The Impute node is used to correct and replace missing values. The node would relate to the complete aspect of data quality. The missing values can be replaced using the mean or average of the values in the data set or the count. The count would be the item with the most frequency and would be used for the missing value.

Figure 4.8: Impute Node

Impute

Interactive Binning Node

The Interactive Binning node allows the creation of bins. For example, an age variable that is specified as an interval (such as 1,2,3,…100,) can be binned into sets of four age bins (or groups) such as 0-25, 26-50, 51-75, 76-100.

Figure 4.9: Interactive Binning Node

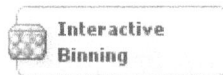

Interactive Binning

Replacement Node

The Replacement node is used to interactively replace missing values for both class and interval variables. The replacement node can be used to update either missing or non-missing data, such as outliers or miscoded data, before running an impute node for missing values.

Figure 4.10: Replacement Node

Replacement

Transform Variables Node

The Transform node is used to generate new variables or Modify existing variables. Transform can be used to standardize variables to improve the overall model fit.

Figure 4.11: Transform Variable Node

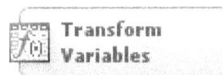

Transform Variables

Experiential Learning Application: Heart Attack Payment Data

Myocardial infarction, more commonly known as a heart attack, occurs when one of the arteries carrying oxygenated blood to the heart becomes blocked and the oxygen-starved heart tissue dies. In the U.S., over

70,000 people have a heart attack each year, with the majority as a first heart attack (CDC, 2017a; WebMD, 2017). About 20-25% of heart attacks occur without warning, others have preceding chest pain, shortness of breath, dizziness, faintness, or nausea. Men and women can also experience different symptoms or rates. Women are more likely to experience heart attack symptoms relating to indigestion, nausea, or fatigue (Davis, 2018). A study in Quebec, Canada, between 1981 and 2014, measured the quantity and duration of snowfall to calculate the relationship between snowfall and myocardial infarction. The results found that an increase in myocardial infarction was found after increased snowfall among men but not among women. The researchers concluded that men might be at greater risk due to their propensity to shoveling snow. In particular, after heavy and continuous snowfalls, shoveling snow is a demanding exercise, with risks increased due to cold temperatures that affect blood pressure and blood flow (Auger et al., 2017).

The state-level data includes payment measures for 30-day episode of care filtered for 2,429 heart attack patients (CMS, 2017b).

Data Set Files:

- 4_EL1_Heart_Attack_Payment_Data.xlsx
- 4_EL1_Heart_Attack_Payment_Missing_Data.xlsx

Variables:

- Provider ID
- Hospital Name
- Address
- City
- State
- ZIP code
- County Name
- Phone Number
- Measure Name
- Measure ID
- Compared to National
- AvgPayment
- Measure Start Date
- Measure End Date

Step 1: Sign in to SAS Solutions On Demand, or Open SAS Enterprise Miner Local Installation.

Step 2. Open the SAS Enterprise Miner Application (click the SAS Enterprise Miner link).

Step 3. Create a New Enterprise Miner Project (click **New Project**).

Step 4: Use the default SAS Server, and click **Next**.

Step 5: Add Project Name *HeartAttackPayment*, and click **Next**.

Step 6: SAS will automatically select your user folder directory that is using SAS Solutions On Demand and click **Next**. (If you are using a local installation version, choose your folder directory, and click **Next**.)

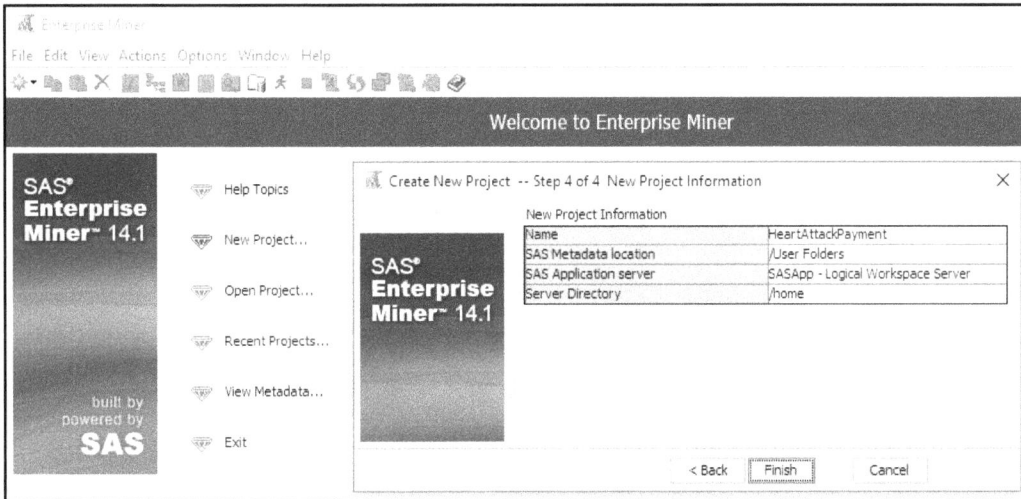

Step 7: Create a new diagram (Right-click **Diagram**).

Figure 4.12: Create New Diagram Step 1 of 2

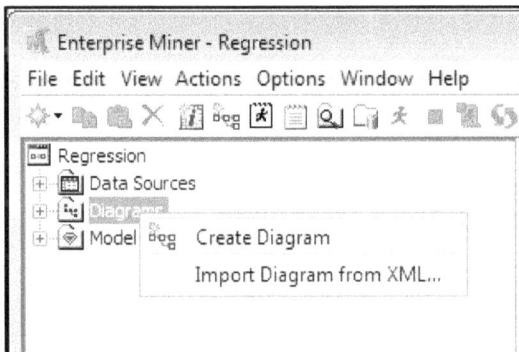

Figure 4.13: Create New Diagram Step 2 of 2

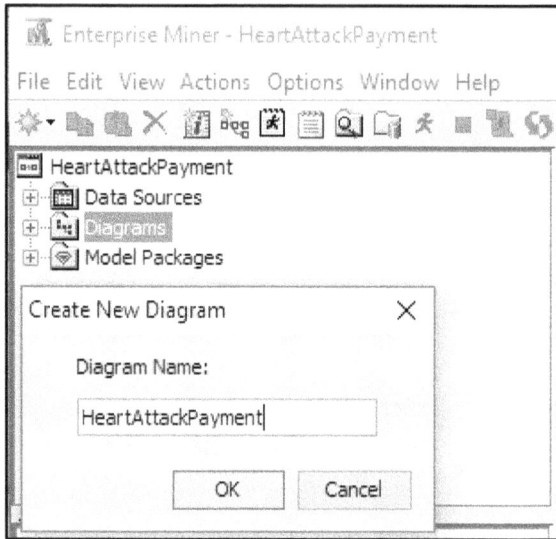

Step 8: Add a File Import node (Click the **Sample** tool tab, and drag the node into the diagram workspace).

Figure 4.14: Add a File Import Node

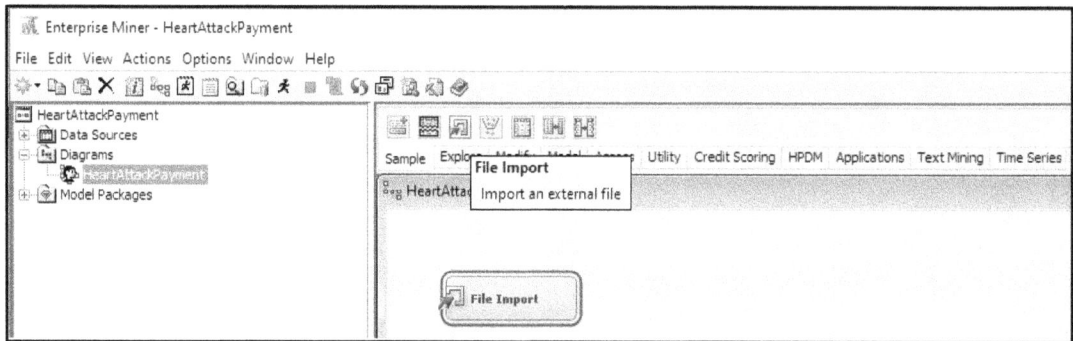

Step 9: Click the File Import node, and review the property panel on the bottom left of the screen.

Step 10: Click **Import File** … and navigate to the Chapter 4 Excel File
4_EL1_Heart_Attack_Payment_Data.xlsx.

Figure 4.15: Browse for File

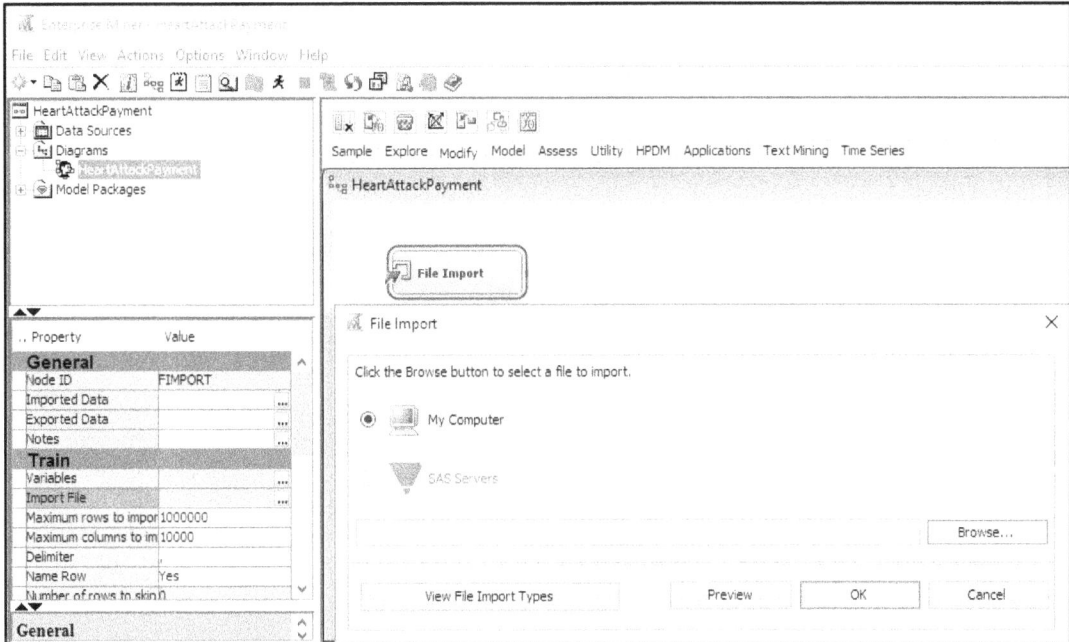

Step 11: Click **Preview** to ensure that the data set was selected successfully, and click **OK**.

Step 12: Right-click the File Import node and click **Edit Variables**.

Figure 4.16: Edit Variables

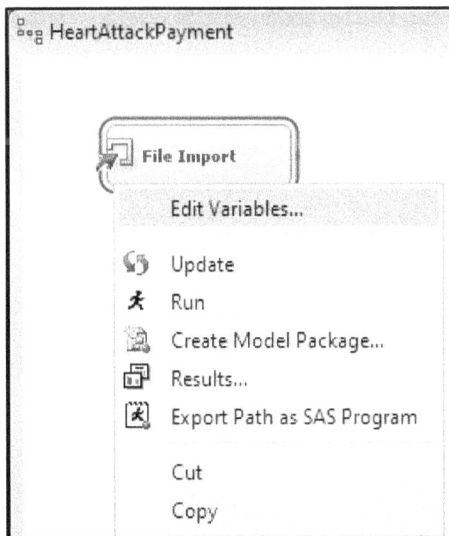

Step 13: Set the Measure_Start_Date and Measure_End_Date to the Time ID Role, and set all other variables to the Input Role. Ensure that the roles and levels for each variable were assigned correctly. You can also click each individual variable and then click **Explore** to review the data further for each variable name. After you have completed your review, click **OK**.

Figure 4.17: Edit Variable Roles

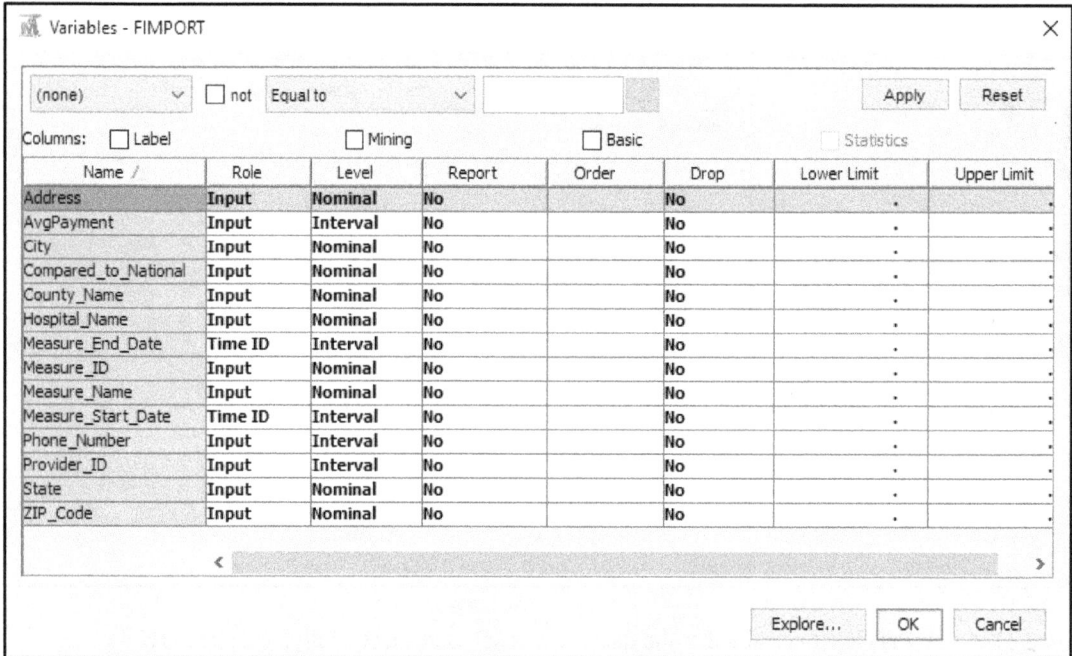

Name /	Role	Level	Report	Order	Drop	Lower Limit	Upper Limit
Address	Input	Nominal	No		No	.	
AvgPayment	Input	Interval	No		No	.	
City	Input	Nominal	No		No	.	
Compared_to_National	Input	Nominal	No		No	.	
County_Name	Input	Nominal	No		No	.	
Hospital_Name	Input	Nominal	No		No	.	
Measure_End_Date	Time ID	Interval	No		No	.	
Measure_ID	Input	Nominal	No		No	.	
Measure_Name	Input	Nominal	No		No	.	
Measure_Start_Date	Time ID	Interval	No		No	.	
Phone_Number	Input	Interval	No		No	.	
Provider_ID	Input	Interval	No		No	.	
State	Input	Nominal	No		No	.	
ZIP_Code	Input	Nominal	No		No	.	

Step 14: Add a StatExplore node (Click the **Explore** tab, and drag the node into the diagram workspace).

Figure 4.18: Add StatExplore Node

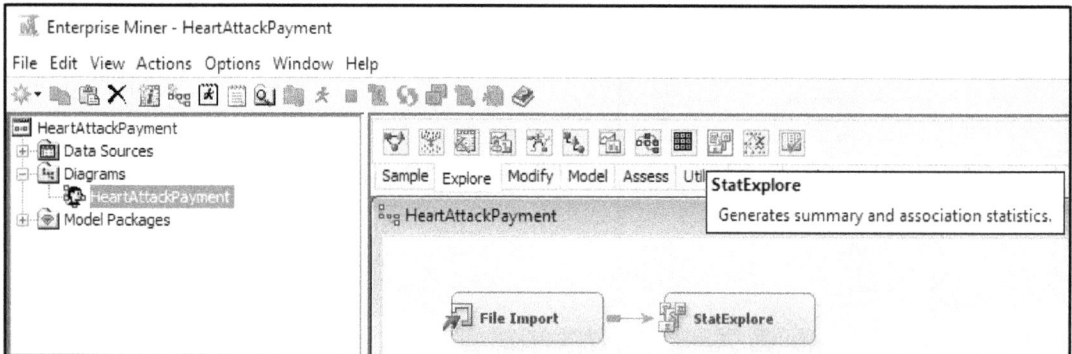

Step 15: Connect the File Import node to the StatExplore node.

Step 16: Right-click the StatExplore node and click **Run**.

Step 17: Expand the Output window results.

Figure 4.19: StatExplore Results

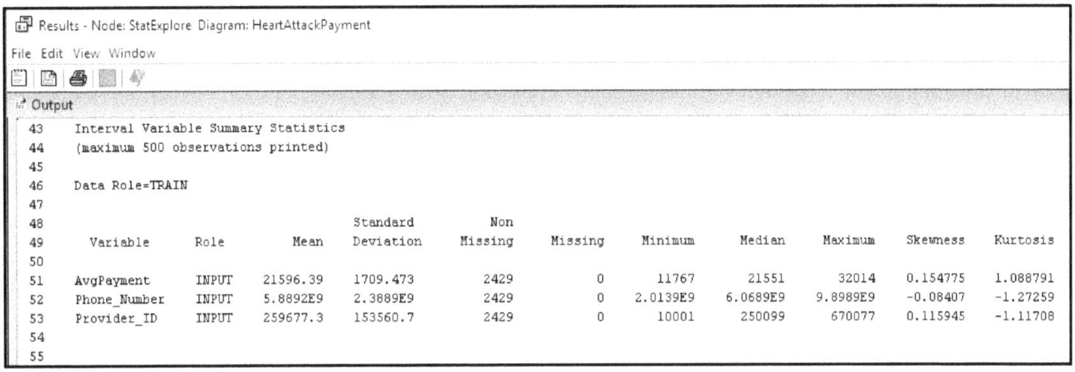

```
Results - Node: StatExplore Diagram: HeartAttackPayment
File  Edit  View  Window

Output
43    Interval Variable Summary Statistics
44    (maximum 500 observations printed)
45
46    Data Role=TRAIN
47
48                              Standard        Non
49    Variable      Role    Mean  Deviation  Missing  Missing  Minimum   Median   Maximum  Skewness  Kurtosis
50
51    AvgPayment    INPUT  21596.39  1709.473    2429        0    11767    21551    32014  0.154775  1.088791
52    Phone_Number  INPUT  5.8892E9  2.3889E9    2429        0  2.0139E9  6.0689E9  9.8989E9  -0.08407  -1.27259
53    Provider_ID   INPUT  259677.3  153560.7    2429        0    10001   250099   670077  0.115945  -1.11708
54
55
```

Step 18: Complete Table 4.2 with the StatExplore descriptive measures.

Table 4.2: Descriptive Measures

Measures	Value
Count of Providers	
AverageHeartAttackPayment	
MedianHeartAttackPayment	
MinHeartAttackPayment	
MaxHeartAttackPayment	
RangeHeartAttackPayment	
StdDevHeartAttackPayment	
VarianceHeartAttackPayment	
Skewness	
Kurtosis	
ProviderProportionwithGreater than the National Average Payment	

Step 19: As most individuals are familiar with Excel for computing descriptive measures, a summary of an Excel output has been provided for comparison. As an option, you can also take the data set and run the Excel formulas to compare the results to SAS Enterprise Miner.

Figure 4.20: StatExplore Excel Comparison

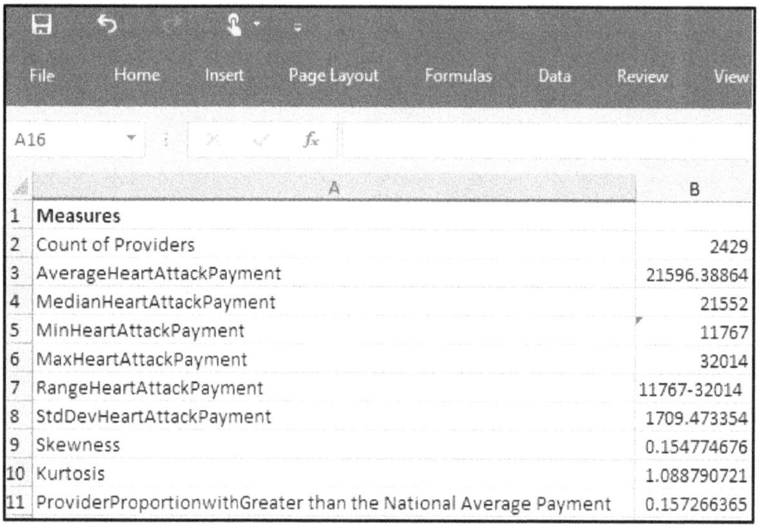

	A	B
1	Measures	
2	Count of Providers	2429
3	AverageHeartAttackPayment	21596.38864
4	MedianHeartAttackPayment	21552
5	MinHeartAttackPayment	11767
6	MaxHeartAttackPayment	32014
7	RangeHeartAttackPayment	11767-32014
8	StdDevHeartAttackPayment	1709.473354
9	Skewness	0.154774676
10	Kurtosis	1.088790721
11	ProviderProportionwithGreater than the National Average Payment	0.157266365

Heart Attack Payment – Missing Data

Step 20: Next, let us import a similar file, but one that has missing data. Add a File Import node (click the **Sample** tab, and drag the node into the diagram workspace). Right-click and rename the node as File Import Missing Data.

Figure 4.21: File Import Node Missing Data

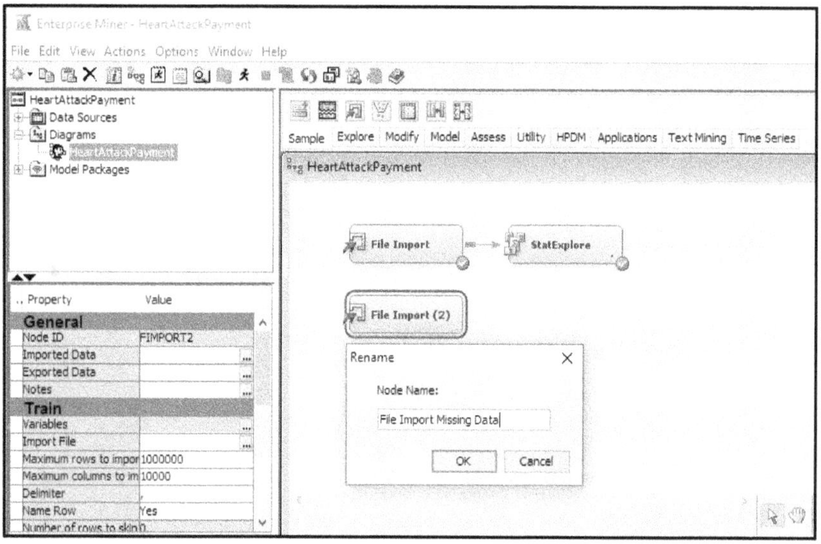

Step 21: Click the File Import node, and review the property panel on the bottom left of the screen.

Step 22: Click **Import File** … and navigate to the Chapter 4 Excel File
4_EL1_Heart_Attack_Payment_Missing_Data.xlsx.

Step 23: Add a StatExplore node. Right-click and rename the node as StatExplore Missing Data.

Figure 4.22: StatExplore Node Missing Data

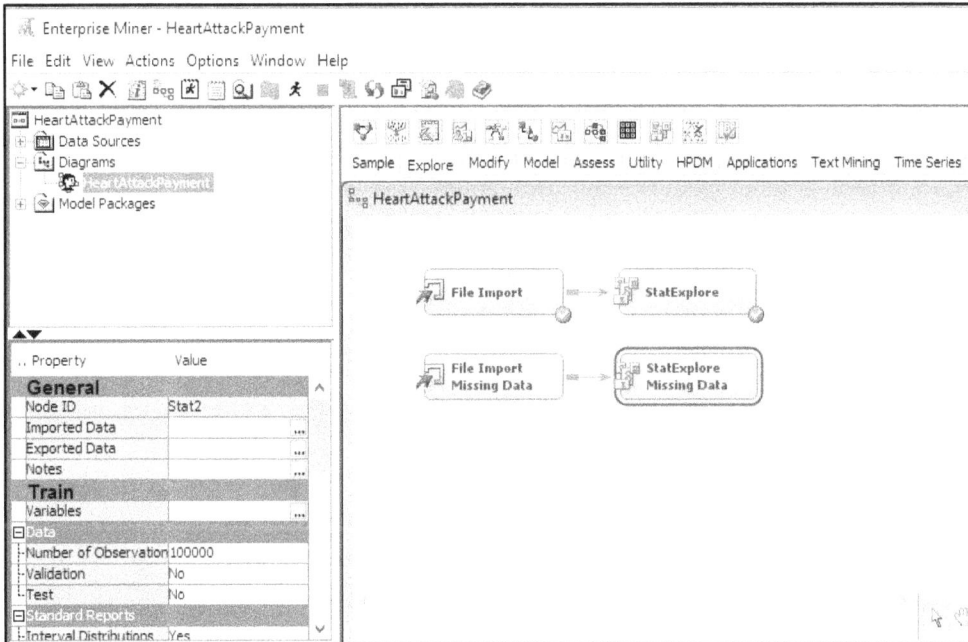

Step 24: Re-run the same file import and StatExplore node and see how the totals change due to the missing data. For the AvgPayment variable, we see that there are 14 missing values. The Skewness and Kurtosis is also high indicating that there might be other data quality issues present such as outliers.

Figure 4.23: StatExplore Results

```
43    Interval Variable Summary Statistics
44    (maximum 500 observations printed)
45
46    Data Role=TRAIN
47
48                                    Standard      Non
49    Variable      Role     Mean     Deviation   Missing   Missing   Minimum   Median    Maximum   Skewness   Kurtosis
50
51    AvgPayment    INPUT   435659.4  20348484    2415      14        11767     21545     1E9       49.14265   2415
52    Phone_Number  INPUT   5.8892E9  2.3889E9    2429      0         2.0139E9  6.0689E9  9.8969E9  -0.08407   -1.27259
53    Provider_ID   INPUT   259677.3  153560.7    2429      0         10001     250099    670077    0.115945   -1.11708
54
```

Step 25: Add a GraphExplore node. Right-click and rename the node as GraphExplore Missing Data. We are adding this node to further Explore the data quality issues.

Figure 4.24: GraphExplore Missing Data

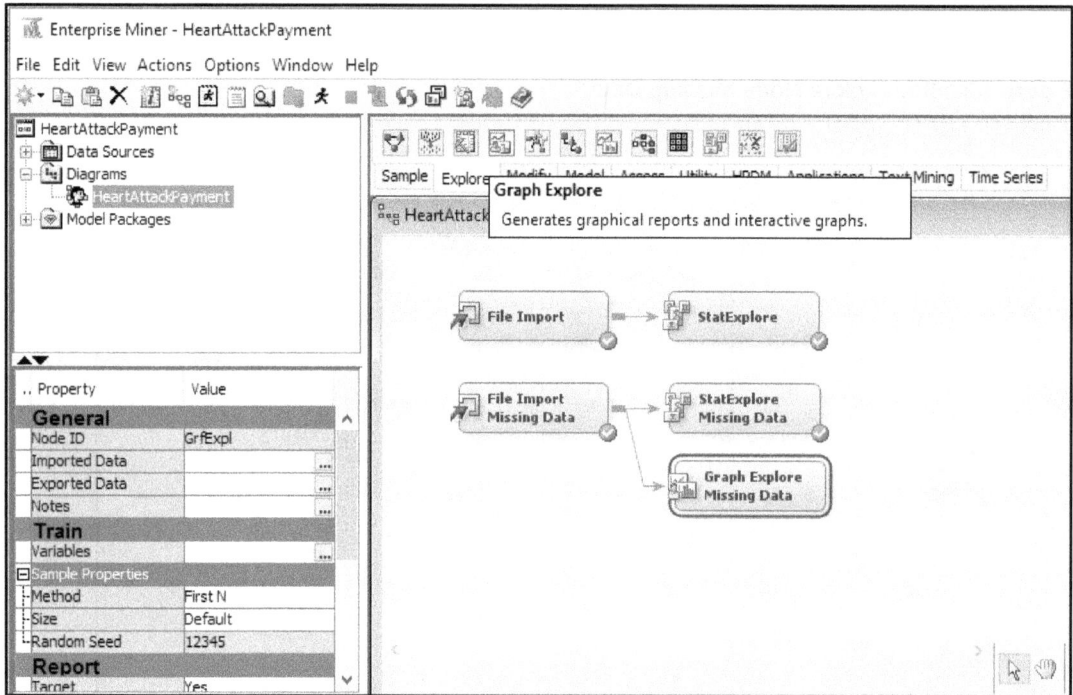

Step 26: Click the Graph Explore Missing Data node. On the bottom left, set the property **Size** to **Max**. The setting will ensure that all the data is selected for the graph to be explored. Otherwise, only a sample subset would be selected.

Figure 4.25: GraphExplore Missing Data Size Property

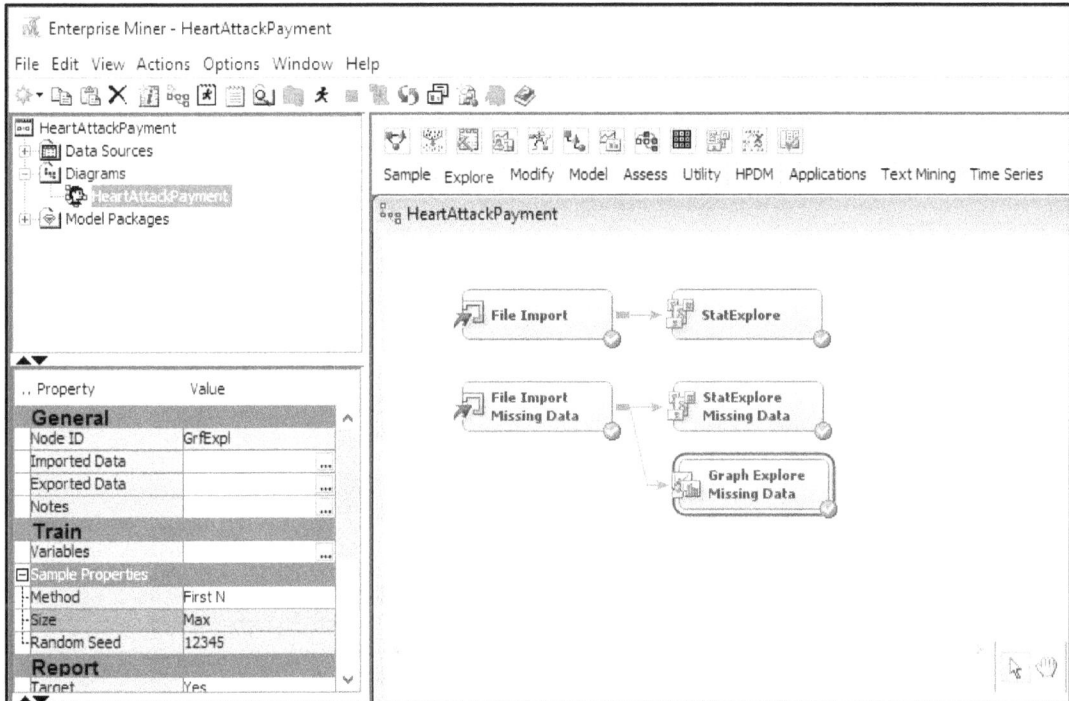

Step 27: Select **View** ▶ **Plot**. We want to create a visualization for the data set.

Figure 4.26: GraphExplore Missing Data Plot

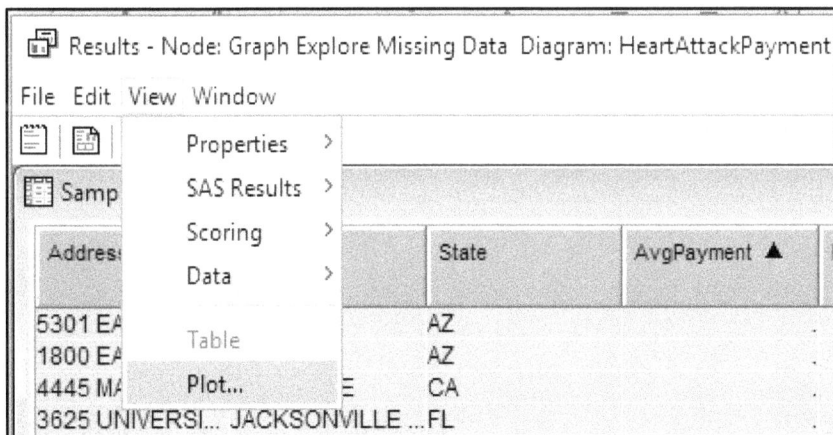

Step 28: Select the Box chart type. This will allow for more easily visualizing any outliers.

Figure 4.27: GraphExplore Missing Data Box Chart Type

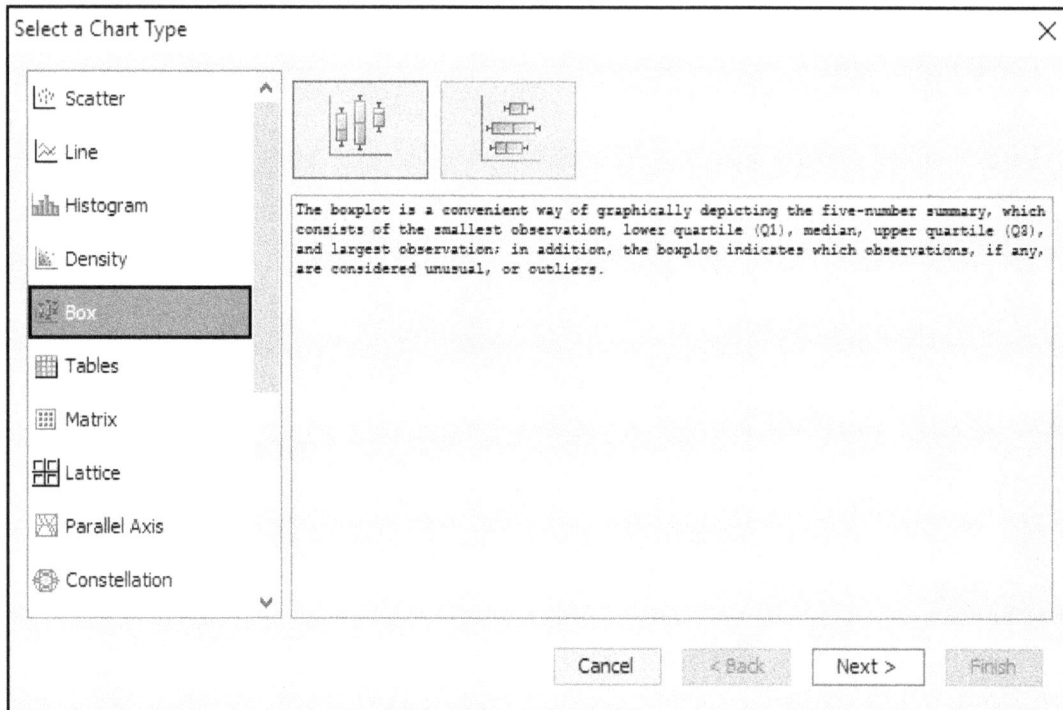

Step 29: Select the role of **Y** for the **AvgPayment** variable. We want to view the box plot for the average payment. Click **Finish**.

Figure 4.28: GraphExplore Missing Data Box Chart Type Role

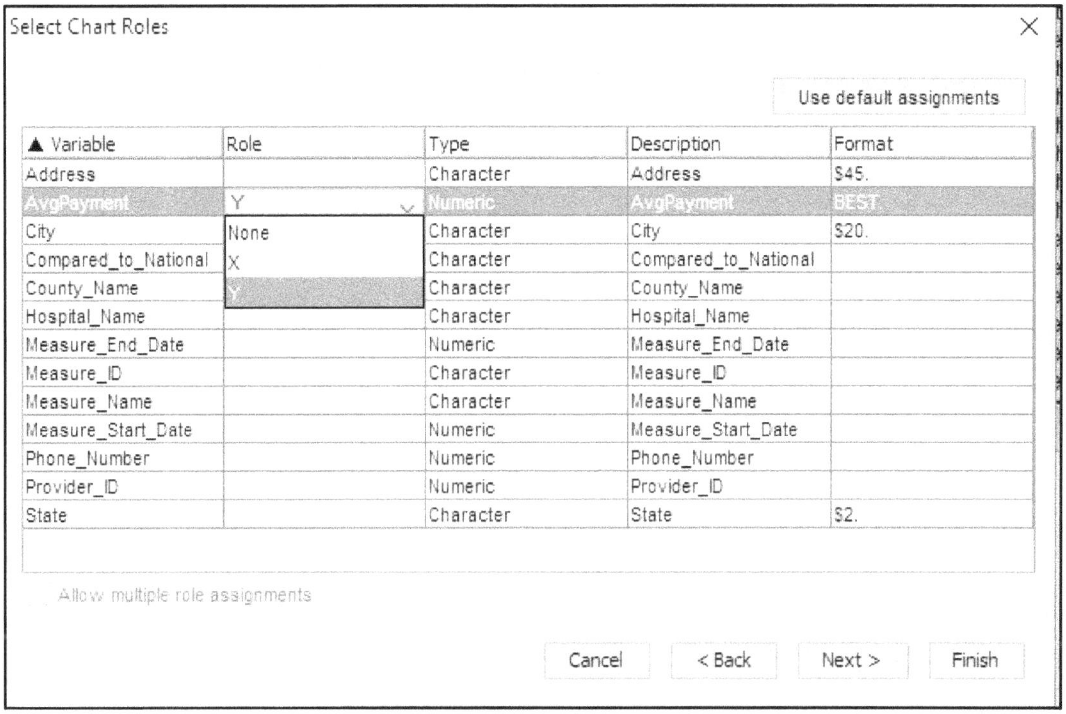
Select Chart Roles
Use default assignments
Variable
Role
Type
Description
Format
Address
Character
Address
$45.
AvgPayment
Y
Numeric
AvgPayment
BEST.
City
None
Character
City
$20.
Compared_to_National
X
Character
Compared_to_National
County_Name
Y
Character
County_Name
Hospital_Name
Character
Hospital_Name
Measure_End_Date
Numeric
Measure_End_Date
Measure_ID
Character
Measure_ID
Measure_Name
Character
Measure_Name
Measure_Start_Date
Numeric
Measure_Start_Date
Phone_Number
Numeric
Phone_Number
Provider_ID
Numeric
Provider_ID
State
Character
State
$2.
Allow multiple role assignments
Cancel
< Back
Next >
Finish

Step 30: Review the results. After review, we see there is one value that is significantly different from the others. If you place your pointer over the outlier value at the top, we see 9s. The outlier might have been a data entry error, a default value, or a conversion error from a system. Close any open windows and return to the main diagram.

Figure 4.29: GraphExplore Missing Data Box Chart Type Role

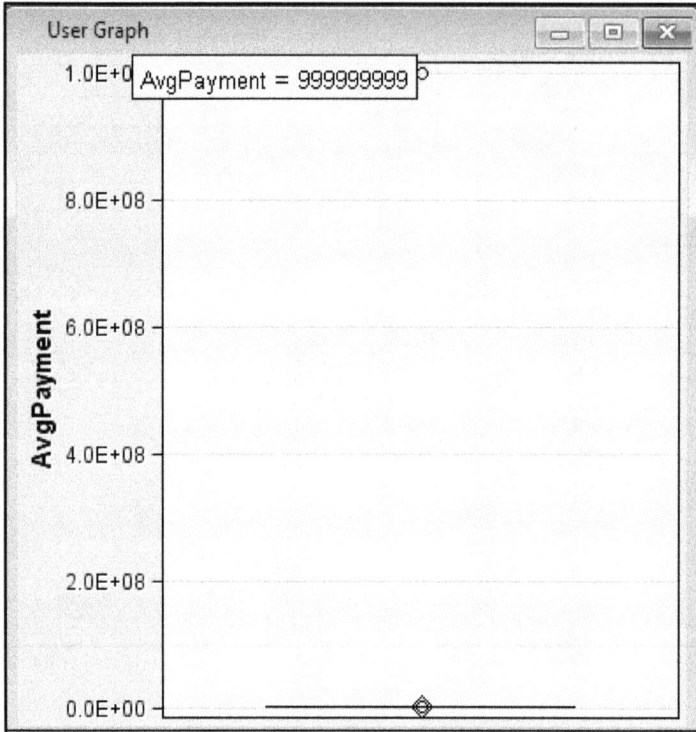

Step 31: Add a MultiPlot node to the diagram.

Figure 4.30: Add MultiPlot Node

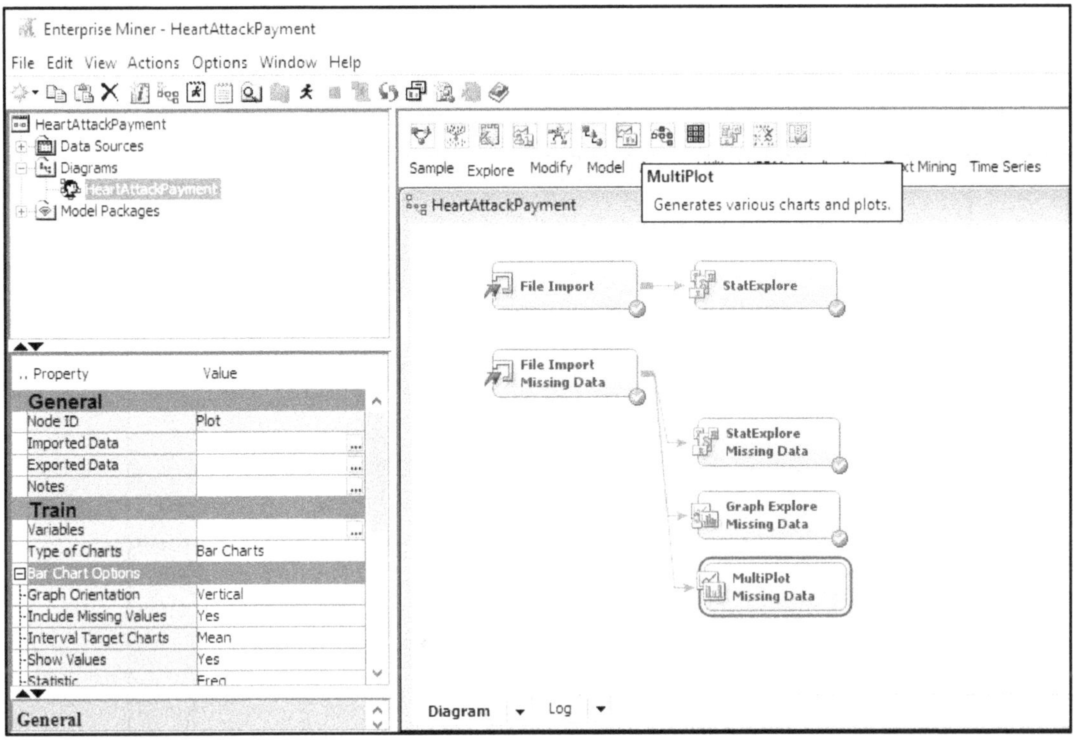

Step 32: Review the results. After review of the MultiPlot for AvgPayment, we see an outlier value on the far right. On the far left, we also see missing data as represented by a period (or dot) in 14 records. Close the windows to return to the main diagram, and we will work to clean up this data set.

Figure 4.31: Add MultiPlot Node Results

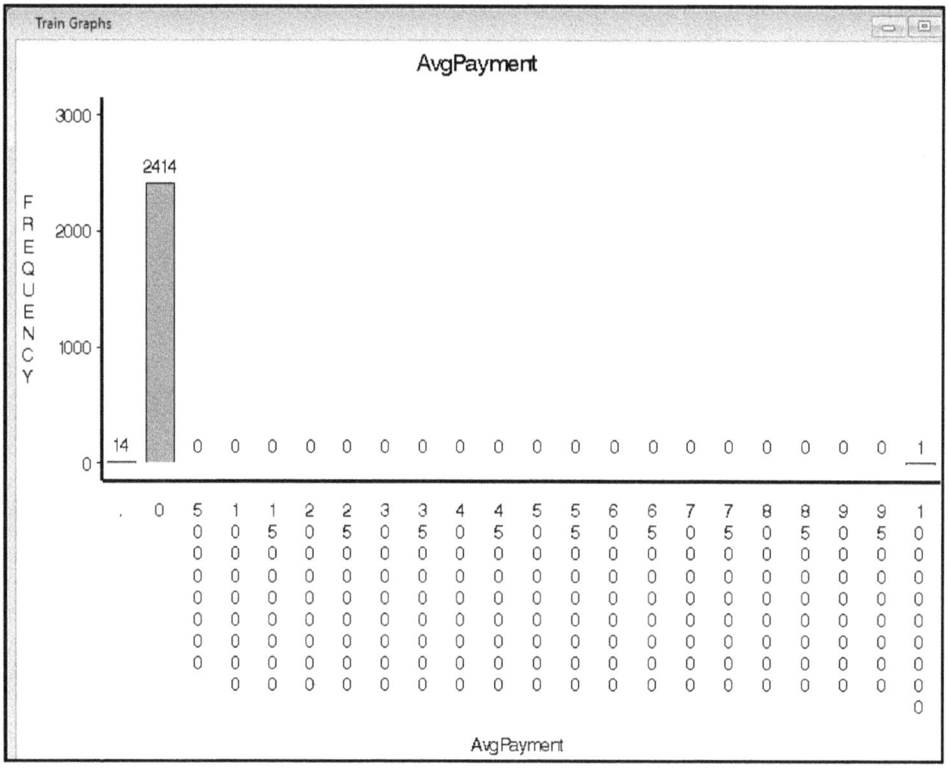

Step 33: Since our focus is on the AvgPayment variable, on the main diagram, right-click the File Import Missing Data node and click **Edit Variables**. Set all variables to the role Rejected, and keep AvgPayment as the only Input.

Figure 4.32: Update File Import Variables Missing Data

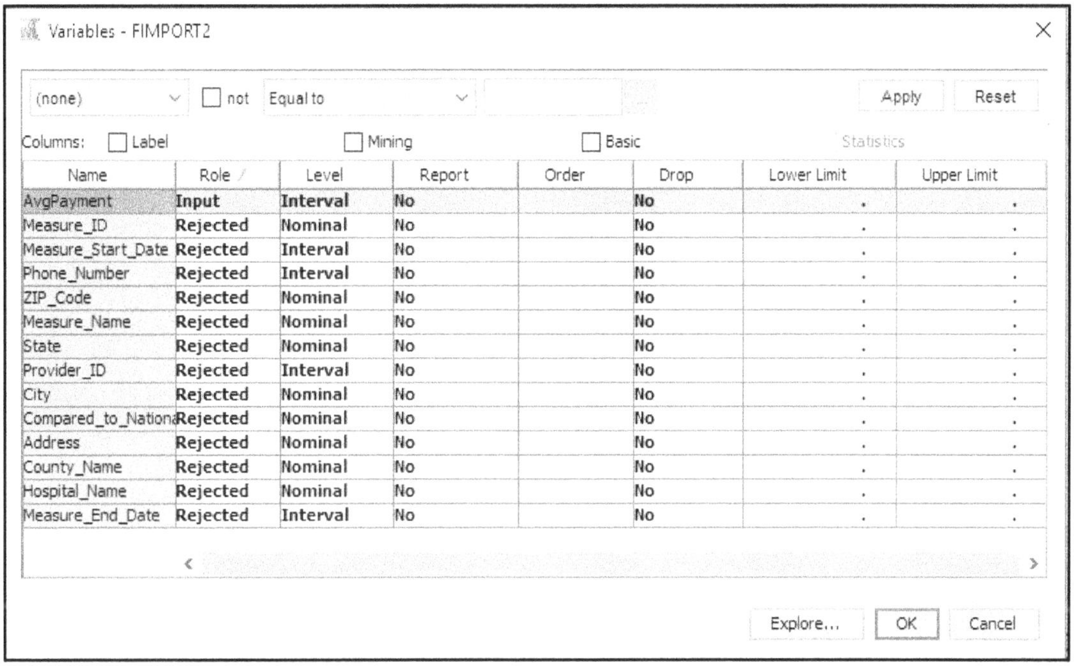

Name	Role	Level	Report	Order	Drop	Lower Limit	Upper Limit
AvgPayment	Input	Interval	No		No	.	.
Measure_ID	Rejected	Nominal	No		No	.	.
Measure_Start_Date	Rejected	Interval	No		No	.	.
Phone_Number	Rejected	Interval	No		No	.	.
ZIP_Code	Rejected	Nominal	No		No	.	.
Measure_Name	Rejected	Nominal	No		No	.	.
State	Rejected	Nominal	No		No	.	.
Provider_ID	Rejected	Interval	No		No	.	.
City	Rejected	Nominal	No		No	.	.
Compared_to_Nation	Rejected	Nominal	No		No	.	.
Address	Rejected	Nominal	No		No	.	.
County_Name	Rejected	Nominal	No		No	.	.
Hospital_Name	Rejected	Nominal	No		No	.	.
Measure_End_Date	Rejected	Interval	No		No	.	.

Step 34: First, let us add the Replacement node. We want to eliminate the extreme outlier value from our data set. In Figure 4.33 notice how the StatExplore Missing Data node and the Graph Explore Missing Data node were moved below Replacement. The nodes were moved below because each of these nodes can run independently. All nodes must follow in sequence after File Import Missing Data, since they are dependent on the file information to run.

Figure 4.33: Add Replacement Node

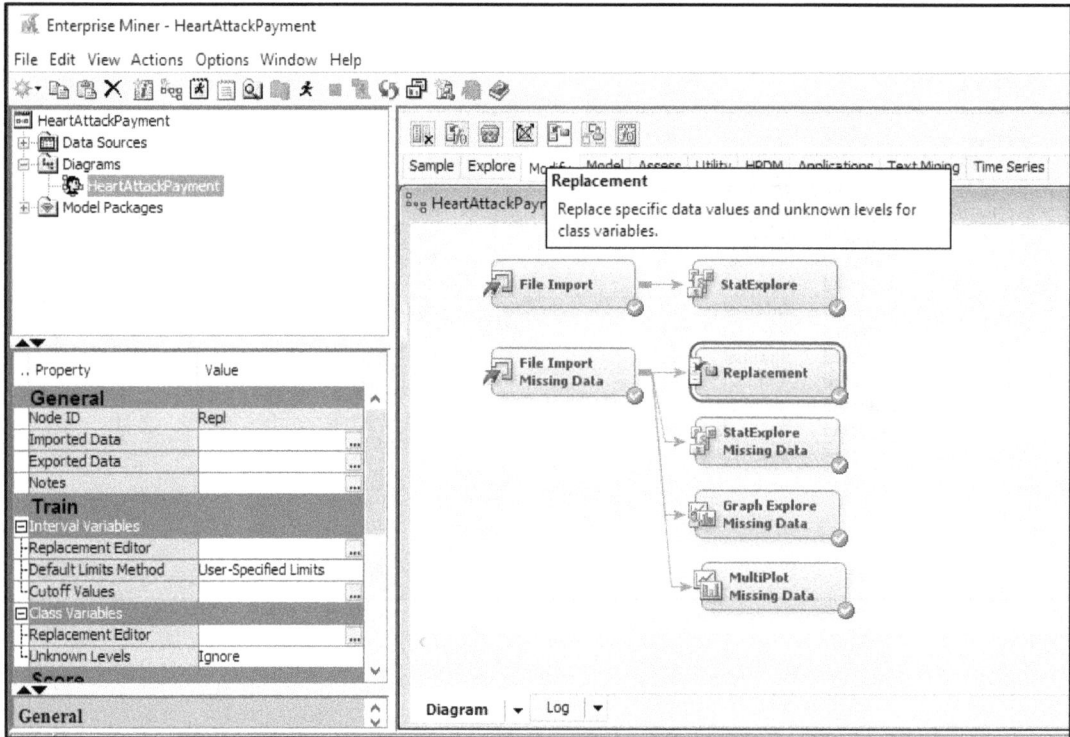

Step 35: Click the Replacement node, and set the **Default Limits Method** to **User-Specified Limits**. Then click the **Replacement Editor** property and click the ellipses. In order to eliminate the extreme outlier value from our data set, set the **Replacement Upper Limit** to **999,999**. The upper limit means anything at or above a $1 million average payment would be replaced. We want to replace this with a missing value, which we will clean up with the next step. Click **OK**.

Figure 4.34: Update Replacement Node Properties

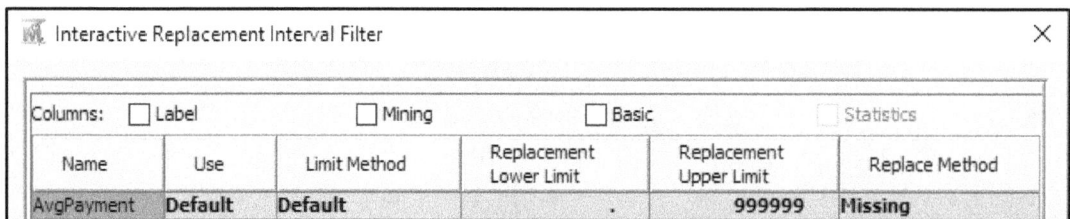

Interactive Replacement Interval Filter					✕
Columns: ☐ Label	☐ Mining		☐ Basic	☐ Statistics	
Name	Use	Limit Method	Replacement Lower Limit	Replacement Upper Limit	Replace Method
AvgPayment	Default	Default	.	999999	Missing

Step 36: Run the Replacement node and view the results to ensure that one record was replaced with a missing value.

Figure 4.35 Replacement Node Results

```
 Output
   44
   45     Replacement Counts
   46
   47     Obs     Variable      Label      Role      Train
   48
   49      1      AvgPayment    AvgPayment  INPUT       1
   50
```

Step 37: Next, let us add and run the Impute node, which will replace missing values.

Figure 4.36: Add Impute Node

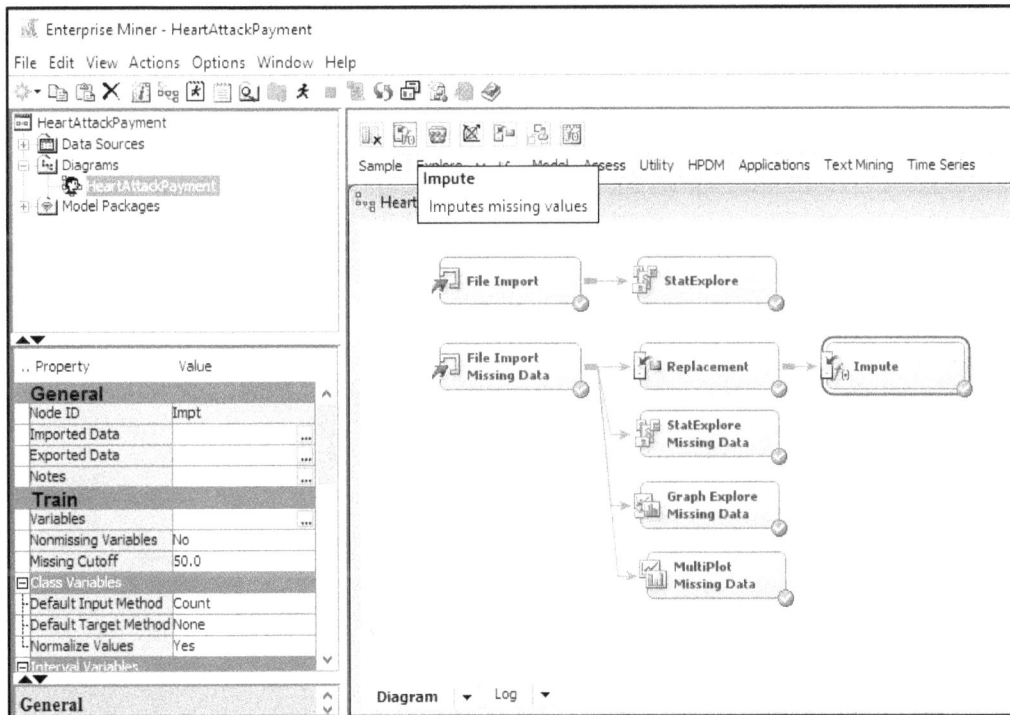

Step 38: Review the Impute node properties. From the properties, we see that there is a 50.0 missing cutoff, which means that if more than 50% of the values of a given variable are missing, the variable will be ignored. Also note the default methods for class variables and interval variables. For **Class Variables**, we see that the default input method is **Count**, which means that the value that has the greatest count will be used to replace a missing value. For **Interval Variables**, we see that the default input method is **Mean**, which indicates that the average (or mean) of the values will be used to replace missing values.

Figure 4.37: Review Impute Node Properties

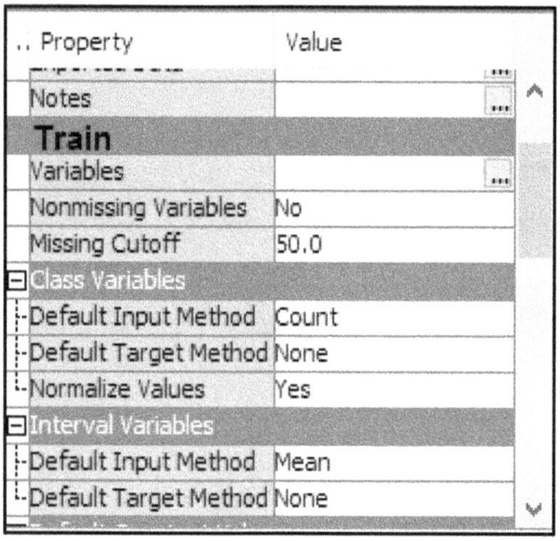

Step 39: Run the Impute node and view the results. From the results, a total of 15 variables were replaced with the average value of 21,589.70. The 15 includes the original 14 missing plus the 1 outlier that is replaced for a total of 15.

Figure 4.38: Review Impute Node Results

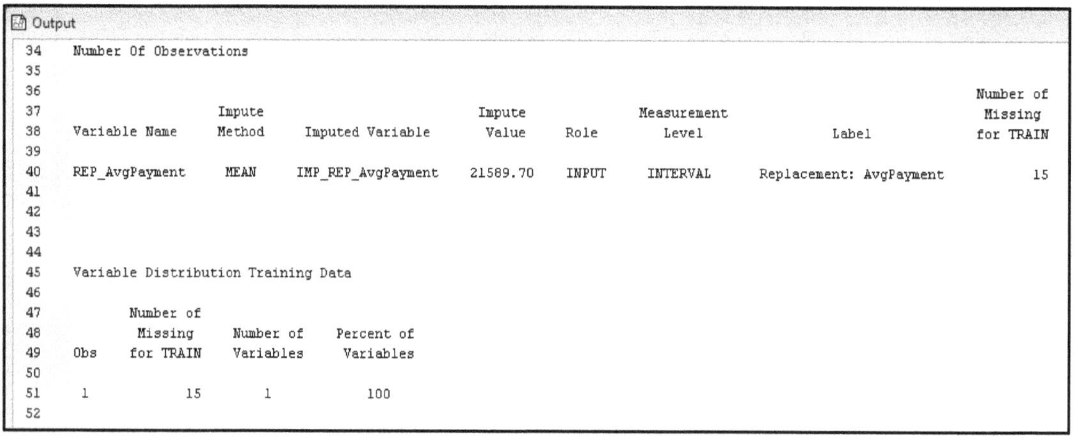

```
Output
Number Of Observations


                         Impute                            Impute             Measurement                              Number of
Variable Name            Method    Imputed Variable         Value    Role      Level        Label                      Missing for TRAIN

REP_AvgPayment           MEAN      IMP_REP_AvgPayment     21589.70   INPUT    INTERVAL      Replacement: AvgPayment          15




Variable Distribution Training Data

          Number of
           Missing     Number of    Percent of
Obs     for TRAIN     Variables     Variables

 1             15         1            100
```

Experiential Learning Application: Data Quality Exploration

The National Committee for Quality Assurance (NCQA) was founded in 1990 and is a private non-profit organization with the aim of improving health quality. Organizations that earn the NCQA seal symbolize quality and pass a rigorous review of over sixty standards with ongoing annual reporting on their performance in over forty areas. In the U.S., NCQA accredited organizations cover 109 million Americans. The Healthcare Effectiveness Data and Information Set (HEDIS) is used by 90% of the health plans in the U.S. to measure performance on healthcare quality and service. HEDIS contains eighty-one measures across five domains of care: effectiveness of care, access/availability of care, utilization, relative resource use, and health plan information. NCQA aggregates HEDIS results from health plans and providers through a questionnaire, which is online through an Interactive Data Submission System (IDSS). HEDIS is used to improve healthcare through performance statistics, benchmarks, and standards. HEDIS data is collected for Medicare, Medicaid, and commercial entities. For your exploration, you'll be using a HEDIS data subset for one measure on medication management compliance for asthma patients. The variables and descriptions are included below (NCQA, 2017a; NCQA, 2017b; NCQA, 2017c; NCQA, 2017d; NCQA, 2017e).

Data Set File: 4_EL2_HEDIS.xlsx

Variables:

- MemberID, the unique identifier of the patient.
- Age, the age in years.
- MemberMonths, the number of months that the member has been enrolled in the health plan for the current year, from 1-12.
- Gender, the gender assigned at birth, F - female, M - Male, O - if the member refused to provide gender. No other options are allowed.
- MedicationManagementandCompliance, The value 1 is used if the member is in compliance with their asthma medication management, and 0 is used if the member is not in compliance, or information about the member is missing.

Figure 4.39: SEMMA Process

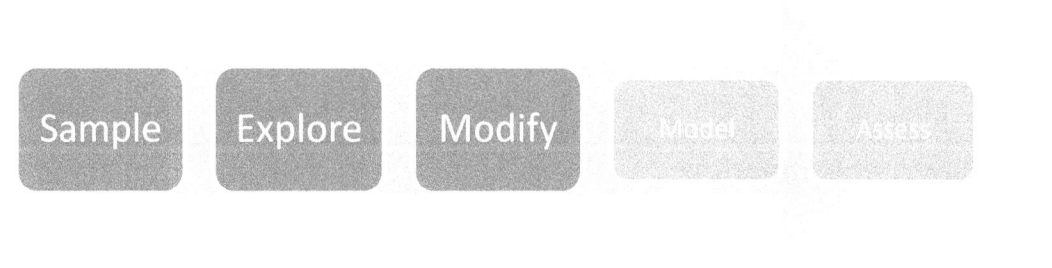

Title	Data Quality Exploration
Introduction	Provide a summary of the business problem or opportunity and the key objective(s) or goal(s). Create a new SAS Enterprise Miner project. Create a new Diagram.

Title	Data Quality Exploration
<u>S</u>ample	Identify data (sources for exploration and model insights). Identify the variables data types, the input, and the target variable during exploration. Add a FILE IMPORT. Provide a results overview following the file import. Input / Target Variables. Generate a DATA PARTITION
<u>E</u>xploration	Provide a results overview following data exploration. Add a STAT EXPLORE. Add a GRAPH EXPLORE Add a MULTIPLOT. Account for the following summary statistics (average, standard deviation, min, max, and so on): <ul><li>Count of Members/Gender/MedicationManagementCompliance</li><li>AverageAge/MemberMonths</li><li>MedianAge/MemberMonths</li><li>ModeAge/MemberMonths</li><li>MinAge/MemberMonths</li><li>MaxAge/MemberMonths</li><li>RangeAge/MemberMonths</li><li>StdDevAge/MemberMonths</li><li>Outliers Age/MemberMonths</li><li>Skewness Age/MemberMonths</li><li>Kurtosis Age/MemberMonths</li></ul>Use descriptive statistics. Account for missing data. Account for outliers.
<u>M</u>odify	Provide a results overview following the modification. Add an IMPUTE.

Learning Journal Reflection

Review, reflect, and retrieve the following key chapter topics only from memory and add them to your learning journal. For each topic, list a one sentence description/definition. Connect these ideas to something you might already know from your experience, other coursework, or a current event. This follows our three-phase learning approach of 1) Capture, 2) Communicate, and 3) Connect. After completing, verify your results against your learning journal and update as needed..

Key Ideas – Capture	Key Terms – Communicate	Key Areas - Connect
Healthcare Quality		
Healthcare Data Quality		

Key Ideas – Capture	Key Terms – Communicate	Key Areas - Connect
Six Sigma Quality		
Value Based Healthcare		
Meaningful Use: Explore Process Step Modify Process Step		
Public Data		
Descriptive Statistics		
Stat Explore Node		
Graph Explore Node		
Impute Node		
Interactive Binning Node		
Replacement Node		
Transform Variable Node		
Heart Attack Payments		
NCQA HEDIS		

Chapter 5: Modeling Patient Data

Chapter Summary

The purpose of this chapter is to develop data modeling skills using SAS Enterprise Miner, with respect to the Model capabilities within the SEMMA process. This chapter builds on previous chapters about the Sample, Explore, and Modify capabilities. This chapter explores patient-generated data sources and analysis with the linear regression model. This chapter also includes experiential learning application exercises on caloric intake, the mobile health (mHealth) heart rate app, and inpatient utilization costs. The Model focus of this chapter is shown in Figure 5.1.

Figure 5.1: Chapter Focus - Model

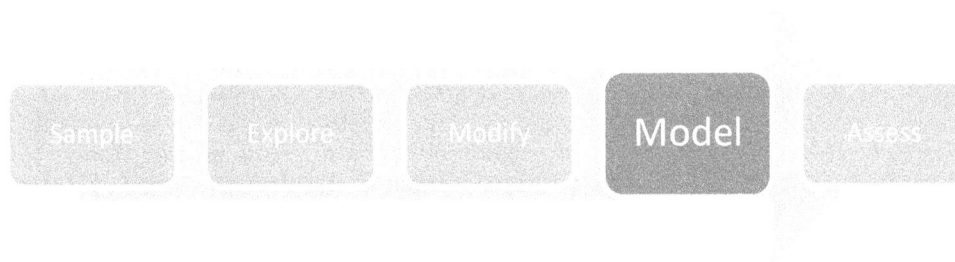

Chapter Learning Goals

- Describe the Model process steps
- Understand patient-level data sources
- Develop data modeling skills
- Apply SAS Enterprise Miner data model functions
- Master the linear regression model

Patients

A patient is an individual who is currently undergoing or awaiting a healthcare treatment, an individual who requires medical or dental care, or an individual who receives services directed at healing an illness or pain. The term, patient, comes from the Latin verb, *patior,* which means "to suffer through and tolerate pain" (MedicineNet, 2016). The term, patient, might also be extended to historical and prospective patients (such as patients who live near a medical facility). There are also many other names used interchangeably for patients, including members, subscribers, consumers, customers, and even guests. Alternative terms have been adopted from other user-focused industries, such as retail and tech, as the healthcare industry continues to evolve and healthcare organizations seek to improve the overall patient experience, patient engagement, and their "bottom line." Some hospital systems have even formed executive positions that are dedicated to this objective, such as Cleveland Clinic's and NYU Langone's Chief Experience Officers (Pearl 2015; MedicineNet, 2016; Whitman, 2017).

Patient Anamatics

Patients are increasingly generating healthcare data versus traditional collection methods such as hospitals, physician offices, or clinical settings. The data generation extends beyond seemingly standard healthcare data of clinical conditions. For example, patients post photos, share what they eat, track exercise progress, monitor health activities, ask healthcare-related questions within networks, among many other methods. Patient data can be analyzed and modeled to understand patterns, and provide data-driven information to patients. Common sources of patient data include personal health records, social media, mobile devices and wearable technology, and employee records and wellness systems.

Patient Data

Personal Health Records

Patients' private information can be stored, maintained, and managed securely through a personal health record (PHR). PHRs can be sponsored and developed by employers, insurance providers, or a third-party business. Proponents of PHRs point to the advantages of consumer-driven healthcare that lead to safer and more efficient care through secure and standardized tools. Healthcare data is typically entered directly by patients to include contact information, emergency contacts, diet plans, diagnosis, medical history, medications, allergies, immunizations, laboratory tests, and insurance, along with provider data (AHIMA, 2005; HealthIT.gov, 2013a; Kim et. al, 2002).

The functionality of PHRs intends to digitize storage of information, allow easy portability, organize and educate the individual patient on personal health information, facilitate decision-making, and allow expansion and adaptation over time. The PHR is compatible with existing health information standards, including vocabulary and standardized code sets, and contains privacy and security settings to ensure a high degree of data integrity (AHIMA, 2005). Early experiences with PHR systems, such as the discontinuation of Google's PHR solution, have created questions about long-term PHR viability, as patients are often more interested in maintaining their Facebook account than their health record account (Andrews, 2011). Facebook also leads us to the next data source, with social media as another common source of patient data.

Social Media

A report by The California Healthcare Foundation discusses the user-generated content movement, with social media technologies providing the platform for users to share health information and get support. Social media sites such as Facebook, Twitter, and YouTube have created a new delivery method for health information. The Centers for Disease Control (CDC) has used social media to allow people to share information about health threats at low cost. The CDC used Facebook, Twitter, YouTube, among others to spread information about upcoming H1N1 and other threats and epidemics (Wagner, 2009; Aikin, 2010). Over 33% of U.S. adults indicated that they use social media to search for health information, and studies have shown that the ability to access health information influences their behavior and health outcomes (Jha and Leesa, 2016).

The next generation of social healthcare is empowering and educating healthcare stakeholders. Social healthcare is defined as the use of social software and its ability to promote collaboration between patients, caregivers, medical professionals, and others in health. With social media, more individuals post stories, share experience, search for treatment information, communicate with physicians, and connect with others in health communities. Most communities have mechanisms to ask questions, add comments, and post user generated content (Yang, et. al, 2016). Individuals with chronic health conditions are sharing their stories for both emotional support and clinical knowledge that they gain, and their collective knowledge rivals that of many organizations. Individuals with chronic conditions can share stories for both emotional and clinical knowledge within a larger group community (Sarasohn-Kahn, 2008). In one reported case from Wisconsin, a woman slipped into a coma shortly after arriving to the emergency room. Physicians were able to find detailed accounts of the medications, symptoms, and conditions for several months on the patient's social media account, which were much more detailed than the current medical records. The patient made a full recovery, and Facebook was attributed to saving her life (Kovell, 2011).

Employee Records and Wellness Systems

Hypertension, diabetes, and cardiovascular diseases have been linked to increased working hours, on-the-job stress, and lack of nutrition and exercise. These factors have led to absenteeism and decreased productivity in the workplace. As a result, employers have realized the importance of employee wellness and have intervened on behalf of their employees in the areas or physical, emotional, and mental health to ensure a productive workforce (Kunte, 2016). Employee data includes personal information, financial information, along with health information completed through online health surveys, fitness trackers, biometric tests, and, in some cases genetic material (Wisenberg, 2016).

Wellness systems are offered through insurance carriers, such as Florida Blue, in an effort to save on healthcare costs and help patients live healthier lives through tools such as health assessments, trackers, symptom checkers, and health education tools (Florida Blue, 2015). Third-party software vendors, such as Limeade, also offer platforms for wellness and well-being in an effort to improve employee engagement and productivity (Limeade, 2016).

Digital Health, Mobile Devices, and Wearable Technology

The World Health Organization (WHO) defines Mobile Healthcare (mhealth) as a component of electronic health services and information delivered through mobile and wireless communication technologies such as cellphones, aimed at improving healthcare outcomes (HIMSS, 2012). At a broader level, digital health includes mHealth, health information technology (HIT), wearable devices, and telemedicine. Patients can use digital health to track their health and wellness and can use greater information access. The combination of people, information, technology, and connectivity is leading to improved healthcare outcomes (FDA, 2018). Historically, patients were largely limited to the single source of their personal physicians to receive information, monitor their conditions, manage their diseases, and create personalized plans of care. Today, healthcare technology has transformed this paradigm with wide accessibility of smart mobile devices along with convenience and cost factors. Patients can now gain access to health data from a variety of sources. In addition, patients can now monitor and manage their own conditions and diseases using mobile devices and software applications (Samsung, 2015). Prior to pervasive health information accessibility through mHealth capabilities, individuals had a 50-50 chance of receiving the recommended care for common illnesses (Mcglynn, 2003). These outcomes are changing with increased access to medical information, where patients have taken a greater role in healthcare management with an opportunity to reduce costs and improve their health. Increases in digital health technologies are driving the demand for a patient-centric and consumer-driven model, similar to music, travel, and publishing industry applications (Woodside, 2016b). One drawback, given the increasing reliance of patients on outside sources, is the inadvertent omission of the expertise of the personal physician (Samsung, 2015). Device manufacturers are also looking to extend health and wellness capabilities further. In one instance, Apple is reportedly developing technology to integrate glucose monitoring into their Apple Watch devices through the main device or alternatively through changeable wristbands (Geller, 2017).

Healthcare Technology Disruption

In the context of patient data, disruptive technologies are altering the healthcare environment. Disruptive technologies are those that are cheaper, simpler, smaller, and more convenient than existing technologies. Disruptive technologies and businesses in their early stages might include smaller margins, smaller target markets, and simpler products and services that at first do not appear to be sufficient substitutes for existing products and services. Due to the smaller market segments and margins, these markets are often ignored by other organizations, creating room for the disruptive organizations to emerge (Christensen, 2017).

In Healthcare, disruptive technologies, such as telemedicine, minute clinics and point-of-sale (POS) systems, are growing. One such disruptive item for patients is health communities in which crowd knowledge and access to online forums and mobile communications for sharing experiences and information prevail over clinical visits. Social media platforms are increasing disruption, where cost is low for both patient access and support, more convenient and simpler due to increasing numbers of users who are familiar with forms of social media from daily use (Woodside, 2012). Another potentially disruptive technology is telemedicine. UnitedHealthcare now allows a virtual doctor visitor to speak with a nurse directly through a mobile app. These methods can be used for minor health issues such as pink eye or sore throat. The physicians are able to diagnose and send prescriptions to local pharmacies (UnitedHealthcare, 2017). Nemours Children's Hospital also offers video doctor visits through a computer or mobile device. The service is available 24/7 in Florida, Maryland, and Pennsylvania and costs $49 per visit with a pediatrician, which might be covered by insurance (Nemours, 2017). The current telemedicine capability fits the definition of a disruptive technology due to its low cost and limited use. However, longer term, this has the potential to disrupt existing physician services within localities. A patient can now seek care from anywhere at any time, not limited to a local area, which might even include international care providers at lower costs. Although legal and industry regulations can prevent the disruption short-term, this might change over time as the technology and quality of care is accepted.

However, disrupting healthcare has proven difficult. Two new generation companies that are trying to disrupt healthcare and aimed at taking a patient-friendly and consumer-friendly approach to the healthcare industry are Oscar and Collective Health. Oscar describes itself as a healthcare provider of the next generation, selling insurance directly to consumers (which has some uncertainty due to potential changes in the ACA). Collective Health primarily focuses on patient employers for self-funded insurance and charges a monthly fee per employee to offer a complete health benefits solution and generate savings through data analytics and machine learning. As part of the solution, employees are offered digital tools, concierge-level support, and integrated access to healthcare services (O'Brien, 2017).

Patient Access Disruption

Patient access is one area that has a potential for disruption. Patient access is a major issue of health reform, and waiting for or not receiving care is one of the components of patient access. As a potential area to improve patient access, over 350,000 cardiac arrests occur within the U.S. each year in locations other than hospitals. As a result, the odds of surviving a cardiac arrest outside a hospital are about 1 in 10. Researchers are now experimenting with drones to deliver automated external defibrillators (AEDs) to save lives. In initial testing, drones were faster to patients than an ambulance by approximately 16.5 minutes, which could save lives with the time between the cardiac arrest and the shock from the AED. In rural areas outside a hospital, where the response time was greater than 30 minutes, the odds of survival from the cardiac arrest were 0%. One of the concerns still to be addressed is whether there would be a qualified medical professional or if bystanders would be able to successfully operate the AED (Columbus, 2017).

Mobile Health Disruption

Mobile health is another area with potential for disruption. The Tricorder XPrize is a $10 million global competition sponsored by Qualcomm that is intended to build a portable device that can monitor health conditions and diagnose thirteen health conditions, while capturing five health vital signs. The required core health conditions include anemia, atrial fibrillation (AFib), chronic obstructive pulmonary disease (COPD), diabetes, leukocytosis, pneumonia, otitis media, sleep apnea, and urinary tract infection, as well as an absence of any of these conditions. Additional monitoring areas include a cholesterol and HIV screen, along with identification of food-borne illness, hypertension, hypothyroidism/hyperthyroidism, melanoma, mononucleosis, pertussis (whooping cough), shingles, and strep throat. The required health vital signs that the device would capture include blood pressure, heart rate, oxygen saturation, respiratory rate, and temperature. The initial XPrize idea was developed in 2010 during a Visioneering summit to discuss and offer incentive grand challenges. This initiative is being driven by consumers with mobile devices who are able to monitor and improve their own healthcare anywhere and at any time. This makes healthcare more convenient, affordable, and accessible. The winning teams for the Human qualification milestone have been announced as Team Dynamical Biomarkers Group (Taiwan) and Team Final Frontier Medical Devices (U.S.). The teams are working toward the development of a multi-functional, portable device that can be used by anyone to help examine living things, diagnose diseases, and collect information about patients (Govette, 2017; Qualcomm, 2017).

Experiential Learning Activity: Personal Health Records

Let's continue the review of patient data with an experiential learning activity on PHRs. Review the description and answer the following questions.

PHR

Description: A personal health record (PHR) is a health technology that individuals can use in improving their own healthcare. Select PHRs are available through providers and payers that are covered under the Health Insurance Portability and Accountability Act of 1996 (HIPAA) Privacy Rule, and these providers and payers are covered under HIPAA are called covered entities. These covered entities must follow the guidelines within HIPAA. PHRs can also be offered by non-covered entities. Although they typically offer the same protections and can be called HIPAA-compliant, they are not required or guaranteed by the HIPAA law and can potentially share information with business partners as part of the typically free PHR platform. For covered entities, an individual's protected health information (PHI) cannot be shared unless permitted by the individual or in certain cases such as obtaining a service from the provider. The HIPPA privacy rule contains many protections, including a requirement to provide patients with the details of privacy practices and individual rights, and allows individuals to receive a detailed log of PHI disclosures by the covered entity up to six years earlier to determine how PHI has been disclosed. It also allows individuals to have updates or corrections made to their PHI and allows individuals to request a copy of their PHI from the covered entity (OCR, 2008).

Are you able access your PHR at your physician's office? What would be the expected process for access?

Describe one advantage and one disadvantage of having access to your PHR:

As a clinician, would you encourage or discourage your patients to provide, access, and update their PHR?

What are the implications and obstacles, if any, for using and accessing a PHR (security, privacy, legal, staffing, and so on)?

SEMMA: Model Process Step

Now that we've covered some of the basics with regard to patient data, we'll continue with our SEMMA process. We've covered the Sample, Explore, and Modify process steps, and now our next step will be to Model our data.

Model Application Examples

Linear regression can be applied to a variety of areas within the healthcare setting. Various international studies have used linear regression to analyze problems on air pollution in the U.S. (Cox, 2013), pulmonary functions in China (Ma, et al, 2013), maternal services in Ethiopia (2015), and factors influencing coronary heart disease in Singapore (Wang, et al, 2016). Linear regression can be used for problems with a continuous target variable (such as age or physical health score) and continuous or categorical input variables (such as income, other patient conditions, and education level).

Model Process Step Overview

During the modeling process step, the data mining model is applied to the data. During the partitioning phase, data is segmented into training and validation data sets. The training data set is used to fit the model, and the validation data set is used to validate the model on a new set of data to demonstrate the reliability of the model. Based on the results, the model can then be tuned to optimal performance. There are many different models that can be selected during this step. The decision to choose each model is based on earlier exploration of data and knowledge of each model. Throughout the text, you will learn several models. We begin this chapter with the model of linear regression.

Model Tab Enterprise Miner Node Descriptions

Regression Node

The Regression node is used for the linear regression model and is associated with the **Model** tab in SAS Enterprise Miner.

Figure 5.2: Regression Node

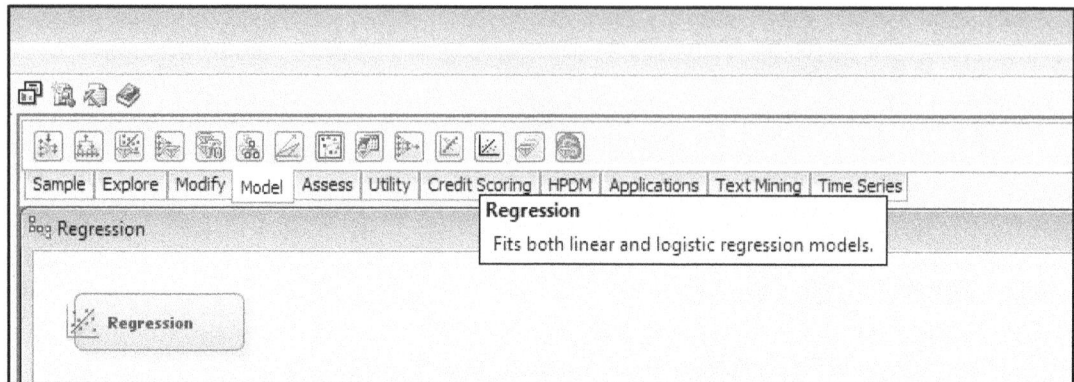

Linear Regression Model Description

Linear regression is a relatively simple, popular, and well-known statistical technique. The applicability to various problems and availability in a wide variety of software applications makes linear regression one of the most universally used techniques. At a basic level, regression is used to explain and measure the relationship between one or more inputs (which is also known as Xs or independent variables) with one target variable (which is also known as Y or a dependent variable). Multiple regression refers to a model that has more than one X (or input) or multiple inputs and one target, whereas simple regression is used to refer to a model with only one input and one target variable. In addition to helping us explain and measure the relationships between the input and target variables, regression can also measure the strength of the relationship, as well as determine whether the relationship is statistically significant and which input variables have the greatest impact on the target variable. Finally, the equation of a regression model can be used to predict results for new data records (Latin, Carrol, and Green, 2003).

- Simple linear regression: 1 Input ▶ 1 Target
- Multiple linear regression: 2+ Inputs ▶ 1 Target

Model Assumptions

Although regression models are effective, they can be affected by outlier values or extreme values. These outlier values are records that are several standard deviations outside of the mean and can be corrected through deletion or imputation of the data, during the exploration and cleaning process to ensure that the records do not affect final regression results. Other causes of outlier values can include data entry errors. Consider the example where a patient weight of 150 pounds accidentally has an extra zero added during data entry, which is now 1500 pounds. These entry errors can be easily corrected through computerized data checks and verification of outliers during the data exploration steps.

One model assumption is that the underlying independent data variables are uncorrelated. If the independent variables explain the same variability of Y, multicollinearity might occur, leading to interpretation difficulty. Multicollinearity is often difficult to detect initially. Although it can be identified by reviewing pairwise plots of input variables and correlation. If the correlation coefficient is greater than 0.70 for two input variables, this can indicate an issue. To address multicollinearity, we can add data, reduce the model by removing input variables, and finally conduct a principal component analysis that will result in uncorrelated inputs (Latin, Carrol, and Green, 2003).

Overfitting can also occur where we measure the model solely on the training sample data, and we attempt to perfect the model for the training data only. This can lead to cases where new data performs poorly, or we are unable to generalize the model beyond the initial data set. To resolve, we can reduce the model to the simplest best model or the parsimonious model. To verify overfitting, we can run the model against a validation set of data. If the error rate for the training data is significantly lower than the validation data, this can also indicate a case of overfitting (Latin, Carrol, and Green, 2003).

Data Preparation

To ensure an adequate sample size, as a general rule of thumb, we want to verify we have at least 25 records for each input variable. For example, if we have three inputs, we should have at least 75 records. The sample size per predictor also takes into account that we'll be splitting our data set into training and validation samples, and 25 records per predictor ensures that we'll have a sufficient number of records in each sample. For example, if splitting our data set into a 60% train and 40% validation sample, 25 records per predictor would give us 15 records for the train sample and 10 records for the validation sample, with 10 records being the minimum recommended per predictor.

Partitioning Requirements

For simple linear regression and multiple linear regression, we partition our data into a training and validation data set. The model is formed based on our training data set, then run through our validation (or hold out) data set to confirm validity of the model. If the model performs well on the training data set and then poorly on the validation data set, this can indicate that we have overfit the model. In other words, we tuned the model too well to the training data set and now the model no longer performs well on a new or general data set. In this case, we might want to remove variables to create a simpler, more general model, which is also known as a parsimonious model.

Model Properties

Selection Model

The model selection can be chosen as a backward selection, forward selection, stepwise, or none. If none is chosen, all inputs are used for the model. If backward selection is chosen, all inputs are used and are removed until the stop criterion is met. If forward is selected, no inputs are used and added until the stop criterion is met. If stepwise is selected, the model starts with the forward selection even though inputs can be removed until the stop criterion is met.

Model Results Evaluation

To evaluate the results of our linear regression model, we can use several items including p-values, R^2, adjusted R^2, errors, lift, and regression equation.

P-Value

Also known as the significance value, generally a p-value below 0.05 is considered significant. For regression, this would indicate that the variable has a significant effect on the model and should be retained, whereas non-significant variables or those with a p-value greater than 0.05 should be removed from the model.

R^2 / Adjusted R^2

Also known as the goodness of fit, they explain the amount of variance in the target based on the inputs. As increasing the number of inputs will increase the R^2, and the adjusted R^2 penalizes the model for each additional input included.

Errors

Errors are calculated from the predicted value minus the actual value, which is also known as the residuals. Common measures of errors are sum of squared errors (SSE) and root mean square error (RMSE). The SSE and RMSE are calculated by taking the square or square root of the errors.

Regression Equation

$Y = b_o + b_1 x_1 + b_2 x_2 + E$, where b_o equals the intercept, x_1 represents our first independent variable, b_1 is the parameter, and E (epsilon) represents the error (or the remaining variance) that is unable to be explained by our input variables.

Now that we have covered the linear regression model, let's continue with an experiential learning application to connect your knowledge of linear regression with a health application on caloric intake.

Experiential Learning Application: Caloric Intake Simple Linear Regression

As an introductory example of a simple linear regression, let's consider the scenario where we want to study whether caloric intake (or the number of calories one consumes) can determine lifespan or longevity. Prior studies suggest that a reduced calorie diet can increase lifespan, and we want to verify the relationship between calorie intake and longevity based on a sample set of data. To start our process, we first want to explore our data and identify our input and target variables. In this example, our input (x) is calories, and our target (y) is lifespan or age. Both our input and target variables are continuous interval level variables. For our linear regression model, the input (or number of calories) will predict the target variable or lifespan. Based on your previous coursework and knowledge, there are often other names for inputs such as predictor, x, and independent variable. There are also other names for the target such as output, response, y, or dependent variable. For this text, we will use the primary terms of input and target.

Data Set File: 5_EL1_Caloric_Intake_Simple_Linear_Regression.xlsx

Variables:

- CaloricIntake
- Age

Step 1: Sign in to SAS OnDemand for Academics, or open SAS Enterprise Miner Local Installation.

Step 2. Open the SAS Enterprise Miner Application (Click the SAS Enterprise Miner link).

Step 3. Create a New Enterprise Miner Project (click **New Project**).

Step 4: Use the default SAS Server, and click **Next**.

Step 5: Add Project Name *Regression*, and click **Next**.

Step 6: SAS will automatically select your user folder directory if you are using SAS OnDemand for Academics (If you are using a local installation version, choose your folder directory), and click **Next**.

Step 7: Create a new diagram (Right-click **Diagram**).

Step 8: Add a File Import node (Click the **Sample** tab, and drag the node onto the diagram workspace).

Step 9: Click the File Import node, and review the property panel on the bottom left of the screen.

Step 10: Click **Import File** and navigate to the Chapter 5 Excel File *5_EL1_Caloric_Intake _Simple_Linear_Regression.xlsx.*

Step 11: Click **Preview** to ensure that the data set was selected successfully, and click **OK**.

Step 12: Right-click the File Import node and select **Edit Variables**.

Step 13: Set Age to the Target variable role, and set Caloric Intake to the Input role. Keep both variables at Interval levels. Click **OK**.

Figure 5.3: Edit Variable Roles and Levels

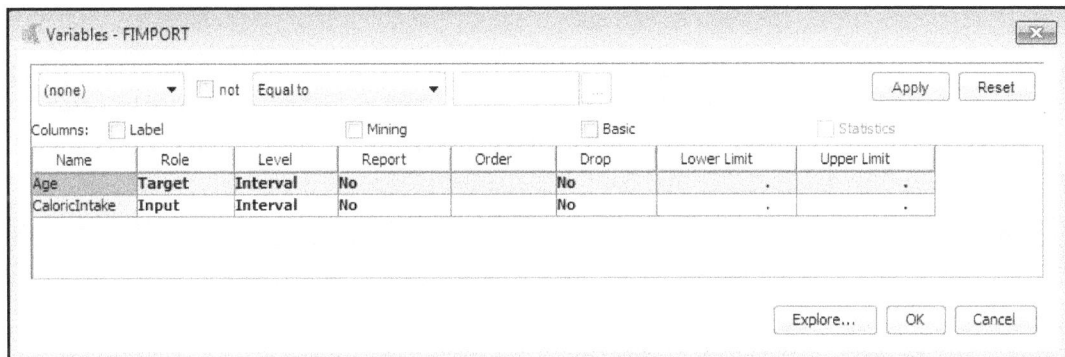

Step 14: Add a Regression node (Click the **Model** tool tab, and drag the node onto the diagram workspace).

Figure 5.4: SAS Model Regression Node

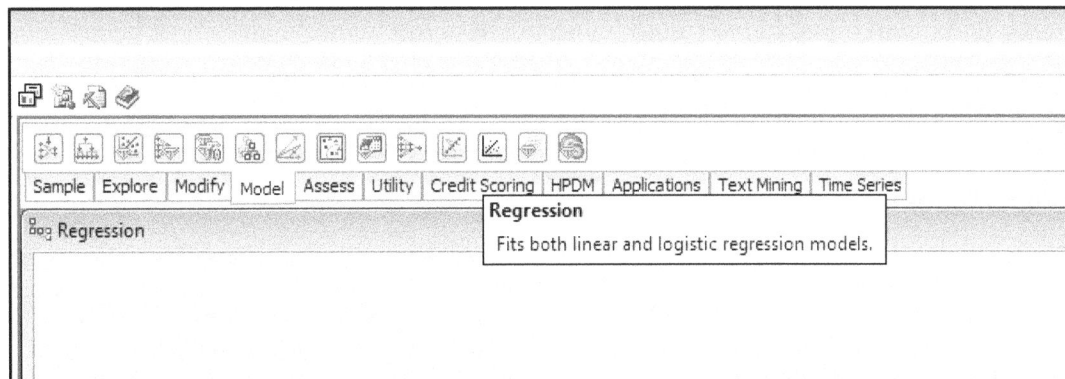

Step 15: Connect the File Import node to the Regression node.

Step 16: Right-click the Regression node and click **Run**.

Figure 5.5: SAS Model Regression Diagram

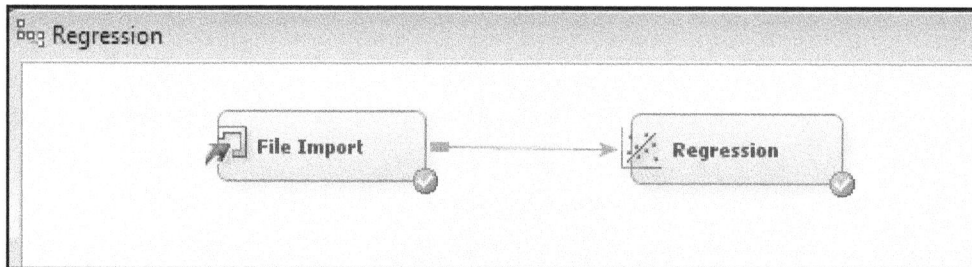

Step 17: Expand the Output window results.

Figure 5.6: SAS Model Regression Output

```
                            Analysis of Variance

                              Sum of
Source                DF      Squares      Mean Square    F Value    Pr > F

Model                  1    7734.309181    7734.309181    177.88     <.0001
Error                 47    2043.609187      43.481047
Corrected Total       48    9777.918367


            Model Fit Statistics

R-Square       0.7910    Adj R-Sq        0.7866
AIC          186.8020    BIC           188.9686
SBC          190.5856    C(p)            2.0000


        Analysis of Maximum Likelihood Estimates

                                Standard
Parameter       DF    Estimate     Error     t Value    Pr > |t|

Intercept        1       120.4    3.8213       31.52     <.0001
CaloricIntake    1     -0.0275    0.00206     -13.34     <.0001
```

Step 18: Review model results.

The output includes an analysis of variance (or ANOVA table), which includes Sum of Squares, Mean Square, F Value (or F statistic), and Pr > F (or significance). In the linear regression output, the first component is the Analysis of Variance (ANOVA), the overall F statistic (Pr>F) is used to test the significance of the overall model. In this case, <.0001 indicates that the model is significant. From our Model Fit Statistics and Adjusted R^2 results, we see that 78.66% of the variance in age can be explained by caloric intake in our simple linear regression model. The remaining 21.44% can potentially be explained by other factors that we have not yet considered in our model such as gender, occupation, lifestyle, income, access to healthcare, location, family history, and so on.

Furthermore, looking at the significance of variables in the model under the Type 3 Analysis of Effects and Analysis of Maximum Likelihood Estimates, which uses the t-statistic $Pr > |t|$, respectively, we find CaloricIntake to be significant based on a p-value < 0.05.

Under the Analysis of Maximum Likelihood Estimates section, the regression equation can also be developed, the standard regression equation is as follows: $Y = b_o + b_1x_1 + E$. Where b_o equals the intercept and x_1 represents our first independent variable and b_1 is the parameter, and E (epsilon) represents the error or the remaining variance that is unable to be explained by our input variables. For our MLR, our regression equation $Y = 120.7 + -0.0275x_1$, where x_1 = CaloricIntake. The regression equation can be used to calculate Age (Y) given CaloricIntake (x_1). For example, calculate expected age based on a 2000 calorie diet.

$$Y = 120.7 + -0.0275x_1$$

Experiential Learning Application: Caloric Intake Multiple Linear Regression

Let's further expand this example. We now add a second input variable of gender, which we believe can help explain additional variance in the model. In addition to calorie intake, we believe gender also has an impact on lifespan. By including a second input (X) variable, we now have a multiple linear regression, and we can measure the significance and impact of both calorie intake and gender on lifespan.

Data Set File: 5_EL2_Caloric_Intake_Multiple_Linear_Regression.xlsx

Variables:

- CaloricIntake
- Age
- GenderMF

Repeat the following steps:

1. Add a File Import node (Click the **Sample** tab, and drag the node onto the diagram workspace).
2. Click the File Import node, and review the property panel on the bottom left of the screen.
3. Click **Import File** and navigate to the Chapter 5 Excel File *5_EL2_Caloric_Intake_Multiple_Linear_Regression.xlsx*.
4. Click **Preview** to ensure that the data set was selected successfully, and click **OK**.
5. Right-click the File Import node and select **Edit Variables**.

Figure 5.7: SAS Model Regression MLR

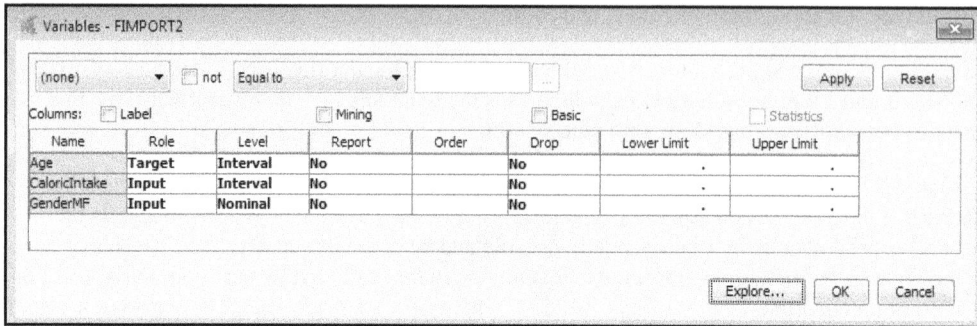

1. Select Age as the Target, CaloricIntake and GenderMF as the Inputs. With Age and CaloricIntake as Interval levels, and GenderMF as Nominal.
2. Add a Regression node (Click the **Model** tab, drag the node onto the diagram workspace).
3. Connect the File Import node to the Regression node.
4. Right-click the Regression node and click **Run**.
5. Expand the Output Window Results.
6. Review the Model Results.

Figure 5.8: SAS Model Regression MLR Output

```
                        Analysis of Variance

                               Sum of
Source                DF       Squares    Mean Square   F Value   Pr > F

Model                  2    7760.087228   3880.043614    88.45    <.0001
Error                 46    2017.831140     43.865894
Corrected Total       48    9777.918367


           Model Fit Statistics

R-Square      0.7936    Adj R-Sq       0.7847
AIC         188.1800    BIC          190.5628
SBC         193.8554    C(p)           3.0000


           Type 3 Analysis of Effects

                         Sum of
Effect              DF   Squares   F Value   Pr > F

CaloricIntake        1   7756.1621  176.82   <.0001
GenderMF             1     25.7780    0.59   0.4472


           Analysis of Maximum Likelihood Estimates

                            Standard
Parameter       DF  Estimate    Error   t Value   Pr > |t|

Intercept        1    120.7     3.8586    31.29    <.0001
CaloricIntake    1   -0.0276    0.00207  -13.30    <.0001
GenderMF     F   1   -0.7315    0.9542    -0.77    0.4472
```

In the linear regression output, the first component is the Analysis of Variance (ANOVA), the overall F statistic (Pr>F) is used to test the significance of the overall model, in this case <.0001 indicates the model is significant. From our Model Fit Statistics and Adjusted R^2 results, we see that that 78.66% of the variance in age can be explained by caloric intake and gender in our multiple linear regression model. This is the same result as earlier, indicating that at least with our sample, the addition of the gender variable does not further help explain the variance in age. Furthermore, looking at the significance of variables in the model under the Type 3 Analysis of Effects and Analysis of Maximum Likelihood Estimates, which use the F statistic Pr>F and t-statistic Pr > |t|, respectively, we find that gender is not significant based on a p-value > 0.05. The results would tell us that our model can use the single input variable of caloric intake to best explain the variance in age. This would be the simplest and best model based on the results (or the parsimonious model).

Under the Analysis of Maximum Likelihood Estimates section, the regression equation can also be developed, the standard regression equation is as follows: $Y = b_o + b_1x_1 + b_2x_2 + E$. Where b_o equals the intercept and x_1 represents our first independent variable and b_1 is the parameter, and E (epsilon) represents the error or the remaining variance that is unable to be explained by our input variables. For our MLR our regression equation $Y = 120.7 + -0.0276x_1 + -0.07315x_2$, where x_1 = caloricintake and x_2 = gender.

Following our multiple linear regression, we can continue to include additional variables in the model to improve overall accuracy or as new inputs are identified. A recent study has found that the percentage of friend requests on Facebook leads to longevity (Elizalde, 2016). As a result of this finding, we could test by adding a third input (X) variable to our model. The updated model would include the inputs of Caloric Intake, Gender, and Facebook Friends, and target or output of Age. We can continue to iterate or refine the model over time.

Model Summary

In summary, the linear regression can take the form of simple regression 1 input and 1 target or multiple linear regression 2 or more inputs and 1 target. The inputs can be interval (continuous) or nominal (categorical), whereas the target must be interval (continuous). To evaluate the linear regression, we can use R^2 or Adjusted R^2, error, and p-value. To apply the linear regression model to future data, we can use the regression equation.

- Model: Regression node
- Simple Linear Regression: 1 input and 1 target variable
- Multiple Linear Regression: 2+ inputs and 1 target variable
- Input: Interval (Continuous), Nominal (Categorical)
- Target: Interval (Continuous)
- Evaluation: R^2, Adjusted R^2, Error, p-value
- Application: $Y = b_o + b_1x_1 + b_2x_2 + E$

Experiential Learning Application: mHealth Heart Rate App

Executive Summary

In the healthcare environment of today, individuals are increasingly connecting to healthcare data using mobile devices such as tablets and smartphones. Providers and patients demand current information at their fingertips during all phases of healthcare delivery to save time, reduce errors, and improve quality outcomes. In 2017, there were already an estimated half-billion mobile healthcare app users (Zapata et al, 2015), and the mobile healthcare market is projected to experience an annual growth rate of 32% and reach nearly $200 billion in value by 2025 (Terry, 2013; Wood, 2017). Mobile Healthcare (or mHealth) can be defined as a component of electronic health services and information delivered through mobile technologies. This is often expanded to include mHealth aimed at improving healthcare outcomes (HIMSS, 2012).

Increasing numbers of mHealth applications are available through vendors such as Apple, IBM, and Samsung. The mHealth technologies include location-based services, beacons, digital photos and camera, push notifications, biometrics, microphone, conferencing, digital signature, and so on. (Apple, 2016; Frost and Sullivan, 2015). In the Apple Watch, the technology uses photo plethysmography (or light reflecting technology), as blood reflects red light and absorbs green light. Apple Watch uses LED lights and light detecting sensors flashing hundreds of times per second to measure the blood flow and calculate the beats per minute. With each heartbeat, blood flows to your wrist and more green light is absorbed, and, between beats, less green light is absorbed (Apple, 2016).

In order to capitalize on the mHealth trend, your organization is offering a new line of wearable heart rate monitors supported by a mobile app. The monitor allows users to take their heart rate using their device and analyze their information through their mobile app. For testing the accuracy of the device and app, the management team is interested in determining whether individual factors influence heart rates as well as determining whether the wearable heart rate monitors are similar to manual methods such as taking your pulse with a finger.

Heart rate is important when exercising to make sure that you are hitting the optimal target heart rate of between 50-85 percent of maximum. For an automated reading, it is recommended to place your finger on the sensor for 10 seconds to capture a reading on the app. For a manual reading, it is recommended that you take your pulse on the inside of your wrist on the thumb side. Using the tops of two fingers, press on your wrists and count your pulse for 10 seconds and multiply by 6 to find the heart rate beats per minute (American Heart Association, 2016).

Regular measurement of heart rate is recommended by the World Health Organization (WHO) and the International Association of Cardiologists (Losa-Iglesias,et al, 2016). The regular measurement is especially recommended for those over 50 years of age and those with cardiovascular disorders. Mobile apps can be used by patients to track their heart rate even though there are often trade-offs between convenience and clinical accuracy (Losa-Iglesias,et al, 2016).

A sample has been provided for you to get started: the Heart Rate App data set includes seven variables and 250 records. Help your management team answer the following questions. Follow the SEMMA process and provide recommendations.

Question 1: Does age, gender, height, BMI, or weight influence heart rate?

Question 2: How does the automated app compare with the manual method for measuring heart rate?

Data Set File: 5_EL3_mHealth_Heart_Rate_App.xlsx

Variables:

- Gender, Male or Female
- Age, in Years
- Height, in Inches
- Weight, in Pounds
- BMI, Body Mass Index
- Method, Manual or App
- Heart Rate, Beats Per Minute (BPM)

Follow the SEMMA process for your experiential learning application. A template has been provided below that can be reused across future projects.

Figure 5.9: SEMMA Process

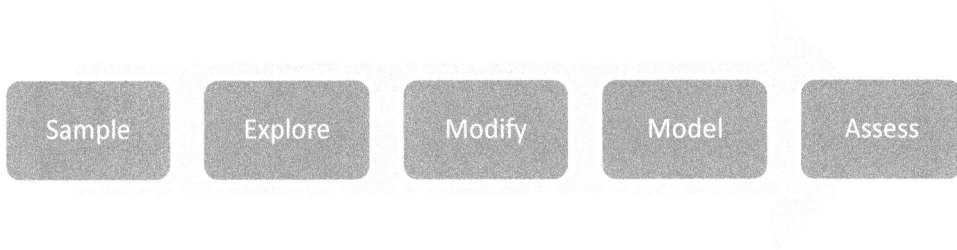

Title	mHealth Heart Rate App
Introduction	Provide a summary of the business problem or opportunity and the key objective(s) or goal(s). Create a new SAS Enterprise Miner project. Create a new Diagram.
Sample	Data (sources for exploration and model insights) Identify the variables data types, the input and target variable during exploration. Add a FILE IMPORT Provide a results overview following file import: Input / Target Variables Generate a DATA PARTITION
Exploration	Provide a results overview following data exploration Add a STAT EXPLORE Add a GRAPH EXPLORE Add a MULTIPLOT Summary statistics (average, standard deviation, min, max, and so on.) Descriptive Statistics Missing Data Outliers

Title	mHealth Heart Rate App
Modify	Provide a results overview following modification Add an IMPUTE
Model	Discovery (prototype and test analytical models) Apply a stepwise regression model and provide a results overview following modeling. Add a REGRESSION Model description Analytics steps Regression results (p-value, R^2, regression equation) Model results (Classification Matrix, Lift, Error, Misclassification Rate) Correlation Matrix (which variables measure the same thing) Selection Model
Assess and Reflection	Provide overall recommendations to business Model advantages / disadvantages Performance evaluation Model recommendation Summary analytics recommendations Summary informatics recommendations Summary business recommendations Summary clinical recommendations Deployment (operationalization plan: timeline, resources, scope, phases, project plan) Value (return on investment, healthcare outcomes)

Experiential Learning Application: Inpatient Utilization - HCUP

Inpatient costs have grown 2% annually over the last decade, accounting for approximately 33% of all healthcare costs. The average hospital stay average is 4.6 days with an average cost of $11,000 based on estimates. The Healthcare Cost and Utilization Project (HCUP) develops a set of databases and products that are sponsored by the Agency for Healthcare Research and Quality (AHRQ). HCUP (or also pronounced "H-CUP") is a set of data, tools, and products from a partnership between federal and state government agencies, and industry. HCUP includes one of the largest collections of hospital care data in the U.S. to allow research into health policies, medical practice, access to care, outcomes, and quality and cost of healthcare. The HCUP data collection includes obstetrics and gynecology, otolaryngology, orthopedic, cancer, pediatric, public, and academic medical hospitals. The data set contains five service lines: Medical, Surgical, Injury, Mental Health, and Maternal/Neonatal. The data set contains data at the discharge level. For example, a patient might be discharged more than once per year and might appear in the data set more than once. Mental health service line includes diagnosis codes such as anxiety, Schizophrenia, and substance-related mental disorders. The injury service line includes diagnosis codes such as fracture, sprains and strains, and poisoning and toxic effects. The surgical service line includes operating room procedures. The medical service line covers procedures not included in the surgical operating room procedures. Costs are based on accounting reports from hospitals and represent actual expenses (such as wages, supplies, utility costs) as well as the amount billed for the hospital stay (Weiss, et al., 2014).

A sample has been provided for you to get started. The HCUP data set includes 5 variables and 376 records. Follow the SEMMA process and provide recommendations.

Data Set File: 5_EL4_HCUP.xlsx

Variables:

- PatientID, unique identifier of the patient
- EncounterID, unique identifier of the episode of care or hospital stay
- ServiceLine, 1=Medical, 2=Surgical, 3=Injury, 4=Mental Health, 5= Maternal/Neonatal
- LengthOfStay, number of days between hospital admission and hospital discharge
- HospitalCosts, total billed costs for hospital stay

Follow the SEMMA process for your experiential learning application. A template has been provided that can be reused across future projects.

Title	Inpatient Utilization - HCUP
Introduction	Provide a summary of the business problem or opportunity and the key objective(s) or goal(s). Create a new SAS Enterprise Miner project. Create a new Diagram.
Sample	Data (sources for exploration and model insights) Identify the variables data types, the input and target variable during exploration. Add a FILE IMPORT Provide a results overview following file import: Input / Target Variables Generate a DATA PARTITION
Exploration	Provide a results overview following data exploration Add a STAT EXPLORE Add a MULTIPLOT Summary statistics (average, standard deviation, min, max, and so on.) Descriptive Statistics Missing Data Outliers
Modify	Provide a results overview following modification Add an IMPUTE

Title	Inpatient Utilization - HCUP
Model	Discovery (prototype and test analytical models)
	Apply a stepwise regression model and provide a results overview following modeling
	Add a REGRESSION
	Model description
	Analytics steps
	Regression results (p-value, R^2, regression equation)
	Model results (Classification Matrix, Lift, Error, Misclassification Rate)
	Correlation Matrix (which variables measure the same thing)
	Selection Model
Assess and Reflection	Provide overall recommendations to business
	Model advantages / disadvantages
	Performance evaluation
	Model recommendation
	Summary analytics recommendations
	Summary informatics recommendations
	Summary business recommendations
	Summary clinical recommendations
	Deployment (operationalization plan: timeline, resources, scope, phases, project plan)
	Value (return on investment, healthcare outcomes)

Reflection

Review, reflect, and retrieve the following key chapter topics only from memory and add them to your learning journal. For each topic, list a one sentence description/definition. Connect these ideas to something you might already know from your experience, other coursework, or a current event. This follows our three-phase learning approach of 1) Capture, 2) Communicate, and 3) Connect. After completing, verify your results against your learning journal and update as needed.

Key Ideas – Capture	Key Terms – Communicate	Key Areas - Connect
Patient Data		
Personal Health Records		
SEMMA		
Model Process Step		
Regression Node		
Linear Regression		
Multiple Linear Regression		
Input Variable		

Key Ideas – Capture	Key Terms – Communicate	Key Areas - Connect
Target Variable		
p-Value		
R^2		
Errors		
Regression Equation		
Caloric Intake Application		
mHealth Heart Rate Application		

Chapter 6: Modeling Provider Data

Chapter Summary

The purpose of this chapter is to develop data modeling skills using SAS Enterprise Miner, and with respect to the Model capabilities within the SEMMA process, this chapter builds on previous chapters with Sample, Explore, and Modify capabilities. This chapter continues modeling using logistic regression. This chapter also includes experiential learning application exercises on hospital acquired conditions and immunizations. The model focus of this chapter is shown in Figure 6.1.

Figure 6.1: Chapter Focus

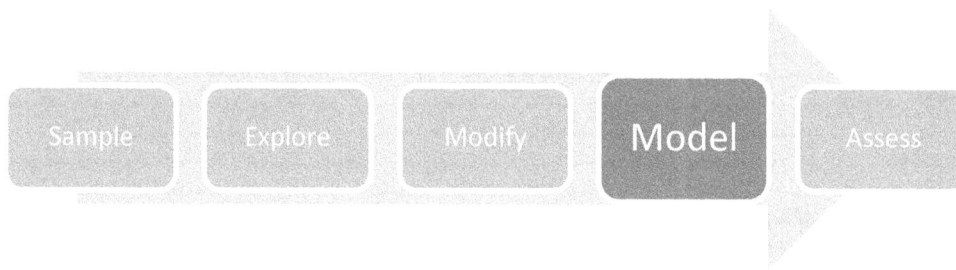

Chapter Learning Goals

- Describe the Model process steps
- Understand provider-level data sources
- Develop data modeling skills
- Apply SAS Enterprise Miner model data functions
- Master the logistic regression model

Providers

Providers are considered the individuals (or professionals) that provide healthcare to patients. Provider facilities are generally broken down into inpatient and outpatient services. Typically, an overnight stay determines the difference between an inpatient and outpatient service. A patient might stay overnight in a hospital, emergency room, or other short-term care center. Providers that offer overnight stays are also called acute care facilities. If a patient has a surgery in a hospital but is discharged, and leaves a few hours later without an overnight stay, this is considered an outpatient service because there is no overnight stay. The number of hospitals has decreased since the 1970s due, in part, to improved surgical techniques and other medical improvements that no longer require overnight stays (Gartee, 2011).

When staying overnight in an inpatient facility, patients follow formal admission and discharge processes. A patient chart is created with information about the services conducted, along with physician, nurse, and medical professional review notes. Once the patient has received appropriate care, the patient is then discharged from the hospital with at-home or follow-up care instructions. Readmission is often tracked through a 30-day readmission frequency. The readmission metric tracks patients who are discharged and return within a 30-day period, possibly indicating that all the underlying causes were not treated. The number of days that a patient stays in the inpatient facility is considered the Length of Stay (LOS). The LOS often drives the cost of the stay and is used for billing purposes. A LOS is calculated from the Discharge Date – Admit Date. For 1-day stays, or, in cases where a patient is admitted and discharged within 24 hours, this counts as a 1-day LOS. If the LOS is less than 30 days. this is considered acute care. If the LOS is greater than 30 days, this is considered long-term care. If a patient requires a long-term stay, but instead of being cared for in a hospital setting, the patient might be transferred to a skilled nursing facility (Gartee, 2011).

Figure 6.2: Provider Settings

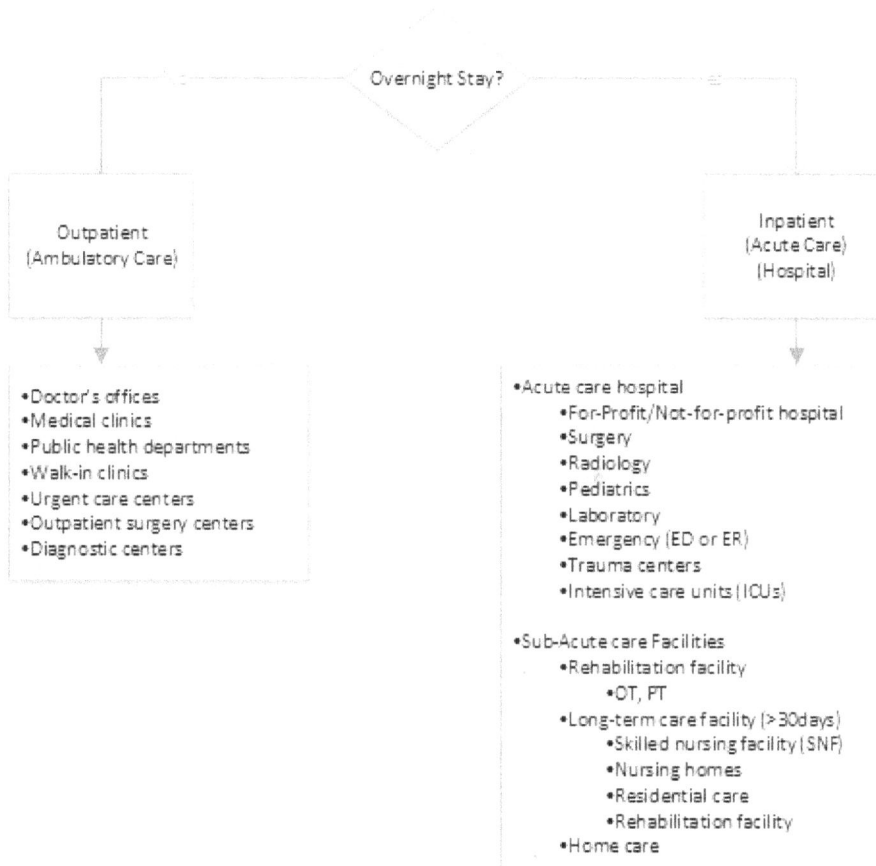

Provider Anamatics

Provider anamatics gives an opportunity to improve patient safety, and financial and operational performance for healthcare. The financial and operational analytics application area includes facilities, staffing, admissions, and reimbursements. The clinical and patient safety area includes evidence-based medicine, clinical decision support, safety, survival rates, and compliance. To date, many providers have focused on descriptive analytics through regulatory reporting and performance improvement program reporting. Predictive analytics have been used for determining admissions volumes and clinical interventions. A five-level pyramid has been proposed to explain provider analytic capability at different stages: 1) analytically impaired, 2) localized analytics, 3) analytical aspirations, 4) analytical companies and 5) analytical competitors. Most providers have achieved level-2 localized analytics. A few providers have achieved level-3 aspirational analytics including Kaiser Permanente, Mayo Clinic, and Cleveland Clinic. The successful achievement of level 3 is attributed to top management support, personnel skill sets, and analytical collaboration between localized groups. Initiatives, such as meaningful use for Electronic Medical Record (EMR) reimbursement and Accountable Care Organizations (ACOs), support the movement toward analytics along with regulatory reporting and compliance through the Centers for

Medicare and Medicaid (CMS) and the Food and Drug Administration (FDA) (Davenport and Miller, 2014).

Provider Data

Provider Systems

From our Chapter 2 review, common provider systems include Electronic Health Record (EHR), Computerized Physician Order Entry (CPOE), Telemedicine, and Pharmacy, with EHRs typically being the starting point for information technology adoption and storage for providers. EHR systems store all the patient history and medical information (such as diagnosis, procedures, and billing) that occurs from a clinical delivery standpoint, and replaces the paper-based processes that many providers have used in the past. EMRs contains medical and clinical data from a provider's office, whereas an EHR also contains data from a complete medical history for the patient. EHRs are accessible across all sites and healthcare providers (HealthIT.gov, 2016a). Given the healthcare market potential, there are a wide variety of EHR vendors available, in the order of hundreds a few years ago and now even over 1,100 EHR vendors to choose from (Shay, 2017). Within hospitals that participated in the Medicare EHR incentive program in 2016, there were 175 EHR vendors for 4,474 hospitals. The top five vendors by reported usage volume included Cerner, MEDITECH, Epic Systems Corporation, Evident, and McKesson (HealthIT.gov, 2016b). Within ambulatory primary care physicians, medical and surgical specialists, podiatrists, optometrists, dentists and chiropractors participating in the Medicare EHR program in 2016, there were 632 EHR vendors for 337,432 providers. The top five vendors by volume reporting include Epic Systems Corporation, Allscripts, eClinicalWorks, NextGen Healthcare, and GE Healthcare (HealthIT.gov, 2016b).

Open versus Closed Systems

For provider systems, there are generally two types of software vendors to choose from: 1) open-source and 2) closed-source. These two types of vendors are common with software vendors across all industries including healthcare. Open-source systems are typically publicly accessible at no cost, and the underlying software source code can be viewed and customized. Examples of open-source EHRs include the Open Medical Record System (OpenMRS) and the Veterans Health Information Systems and Technology Architecture (VistA) in which the open-source code and free software was made available through the U.S. Freedom of Information Act (OpenMRS; Plone, 2017). Closed-source systems are typically purchased for a license fee, and the source code is prohibited from viewing. Customization can take place through approved methods such as application programming interfaces even though the base product cannot be customized directly. The majority of the top EHR systems in use by providers are closed-source including Cerner and MEDITECH.

Cloud versus On-Premise Systems

EHR vendors can also be cloud-based or on-premise, similar to analytics platforms such as SAS OnDemand for Academics, which was discussed in Chapter 2. Cloud-based systems are accessible through a standard internet connection and web browser and do not require specialized hardware or software installation. Cloud-based systems also typically offer a fast set-up time since there is no on-site installation required. Cloud-based systems typically offer a pay-as-you go or pay-for-usage model on a monthly basis. On-premise systems are locally installed and typically have minimum hardware requirements for installing the software application, and, many times, a continuous internet connection is not required. Typically, for on-premise software installations, the number of users and usage requirements (such as storage space and processing power) are determined in advance on an annual basis. As a result of the demand and growth for cloud-based systems, many vendors are transitioning to offer the cloud-based capabilities over on-premise installations. Epic Systems, one of the largest EHR providers in the world, has been using a hybrid approach of both on-premise for existing customers and cloud-based software for new customers as the transition occurs. Cerner also offers a cloud-based solution as a result of the many community hospitals lacking the expertise of on-site programmers, administrators, and engineers required to run an on-premise system. The cloud-based systems are also sometimes referred to as remote-hosted or software as a service (SaaS) offerings even though they carry the same functionality (Sullivan, 2014). Highlighted as the number one cloud-based EHR in the U.S. by SK&A, Practice Fusion was originally developed solely as a cloud-based software platform. Although Practice Fusion is a closed source, the company previously offered the software free of charge, supported through advertisements, care coordination, and data analysis (Practice Fusion, 2017). The company has since transitioned to a monthly subscription-based model, following its $100 million acquisition by Allscripts (Monica, 2018).

To learn more about successful EHR systems, we will review two well-known EHR implementations, followed by the factors ensuring successful implementation of EHRs.

EHR Implementations

Mentioned as an open-source platform example, one of the most cited implementations of EHR is through the Department of Veteran's Affairs (VA). The Veterans Health Information Systems and Technology Architecture (VistA) expands beyond electronic health records and includes electronic patient records, imaging, laboratory and computerized order entry, which has been developed over several decades. In one study, researchers estimated the costs against the benefits for the VA through the year 2007. The total costs (or investment) by the VA in VistA was estimated at $4.07 billion, with the total benefits estimated at $7.16 billion. The savings were achieved by eliminating redundancies, reducing usage, improving quality through reduction of errors, reduced work, and reduced cost of space. The findings showed a net resulting benefit of $3.09 billion to the VA even though this result took a large commitment and many years of effort from the 1990s through 2004 to achieve a full adoption, and to break-even with the investment and to begin to achieve net positive benefits as a result of VistA (Byrne et al., 2010).

Founded in 1945, Kaiser Permanente is the largest non-profit healthcare system in the U.S., covering approximately 12 million members with annual revenue of $73 billion. In addition, Kaiser contains approximately 40 hospitals, 700 medical offices, 22,000 physicians, 57,000 nurses, and 210,000 employees. Kaiser Permanente combines a health insurance plan, medical facilities, and physician groups to promote timely interventions and collaboration to improve healthcare. A key component of this initiative was the roll out of their Kaiser Permanente (KP) HealthConnect system, which includes an integrated EHR with online member access (McCarthy et. al, 2009; Kaiser Permanente, 2018). An example of a closed-source platform, Kaiser Permanente has been using information technology for several decades, and Kaiser Permanente's usage of EHRs began in the 1990s. In 2003, Kaiser Permanente formally launched a $4 billion initiative and system called KP HealthConnect. The implementation was completed in 2008 and was the largest civilian implementation of EHRs in the U.S., to date. The underlying technology was purchased from Epic Systems Corp, and includes features such as laboratory, medication, imaging, Computerized Physician Oder Entry (CPOE), population health management, chronic condition management, decision support, preventative care, electronic referrals, reports, performance management, and patient billing features. Five key attributes in the healthcare delivery system were identified as requirements for the EHR (McCarthy et. al, 2009).

1. Information Continuity – the availability of information to all providers through the EHR. Kaiser Permanente has a web portal for online access to medical records, visits, appointments, prescriptions, lab results, and messaging. Medical staff is focused on patient sign-up to improve engagement to monitor their preventative care, schedule appointments, refill prescriptions, and communicate with their medical team. In the Hawaii region, patient visits decreased by 26%, and overall patient contacts decreased by 8%, facilitated by the EHR efficiency and online access.

2. Care Coordination / System Accountability – the coordinated care between all providers. Care teams track and improve preventative medicine as needs arise. Screenings improved over national levels. For example, breast cancer screening increased to as high as 86% for Medicare members and exceeded national averages of 67%, and cholesterol screening increased from 55% to 97%.

3. Peer Review and Teamwork – the teamwork of providers, nurses, and care personnel all support and review one another for high-quality output. Unblinded performance data is shared among the group. Internal transparency was further improved through information system tracking and physicians were eligible for a performance incentive up to 5% based on quality, patient satisfaction, workload, and team contributions.

4. Continuous innovation – the innovation and learning to continuously improve health delivery. An innovation center is used as a learning lab for simulation and evaluations of care. As part of continuous improvement, Kaiser Permanente has an in-house Permanente Journal, innovation awards, workshops, the Garfield Innovation Center. and a Care Management Institute to investigate and identify best practices for patient outcomes. In one application, development of an osteoporosis disease management program led to a 37% reduction of hip fractures. In a separate effort, the use of generic equivalent cholesterol medications led to annual cost savings of over $150 million.

5. Access to care – multiple entry points and providers are responsible and available for care. Patient modules allow translations to native languages along with a bilingual staff. KP HealthConnect entry points included 24-hour call centers and nurse lines, web-based appointment booking, urgent care after hours, phone appointment booking, and electronic messaging. In Northern California, the combination of efforts led to a 32% decrease in emergency department visits.

In the EHR implementation at Kaiser Permanente, a shared partnership and teamwork among managers, physicians, and employees was seen as a key to success. Sharing a corporate culture dedicated to quality care and continuous improvement was an important aspect of the overall success. Next, we will discuss in further detail additional success factors and implementation process for EHRs.

EHR Implementation and Success Factors

During the 1990s, information technology investment across most industries in the U.S. was eight times higher per worker than in the healthcare industry, and the healthcare sector contained numerous issues relating to costs, errors, and efficiency of care. To address these issues, a 2004 presidential executive order was established, calling for widespread and nationwide adoption of Electronic Health Records (EHRs) over the next decade. Barriers to implementation of EHRs include up-front costs and savings that benefit third-party payers. Globally, EHR adoption has been spurred by governmental subsidies to provider technology. Studies have shown the relationship between EHR adoption and provider size in which the adoption rate decreased as provider size in terms of full-time physicians decreased. The total cost of an EHR (which includes purchase of hardware, software, implementation, maintenance, training, customization and support for operational changes) remains high. Resources (such as insufficient capital, return on investment (ROI), data use concerns, consumer education, and costs) were listed as the top limiting factors and barriers to EHR adoption (Woodside, 2007a).

Providers currently assume the majority of the up-front cost and risk of EHR investment, and they often do not receive a full return on benefits, as a result of software licensing, clinical workflow changes, and ability to organize beyond a local area to achieve economies of scale. The costs and benefits factors can include practice size, specialty, geographic location, operational efficiency, affiliations, IT support, and market incentives. An agreed upon level of incentives is one that compensates the additional cost of obtaining data and is fair, equitable, attainable, and reviewed with increases. Incentives are intended to provide adoption momentum in the market. Past implementations have shown that physicians are often resistant to practice changes required by EHR. Given potential resistance, incremental approaches to adoption have been suggested. To encourage information technology adoption, some advocate for direct or indirect incentives. Examples of direct rewards include regional grants and contracts, low-rate loans, pay-for-use, and an EHR system in exchange for data used for third parties. Examples of indirect rewards are pay-for-performance. Regional grants and contracts would be used to promote EHR at local levels, with the hope of creating local and regional data exchanges. Some solutions intended to encourage adoption, such as pay for performance, do not apply universally and require localized incentives. Federal low-rate loans could also reduce the entry barrier for EHRs. Pay-for-use explores ways to provide reimbursement based on new codes and modifiers or through direct incentives. Pay-for-performance would provide incentives for those practices with the highest quality but not quantity, which is expected to be enhanced through EHR adoption. Once large providers adopt IT, smaller providers might follow. The idea is to move the adoption curve forward until the market is saturated and then to require remaining providers to implement EHRs. Larger practices or those with larger population centers have a higher percentage of adoption (Woodside, 2007).

Hospitals have now achieved near universal adoption of EHRs, representing a nine times increase in EHR adoption between 2008 and 2016 (Henry et al, 2016). Although adoptions rates for EHR systems have continued to increase for hospitals, full adoption still has not been achieved by physician practices. Providers of care are not adopting at as high a rate as a result of the cost-benefits, business process re-engineering, and potential legal barriers. A survey showed that those physicians without an EHR were more likely to work in a smaller, lower-income practice. Age also had an impact. For example, one physician planned to retire in a few years and decided to avoid an EHR implementation (Mazzolini, 2013) because a typical ROI cycle for an EHR required about three years. As further evidence of opposition, there is even an entire Twitter hashtag dedicated to opposing EHRs at #EHRbacklash (Twitter, 2018).

Increasingly, providers are being measured on quality, and that quality is tied to pay-for-performance or other programs that directly affect the payment to the provider. EHR systems make quality measures available for reporting and compliance requirements. The Veterans Health Administration showed improvements in employee-patient ratios and cost-per-patient decreases as compared with the U.S. consumer price index increase. Quality was improved through preventative screening measures and disease management. The Illinois Department of Healthcare identified a medical home model to improve chronic conditions. The model used electronic health records to link all members of the patient care team and added pay-for-performance incentives that were based on national measures. Further analysis showed that care following hospital discharge (in terms of follow-up physician visits, medication adherence, and ongoing testing) was not always provided. Post-discharge care was identified by the Illinois Department of Healthcare as an action item for quality improvement (Woodside, 2013b). It has been shown that quality and efficiency can improve through an EHR, which leads to reduced medical errors and usage. The potential creates an issue as to whether the market supports technology in terms of societal benefit. In an effort to further improve EHR adoption, the Meaningful Use program from 2011 through 2014 was developed by the Centers for Medicare and Medicaid Services (CMS) to encourage EHR adoption through incentives. The Meaningful Use program has completed payments of over $20 billion and 370,000 providers received incentives. By early 2017, 67% of all providers used an EHR. Total costs for EHRs vary though annual spending is estimated near $37 billion, with more than half of providers incurring costs over $100,000 annually for EHRs and identified EHRs as one of their greatest increases in IT spending. Despite the spending, nearly 70% of providers indicated their functionality of the EHR system was lacking (Shay, 2017).

EHR system implementations are no different from other information system implementation such as Enterprise Resource Planning (ERP) systems. These types of system implementations are prone to high failure rates due to the complexity and scope of changes and user resistance. Some estimates show that nearly 80% of IT projects fail in terms of meeting the goals of the original implementation: 1) implemented on time, 2) implemented on budget, 3) implemented with agreed features and functionality, 4) achieved quality outcomes, and 5) in the case of EHR achieved of Meaningful Use.

Some of the key success factors in any implementation include 1) collaborative culture, 2) customization, 3) project management, 4) resources, 5) top management support, 6) systems integration, and 7) training. Collaborative culture includes the teamwork aspect of implementation and the close cooperation between all department areas and business and technical teams, including top management, consultants, end users, and vendors in support of a common objective and set of goals. Communication should be started early in the form of multiple announcements and organizational newsletters, including email, meeting, and intranet announcements. Kick-off meetings with key personnel and staff resources should also take place, along with regularly recurring meetings and weekly communications to key stakeholders. Customization is the ability of an organization to adapt their business processes to fit within the existing EHR application. Many times, organizations try to adjust the software to match their business processes. However, this is often costly, time consuming, and prevents future upgrades and support. Some EHR platforms have more flexible architectures that support and allow customizations such as modifying field names to match an organization's terminology.

Figure 6.3: Implementation Success Model

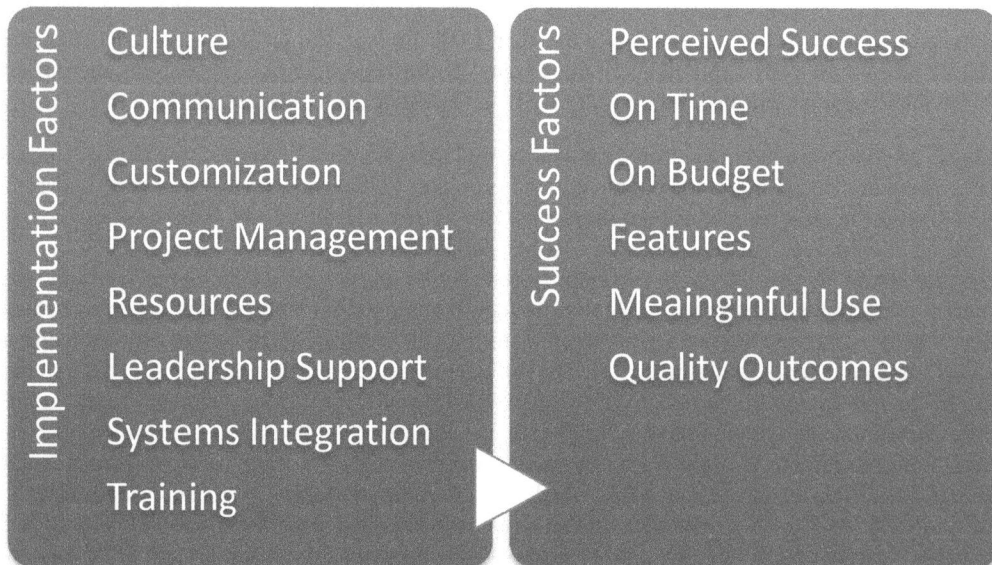

Implementation should follow a formal project management methodology. Many projects fail to adequately account for organizational requirements, resources, and funding that is necessary to support a successful implementation. Project Management includes coordinating, scheduling, scope, and monitoring activities and resources in line with the project objectives. Agile project methodologies allow the implementation to proceed iteratively, allowing improved deployment speed and continuous feedback. Resources can include financial, people, hardware, software, and time for project completion. Consultants are often required due to a knowledge gap and complexity of new systems. User involvement can occur through requirements gathering, implementation participation, and continued use after a go-live date. Dedicated department or area-based resources should be assigned for local subject-matter experts and knowledge diffusion. Top management provides the required resources in a direct or indirect manner through financing as well as the power and support. Top management is also responsible for setting a clear direction, overall project objectives, project guidance, representation, and establishing these throughout the organization. Sponsorship across the entire management team allows others in the organization to support the project through reducing political resistance and facilitating participation.

During most implementations, multiple-vendor solutions are required for purchase because a single vendor cannot provide a fully integrated solution. As an alternative, other companies choose a best-of-breed approach based on vendor offerings. Today, several vendors offer completely integrated solutions with equitable offerings across services. The use of a single vendor also improves delivery time through ease of installation and avoids integration issues that commonly arise when using multiple vendor solutions. Training end users is important toward gaining knowledge and appropriate use of the system. Training modules and materials should be developed before the initial go-live date, along with governance plans (such as best practices, content and technical standards, and policies and procedures for training purposes). In addition, proactive monitoring of training issues should be immediately addressed to avoid long-term paradigm creation in early stages. New users should be required to take established training, and existing users, on an annual basis, should be directed to training materials as questions arise (Woodside, 2011).

EHR Implementation Process

To assist with a successful EHR implementation, an EHR implementation process can follow six steps:

1. Assess Practice Readiness,
2. Plan Your Approach,
3. Select or Upgrade to a Certified EHR,
4. Conduct Training and Implement EHR,
5. Achieve Meaningful Use,
6. Continue Quality Improvement (HealthIT.gov, 2013b).

As you might note, in this textbook, there are a variety of process steps, including SEMMA, that are followed. These process steps allow repeatable formalized methodologies that are based on best practices to ensure successful results.

Figure 6.4: Implementation Steps

Assess Readiness › Plan Your Approach › Select EHR › Train and Implement › Meaningful Use › Improve Quality

Assess Readiness

In the first process stage of Assess Practice, and Readiness, current practices should be reviewed

For example, are all processes organized, efficient, and documented? Are data collection processes established and documented? Is high-speed internet available, and are employees ready to adopt new technology? In addition, future state and goals should be reviewed. For example, what is the expected result for patients, providers, and staff? What are the specific goals that will be used to determine the success of the project.

Plan Approach

In the second process stage of Plan Your Approach, the work environment should be reviewed and priorities should be set. In the planning phase, current processes should be documented, workflow should be mapped, process improvement should be reviewed, backup and disaster recovery plan should be developed, project plans should be developed, patient chart conversion and data migration should be planned, and privacy and security protocols should be developed.

Select EHR

In the third process stage of Select or Upgrade to a Certified EHR, vendor products should be evaluated according to the planning requirements. Pricing should be calculated, and account for hardware, software, maintenance, implementation, interfaces, and customizations. Roles and responsibilities should be defined, and the delivery of systems, integration capabilities, privacy and security, vendor reliability, vendor market share, and legal review should be conducted. Table 6.1 includes a set of common features that are typically included in a technical system (such as EHR or similar enterprise level systems), and that comprise the platform. These features are categorized within Functional Quality, Architectural Quality, and Vendor Quality (Woodside, 2013d). For a selection process, the features are identified and these features are ranked according to organizational need. Next, the software vendors are evaluated based on the set of features in order to offer an objective selection process. After the top two to three vendors are selected, contract negotiation occurs to add a cost value review to the features.

Table 6.1: Software Platform Features

Software Platform Category	Software Platform Features
Functional Quality (Functionality)	Meaningful Use, Online Analytical Processing (OLAP), Data Mining, Predictive Analysis, Business Analysis, Statistical Analysis, Geospatial Analysis, Scorecard, Strategy Maps, Key Performance Indicators, Querying, Reporting, Multiple report output formats, Customization, Interface type, Ad Hoc programming languages, Ease of Use, Automation, Knowledge Management, Portal, Dashboard, Charting, Data Warehouse, Content Management, Search

Software Platform Category	Software Platform Features
Architectural Quality (Reliability)	Availability, Performance, Database standards, Communication standards, Security levels, Number of independent modules, Number of workstations, Vendor hosted, Compatibility, Source code availability, Operational system integration
Vendor Quality	Popularity / Reputation, Technical support, User training, Experience with vendor products, Business focus, References, Past experiences with vendor, Consulting, Licensing costs

Train and Implement

In the fourth process stage of Train and Implement, user training, practice implementation, and testing is completed. In this stage, the data or charts are migrated to the EHR system, policies and procedures including the privacy, security, and backup plans are developed and put in place. Communication to patients, providers, and staff is ongoing to ensure a successful implementation.

Meaningful Use

The fifth process stage of Meaningful Use occurs after implementation and is one of the key goals of the EHR. In this stage, the goals are to improve quality, safety, and efficiency, engage patients, improve care coordination, improve population health, and ensure privacy and security. The ultimate goal is providing a fully electronic environment that improves the outcomes of healthcare. Achievement of Meaningful Use is also an important component to receiving incentives.

Improve Quality

In the sixth and final stage of Improve Quality, ongoing continuous quality improvements are happening. The stage includes further improvements to workflows, additional training, data capture, customizations, definitions of roles and responsibility, technology reliability, speed, upgrades. These are all done to ensure continued efficient operation and ongoing patient and staff satisfaction (HealthIT.gov, 2013).

Experiential Learning Activity: Electronic Health Records

EHR

Description: For our experiential learning activity, you will connect your knowledge of EHRs and the implementation process.
- Research and search the internet to find three different EHR vendors.
- List a set of three features or factors that you will use to evaluate the systems such as cost, quality, performance, and so on. In an earlier study, one of the top criteria (or factor) for selection was whether the EHR system achieved the criteria for Meaningful Use (Shay, 2017)
- Compare and provide a description of the three systems using the set of three factors above.

EHR

- Provide a recommendation for a vendor system for a small physician practice with 10 employees and provide a recommendation for a major hospital system with 30,000 employees.

Vendor Name	*Feature/Factor 1*	*Factor 2*	*Factor 3*
1.			
2.			
3.			

Physician Practice and Hospital:

Following EHR implementation, a wide variety of clinical data that is captured electronically can now be used for analysis. We'll continue with our SEMMA modeling process using data available through EHR systems and review the logistic regression model.

SEMMA: Model

Model Application Examples

Logistic regression can be applied to a variety of areas with the healthcare setting. Logistic regression has been used in medical and biomedical research such as medical risk and treatment compliance. A global study used logistic regression to identify treatment failure for the infectious disease of tuberculosis. Researchers found that gender, age, weight, and nationality were among the significant factors that predicted successful treatment completion (Kalhori, 2010). A study in Taiwan reviewed predictors of health information usage between 2,741 internet users between ages of 20-65. Using a logistic regression analysis, the study found that higher educational level (odds ratio 3.6, p <0.01), living alone (odds ratio 1.77, p <0.05, exercise (odds ratio 2.41, p<0.01), living in a city (odds ratio 1.28, p<0.05), high perceived health (odds ratio 1.34, P<0.05), and use of Western medicine services (odds ratio 1.51, P<0.05) all had an impact in health information usage for male internet users. By contrast, for female internet users being married (odds ratio 1.68, p<0.01) and higher personal income (odds ratio 1.56, P,0.01) had an impact. Predictors similar for both groups included higher educational level, and exercise that led to an increase in the odds for use of online health information (Koo et al., 2016).

Model Process Step Overview

During the modeling process step, the data mining model is applied to the data. During the partitioning phase data is segmented into training and validation data sets. The training data set is used to fit the model, and the validation data set is used to validate the model on a new set of data to demonstrate the reliability of the model. Based on the results, the model can then be tuned to optimal performance.

There are many different models that can be selected during this step, and the decision to choose each model is based on earlier exploration of data and knowledge of each model. We will continue this chapter with the model of logistic regression.

Model Tab Enterprise Miner Node Descriptions

Regression Node

The Regression node is used for the linear regression model and is associated with the **Model** tab in SAS Enterprise Miner. Logistic regression is a nonlinear extension of linear regression, where the input variables are interval (continuous) or nominal (categorical), and the target variable (also known as Y or dependent variable) is binary (such as 0/1 or No/Yes). In special cases, logistic regression can also be used for nominal target variables such as Yes/No/Maybe. In the case of a binary target variability, the regression can be interpreted as a probability. If the probability of belonging to a 0/1 class is above or below a certain threshold, commonly 0.5, then we predict the corresponding class. If the probability of belonging to class 1 is 0.7, we predict 1, and if the probability is 0.3, then we predict 0 (Klimberg and McCullough, 2013).

As a recap, linear regression is used for interval targets, whereas logistic regression is used for binary targets.

Linear Regression: Interval and Nominal Inputs ▶ Interval Target

Logistic Regression: Interval and Nominal Inputs ▶ Binary Target

Model Assumptions

Similar to linear regression, one model assumption about logistic regression is that the underlying data variables are uncorrelated. If the independent variables explain the same variability of Y, multicollinearity can occur, leading to interpretation difficulty. Multicollinearity is often difficult to detect initially, although it can be identified by review of pairwise plots of input variables and correlation. To address multicollinearity, we can add data, reduce the model by removing input variables, and finally conduct a principal component analysis, which will result in uncorrelated inputs (Latin, Carrol, and Green, 2003). Like linear regression, logistic regression can be affected by outlier values and can be corrected by methods such as deleted or imputed during the data exploration and cleaning process. Overfitting the model based on the training data can also occur. This can be addressed through choosing a simpler model that can be generalized for future applications and data sets. A check for overfitting the model can be run against a validation set of data. Also, if the error rate for the training data is significantly lower than the validation data, this might indicate a case of overfitting (Latin, Carrol, and Green, 2003).

Data Preparation

To ensure an adequate sample size, as a general rule of thumb, we might want to verify that we have at least 25 records for each input variable. If we have three inputs, we should have at least 75 records. The additional records also ensures that we'll have a sufficient number of records in each sample after splitting our data set into training and validation samples. If splitting our data set into a 60% train and 40% validation sample, 25 records per predictor would give us 15 records for the train sample and 10 records for the validation sample.

Partitioning Requirements

For logistic regression, we partition our data into a training and validation data set. The model is formed based on our training data set and is then run through our validation data set or hold-out data set to confirm validity of the model. If the model performs well on the training data set but poorly on the validation data set, this can indicate that we have overfit the model. In other words, we tuned the model too well to the training data set, and now it no longer performs well on a new or general data set. In this case, we might want to remove variables to create a simpler, more general model.

Model Properties

Selection Model

The model selection can be chosen as a backward selection, forward selection, stepwise, or none. If none is chosen, all inputs are used for the model. If backward selection is chosen, all inputs are used, and it removes inputs until the stop criterion is met. If forward is selected, no inputs are used, and it adds inputs until the stop criterion is met. If stepwise is selected, the model is started with the forward selection, although inputs can be removed until the stop criterion is met.

Model Result Evaluation

To evaluate the results of our logistic regression model, we can use several items including p-values, errors, lift, misclassification, and odds ratio. For the differences between logistic regression and linear regression, logistic regression contains a new evaluation measure of odds ratio and no longer includes a regression equation and R^2 as with linear regression. The lack of the regression equation is due to the fact that the output of logistic regression is binary, such as 0/1, and therefore would not plot along a continuous line as with the linear regression variable output, such as height.

P-value

Also known as the significance value, generally p-value below 0.05 is considered significant. For regression, this would indicate that the variable has a significant effect on the model and should be retained, whereas non-significant variables (or those with a p-value greater than 0.05) should be removed from the model.

Errors

Errors are calculated from the predicted value less the actual value, which is also known as the residuals. Common measures of errors are sum of squared errors (SSE) and root mean square error (RMSE). The measures are calculated by taking the square or square root of the errors.

Lift

Lift is a measurement between a random or baseline model against the analytical model. A higher lift or outperformance of the random selection is best.

Misclassification Rate

For models with nominal or binary targets, the percentage of total records misclassified as false positive or false negative.

Odds Ratio

Odds ratio is the relationship between probabilities. If the odds ratio is 2, for each unit increase in the input, the odds of belonging to the class increase by a factor of 2. The odds of an event can be measured by the odds of the true case divided by the odds of the false case. An odds ratio less than 1 can be interpreted as for each unit increase in the input, the odds of belonging to the class decrease.

Now that we have covered the logistic regression model, let's continue with an experiential learning application to connect your knowledge of logistic regression with a health application about hospital acquired conditions.

Experiential Learning Application: Hospital-Acquired Conditions

Past studies have attempted to review factors associated with hospital-acquired conditions, and have found that these hospital-acquired conditions can affect any age, gender, or region. In a Veterans Affairs (VA) hospital study, researchers examined hospital acquired acute kidney injury (HA-AKI) in order to identify high risk patients and prevent a potential cause of death. Acute kidney injury (AKI) occurs in 1-5% of hospitalized patients with mortality rates starting at 15% and is associated with other conditions such as myocardial infarction, chronic kidney disease, and end-stage renal disease. For the original study, over 1.6 million patients and over 116 VA hospital's EHR data was collected. Risk factors examined included medications, diagnosis, and demographic factors. Many existing risk models for identifying AKI use logistic regression (Singth et al., 2013; Cronin et al., 2015).

As an example of logistic regression, let's use a similar study to determine the risk factors that led to hospital acquired acute kidney injury (HAAKI). To start our process, we first want to identify our input and target variables. In this example, our inputs (x) are demographics, medications, and diagnosis, and our target (y) is hospital acquired acute kidney injury (HAAKI). Based on the model parameters for logistic regression, our inputs can be continuous, nominal, or binary, and our output will be binary (yes/no, or 0/1). In this application, the goal or target variable will be to determine whether a patient is at risk for HAAKI. For identified at-risk patients, interventions can be completed to improve mortality outcomes.

Data Set File: 6_EL1_HAAKI.xlsx

Variables:

- PatientID, unique patient identifier
- Admit_Age, in years
- BMI, body mass index
- Gender, male (M) or female (F)
- Race, (American Indian-Alaskan, Asian-Pacific Islander, Black, White)

- Alcoholism, condition indicator (1=True, 0=False)
- Anemia, condition indicator (1=True, 0=False)
- Cancer, condition indicator (1=True, 0=False)
- COPD, Chronic Obstructive Pulmonary Disease or lung disease condition indicator (1=True, 0=False)
- Dyslipidemia, cholesterol condition indicator (1–True, 0–False)
- Hepatitis, condition indicator (1=True, 0=False)
- HIV, Human Immunodeficiency Virus condition indicator (1=True, 0=False)
- Dementia, condition indicator (1=True, 0=False)
- RA, Rheumatoid Arthritis condition indicator (1=True, 0=False)
- Hemiplegia, paralysis condition indicator (1=True, 0=False)
- HAAKI, Hospital Acquired Acute Kidney Injury condition indicator (1=True, 0=False)

Step 1: Sign in to SAS OnDemand for Academics.

Step 2. Open SAS Enterprise Miner (Click the SAS Enterprise Miner link).

Step 3. Create a New Enterprise Miner Project (click **New Project**).

Step 4: Use the default SAS Server, and click **Next**.

Step 5: Add Project Name *HospitalAcquiredConditions*, and click **Next**.

Step 6: SAS will automatically select your user folder directory (If you are using desktop version, choose your folder directory), and click **Next**.

Step 7: Create a new diagram (Right-click **Diagram**).

Step 8: Add a File Import node (Click the **Sample** tab, and drag the node onto the diagram workspace).

Step 9: Click the File Import node, and review the property panel on the bottom left of the screen.

Step 10: Click **Import File** and navigate to the 6_*EL1_HAAKI.xlsx* Excel File.

Step 11: Click **Preview** to ensure that the data set was selected successfully, and click **OK**.

Step 12: Right-click the File Import node and click **Edit Variables**.

Step 13: Set HAAKI to the Target variable role, set PatientID to the ID role, and all other variables to the Input role. Set the Admit_Age, BMI, and PatientID variables at Interval levels. Set the Gender and Race variables to Nominal levels. Set all remaining condition variables to Binary levels. To review an individual variable to verify its role and level assignment, click the variable name and click **Explore**. After you have finished setting all variables, click **OK**.

Figure 6.5: Edit Variables

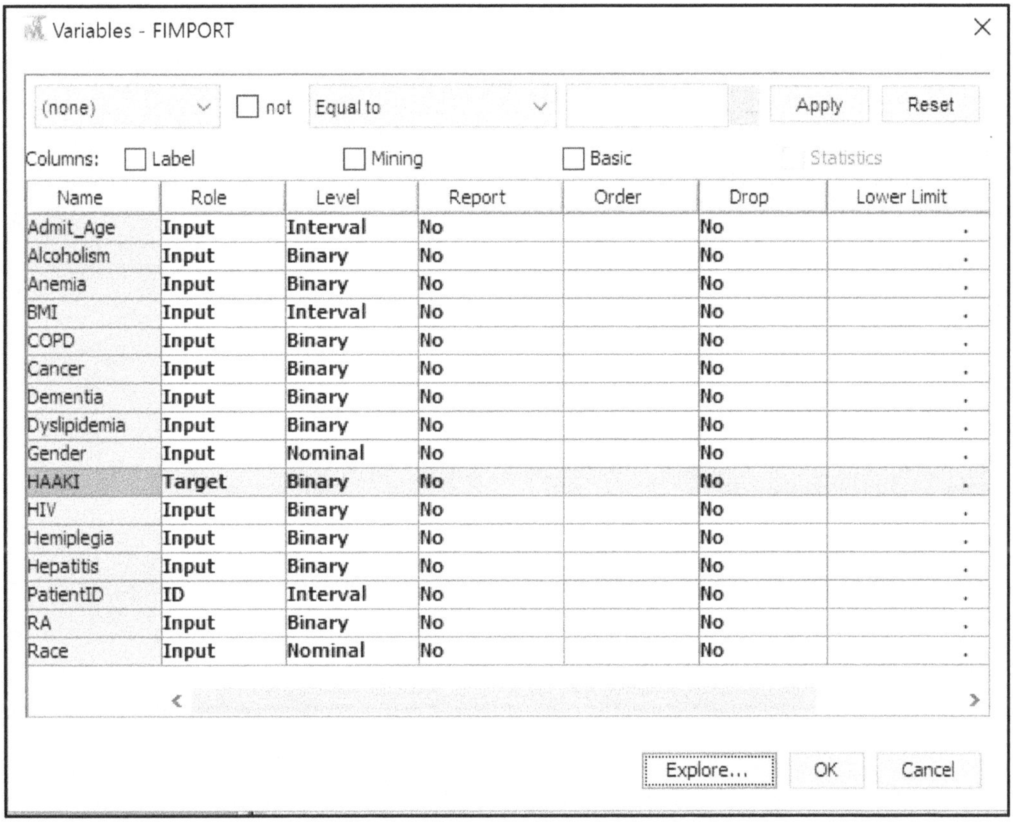

Name	Role	Level	Report	Order	Drop	Lower Limit
Admit_Age	Input	Interval	No		No	.
Alcoholism	Input	Binary	No		No	.
Anemia	Input	Binary	No		No	.
BMI	Input	Interval	No		No	.
COPD	Input	Binary	No		No	.
Cancer	Input	Binary	No		No	.
Dementia	Input	Binary	No		No	.
Dyslipidemia	Input	Binary	No		No	.
Gender	Input	Nominal	No		No	.
HAAKI	Target	Binary	No		No	.
HIV	Input	Binary	No		No	.
Hemiplegia	Input	Binary	No		No	.
Hepatitis	Input	Binary	No		No	.
PatientID	ID	Interval	No		No	.
RA	Input	Binary	No		No	.
Race	Input	Nominal	No		No	.

Step 14: Add a Data Partition node (Click the **Sample** tab, and drag the node onto the diagram workspace). Set the Data Partition Property **Data Set Allocations** to 60.0 for Training, 40.0 for Validation, and 0.0 for Test.

Figure 6.6: Data Partition Node

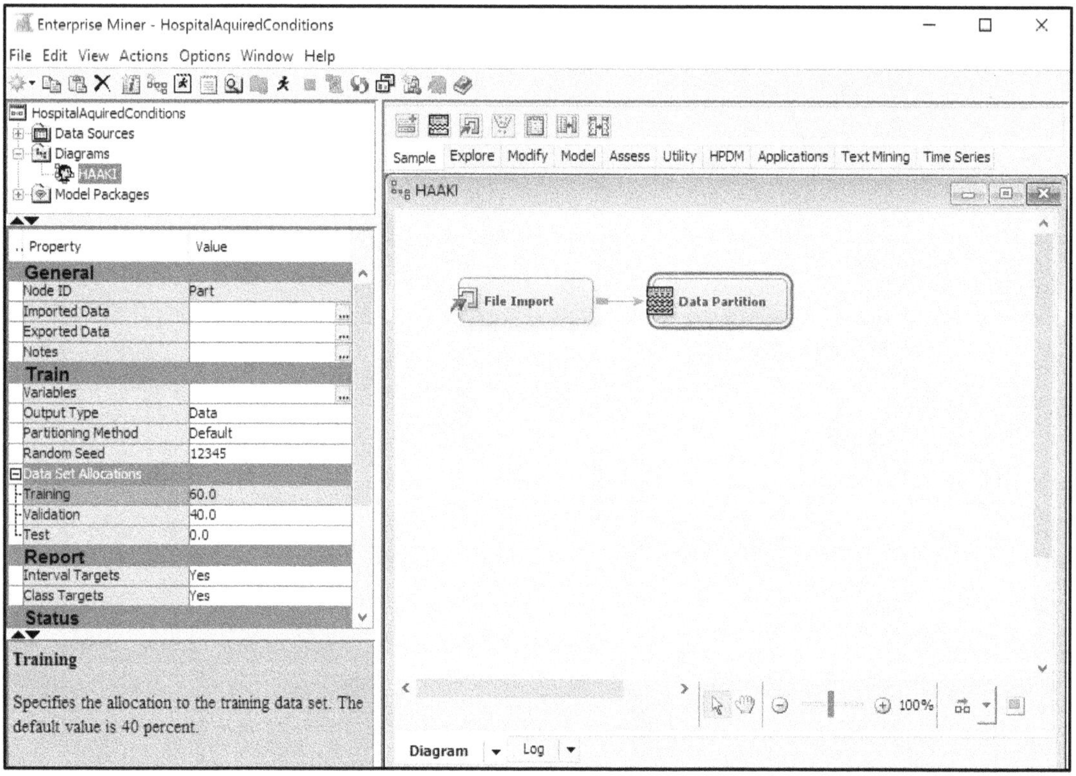

Step 15: Review the Data Partition results, and verify the number of observations in the original data (of 16000 records), and the split between Train of 9599 and Validate of 6401.

Figure 6.7: Data Partition Results

```
 Output

 25
 26     Partition Summary
 27
 28                                     Number of
 29     Type            Data Set        Observations
 30
 31     DATA         EMWS3.FIMPORT_train      16000
 32     TRAIN        EMWS3.Part_TRAIN          9599
 33     VALIDATE     EMWS3.Part_VALIDATE       6401
 34
 35
 36     *--------------------------------------------------------------*
 37     * Score Output
 38     *--------------------------------------------------------------*
 39
 40
 41     *--------------------------------------------------------------*
 42     * Report Output
 43     *--------------------------------------------------------------*
 44
 45
 46
 47
 48     Summary Statistics for Class Targets
 49
 50     Data=DATA
 51
 52                  Numeric    Formatted    Frequency
 53     Variable      Value       Value        Count       Percent    Label
 54
 55      HAAKI          0           0          15200          95       HAAKI
 56      HAAKI          1           1            800           5       HAAKI
 57
 58
 59     Data=TRAIN
 60
 61                  Numeric    Formatted    Frequency
 62     Variable      Value       Value        Count       Percent    Label
 63
 64      HAAKI          0           0           9119       94.9995     HAAKI
 65      HAAKI          1           1            480        5.0005     HAAKI
 66
 67
 68     Data=VALIDATE
 69
 70                  Numeric    Formatted    Frequency
 71     Variable      Value       Value        Count       Percent    Label
 72
 73      HAAKI          0           0           6081       95.0008     HAAKI
 74      HAAKI          1           1            320        4.9992     HAAKI
```

Step 16: Add a Stat Explore node, Graph Explore node, and MultiPlot node (Click the **Explore** tab, and drag the nodes onto the diagram workspace). Set the Graph Explore Property Sample Size to **Max**.

Figure 6.8: StatExplore, Graph Explore, and MultiPlot Nodes

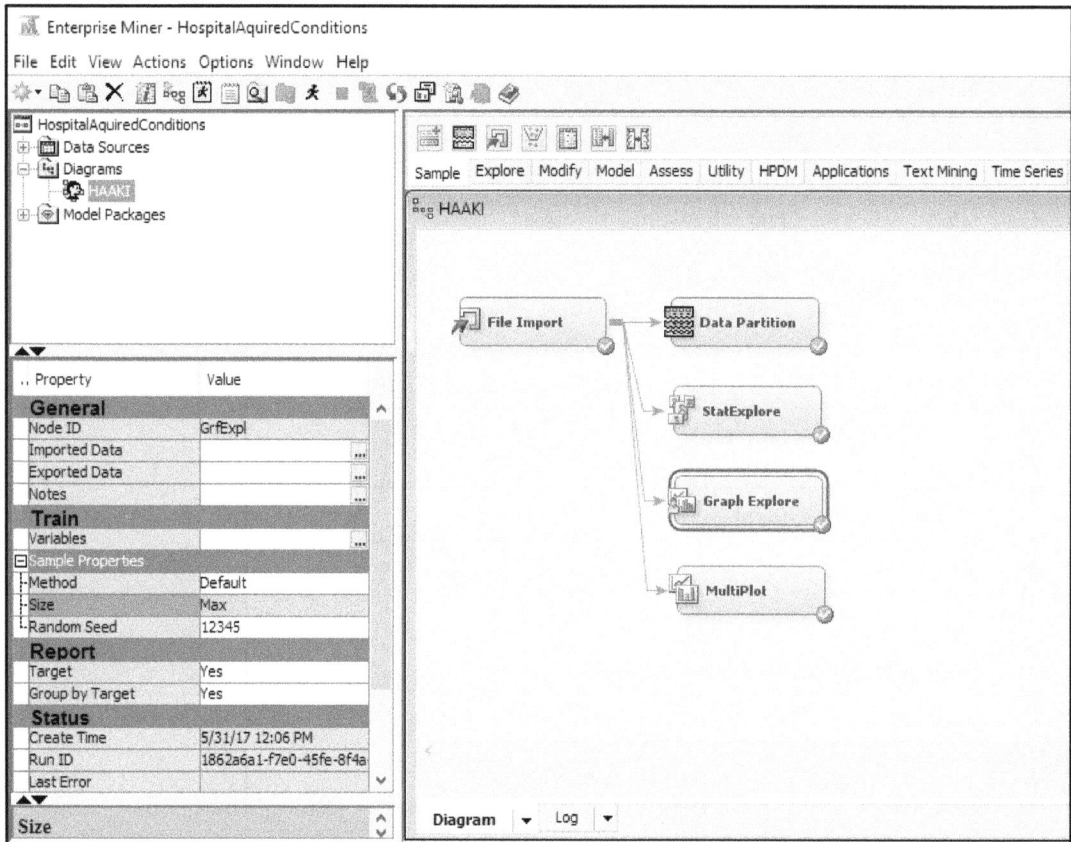

Step 17: Review the Stat Explore results for missing data and outliers. From the data, the BMI and Race variables are the most commonly missing value, with 642 and 467 records missing, respectively. Note the BMI and Race variables are included in two different sections of the output since BMI is an interval variable and Race is a nominal variable with five levels or categories. As this data set is simulated from VA study data, many times, the forms require some information and leave other information, as optional. There might also be cases where the form might not include all options if only a limited selection of choices, such as for race, are available, thereby causing individuals to bypass or skip the data entry. There might also be a case where individuals might not know or be able to easily calculate their BMI. For example, most individuals know their age, height, and weight. However, converting this to a BMI number is typically something most individuals would not be expected to know. In this instance, the system or electronic form could automatically compute this value, which would help improve the overall data quality and also would improve the efficiency of form completion. The automatic filling or calculation in forms would be ideal. For example, some forms have date of birth and age. However, age can be easily calculated from date of birth and can be considered duplicate or double-entry. As another example, a patient enters an ID on an electronic form, and the form then would autocomplete all demographic and medical information to allow the patient to complete only the additional necessary sections, saving time and improving data quality.

Figure 6.9: Stat Explore Results

```
Output
31   HAAKI       TARGET        2
32   PatientID   ID        15930
33
34
35
36   Class Variable Summary Statistics
37   (maximum 500 observations printed)
38
39   Data Role=TRAIN
40
41                            Number
42   Data     Variable          of                          Mode                Mode2
43   Role     Name        Role  Levels  Missing  Mode   Percentage  Mode2   Percentage
44
45   TRAIN    Alcoholism  INPUT    2       0      0        79.51      1        20.49
46   TRAIN    Anemia      INPUT    2       0      0        71.91      1        28.09
47   TRAIN    COPD        INPUT    2       0      0        65.75      1        34.25
48   TRAIN    Cancer      INPUT    2       0      0        74.10      1        25.90
49   TRAIN    Dementia    INPUT    2       0      0        94.08      1         5.92
50   TRAIN    Dyslipidemia INPUT   2       0      1        57.72      0        42.28
51   TRAIN    Gender      INPUT    2       0      M        96.11      F         3.89
52   TRAIN    HIV         INPUT    2       0      0        98.67      1         1.33
53   TRAIN    Hemiplegia  INPUT    2       0      0        95.78      1         4.22
54   TRAIN    Hepatitis   INPUT    2       0      0        89.84      1        10.16
55   TRAIN    RA          INPUT    2       0      0        96.92      1         3.08
56   TRAIN    Race        INPUT    5      467    White     76.17     Black     18.54
57   TRAIN    HAAKI       TARGET   2       0      0        95.00      1         5.00
58
59
60
61   Distribution of Class Target and Segment Variables
62   (maximum 500 observations printed)
63
64   Data Role=TRAIN
65
66   Data     Variable                  Frequency
67   Role     Name        Role  Level     Count   Percent
68
69   TRAIN    HAAKI       TARGET    0      15200      95
70   TRAIN    HAAKI       TARGET    1        800       5
71
72
73
74   Interval Variable Summary Statistics
75   (maximum 500 observations printed)
76
77   Data Role=TRAIN
78
79                           Standard       Non
80   Variable   Role   Mean  Deviation  Missing  Missing  Minimum  Median  Maximum  Skewness  Kurtosis
81
82   Admit_Age  INPUT 67.57881 5.776634   16000      0       58       68       77    -0.01876  -1.20072
83   BMI        INPUT 27.99114 2.591364   15358     642       24       28       32     0.004871 -1.23837
```

Step 18: Review the Stat Explore results variable worth. The graph gives an early indication of the variables that might prove most important in the model. For this model, the three variables that have the most worth with regard to the HAAKI target variables, in order of worth, are HIV indicator, Admit_Age, and Cancer indicator. From the graph, we also see Gender has the least worth. In attempting to describe the data results, you might review why gender has less worth for the target. In looking back at our data set description, the data set represents patients at a veteran's hospital (or for military personnel), and the gender is predominantly male at nearly 97%. Since the majority of the records contain a single value for gender, the ability of the model to use gender to determine HAAKI is limited. Expanding the data set to include additional hospitals with female patients might increase the gender variable importance.

Figure 6.10: Stat Explore Results Variable Worth

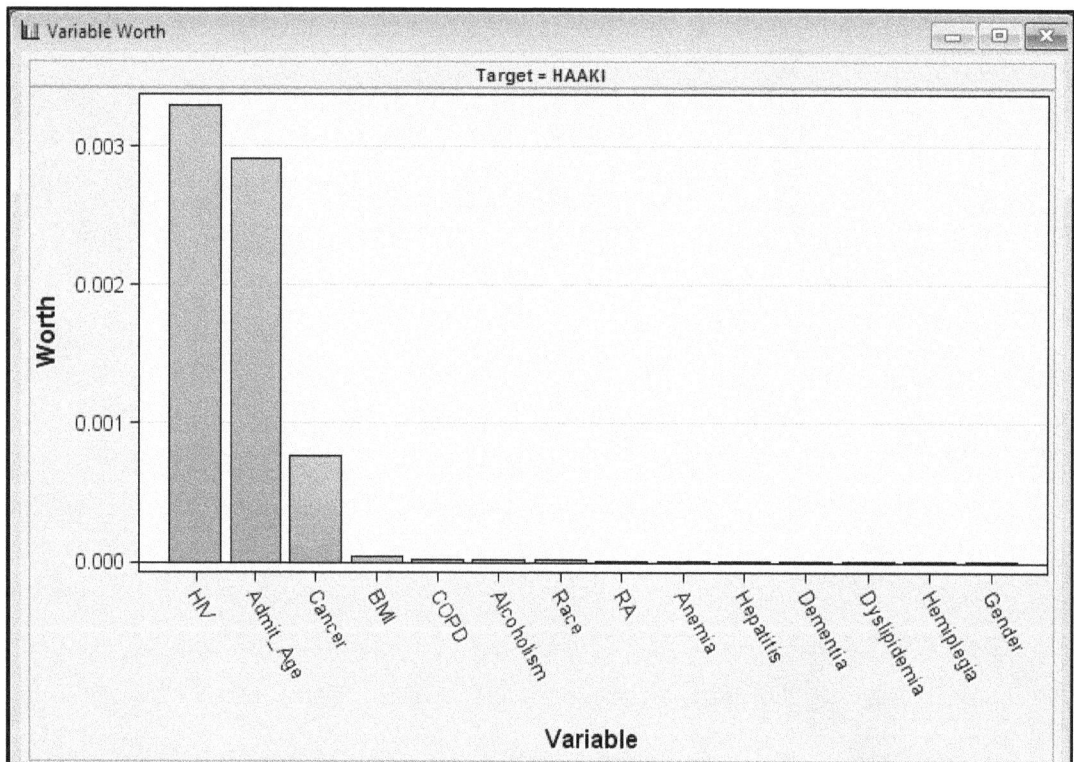

Step 19: Review the Graph Explore results. Place your pointer over the bars to see the values as tooltip text. HAAKI = 1 or True occurred in 800 records.

Figure 6.11: Graph Explore Results

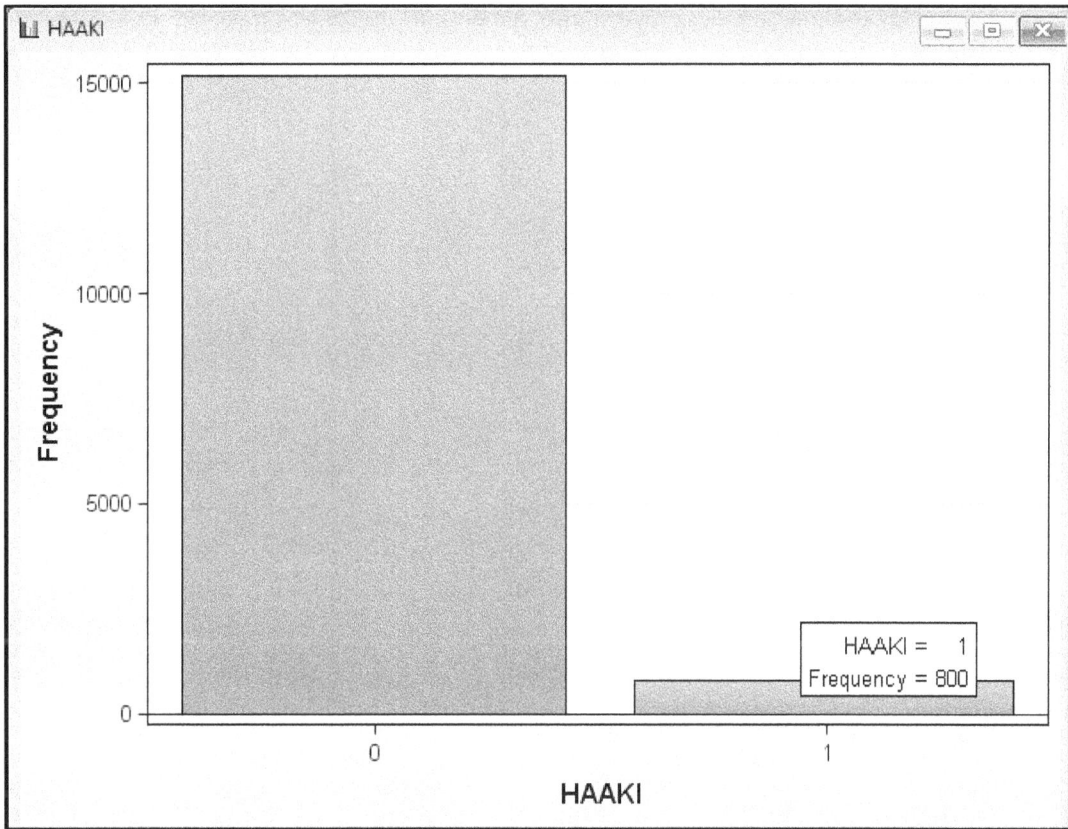

Step 20: Review the MultiPlot results. Scroll through the variables, and note the gender variable breakdown. We will revisit this in a later step.

Figure 6.12: MultiPlot Results

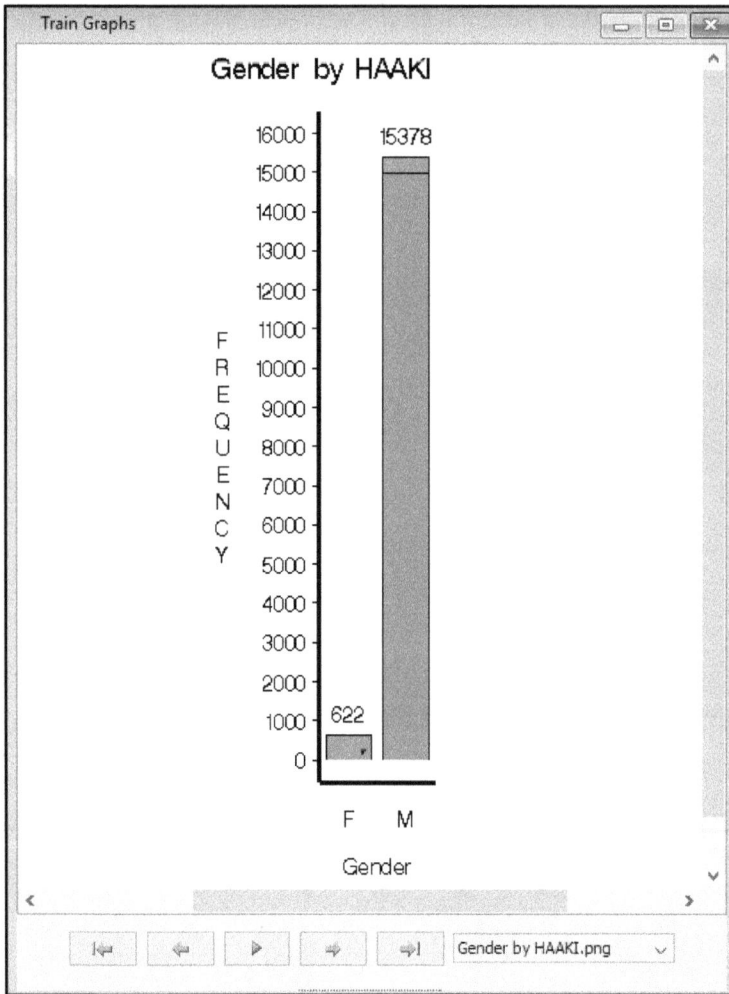

Step 21: Add an Impute node (Click the **Modify** tab, and drag the node onto the diagram workspace). Verify that the Impute Property is set to Count for Class variables and that Mean is set for Interval variables.

Figure 6.13: Impute Node

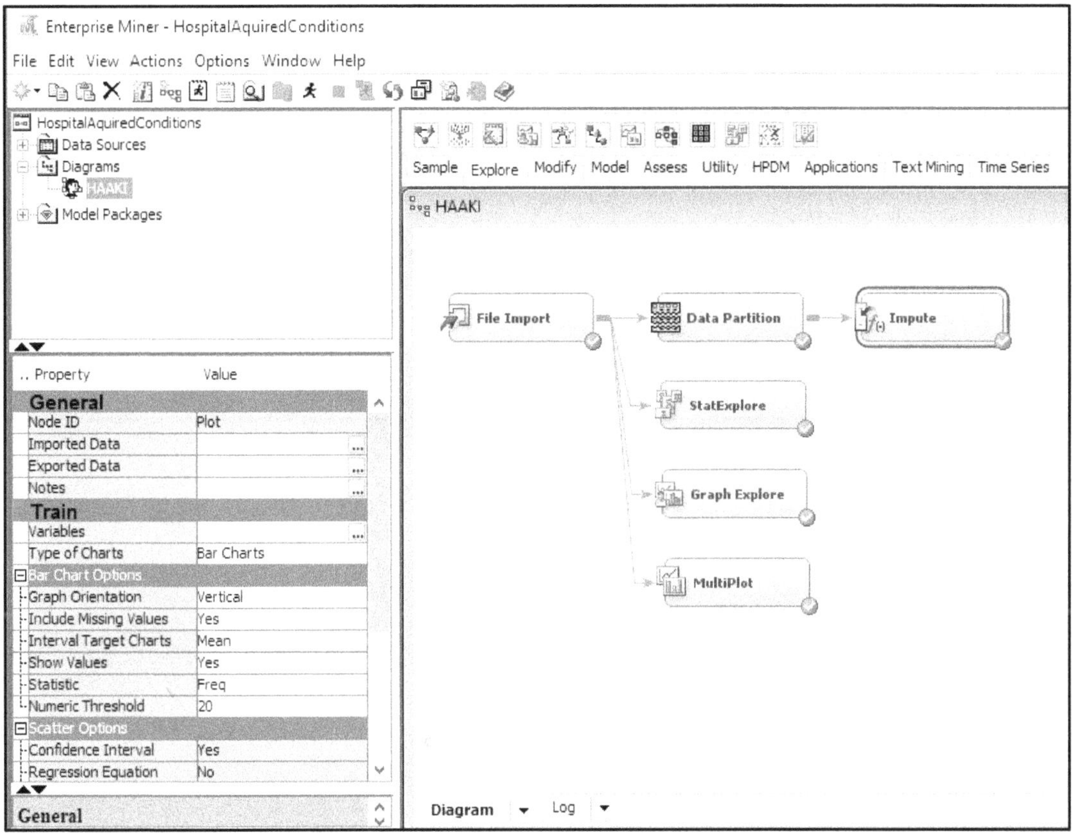

Step 22: Review the Impute results. The BMI and Race variables were replaced. Since these are different types of variables, as indicated in the measurement level, either the mean or count was used. For BMI, since this is an interval variable, the average or mean of 27.99 was used to replace any missing values. For Race, since this is a nominal variable, the value with the highest count in the data set of White was used to replace the missing values. Note that because the Impute node new variables are created in the data set, we now have IMP_BMI and IMP_Race. The two variables are added to the end of the data set with the original variables preserved.

Figure 6.14: Impute Results

```
Output
34    Imputation Summary
35    Number Of Observations
36
37                                                                                      Number of
38    Variable    Impute    Imputed                              Measurement              Missing
39      Name      Method    Variable    Impute Value    Role        Level       Label    for TRAIN
40
41    BMI         MEAN      IMP_BMI     27.994900727    INPUT     INTERVAL      BMI         382
42    Race        COUNT     IMP_Race    White           INPUT     NOMINAL       Race        283
```

Step 23: Add a Regression node (Click the **Model** tab, and drag the node onto the diagram workspace). Note that in the property panel for class targets, the Logistic Regression model is automatically selected. The model selection is based on the Target variable Role in the Edit Variables under the File Import, which we completed in the earlier steps.

Figure 6.15: Regression Node

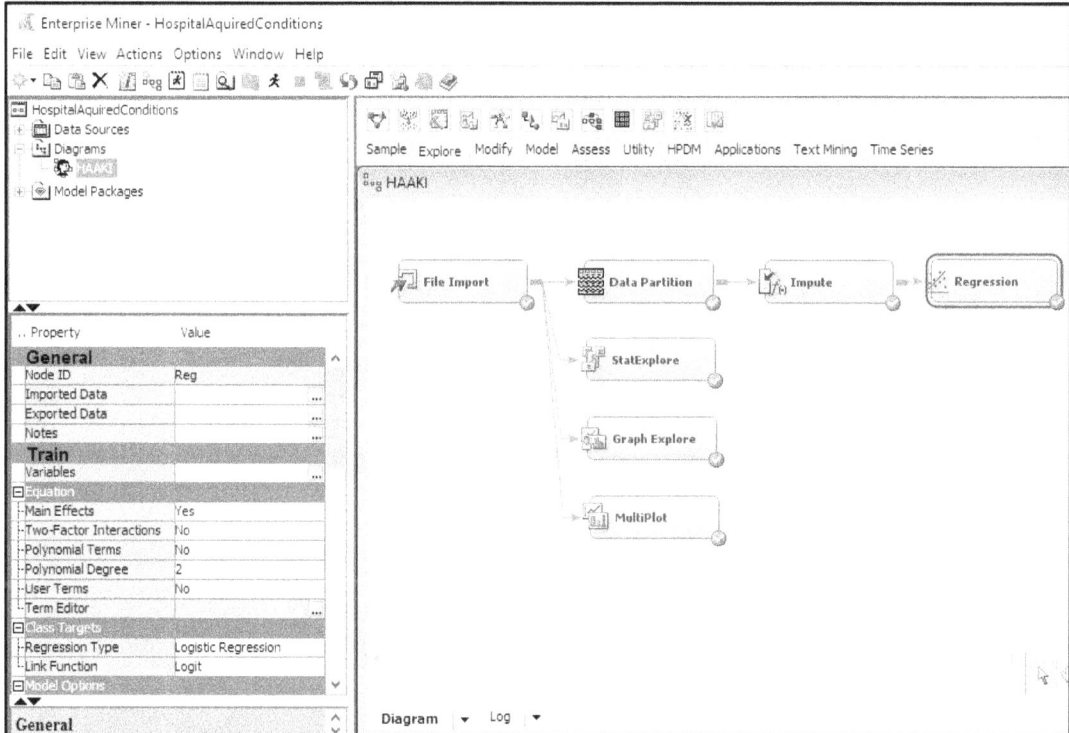

Step 24: Right-click the Regression node, and click **Run**.

Step 25: Expand Output Window Results.

Step 26: Review Analysis of Maximum Likelihood Estimates Model Results. The output includes an analysis of maximum likelihood estimates, which includes parameter and Pr > ChiSq (or significance).

Figure 6.16: Analysis of Maximum Likelihood Estimates

```
Output
148
149                         Analysis of Maximum Likelihood Estimates
150
151                                          Standard      Wald                 Standardized
152   Parameter                     DF   Estimate   Error  Chi-Square  Pr > ChiSq   Estimate   Exp(Est)
153
154   Intercept                      1    -7.9185  0.8801      80.94     <.0001                  0.000
155   Admit_Age                      1     0.0969  0.00905    114.47     <.0001      0.3103       1.102
156   Alcoholism    0                1     0.0926  0.0615       2.27      0.1322                  1.097
157   Anemia        0                1     0.0415  0.0554       0.56      0.4533                  1.042
158   COPD          0                1    -0.0320  0.0506       0.40      0.5265                  0.968
159   Cancer        0                1    -0.4292  0.0492      76.22     <.0001                   0.651
160   Dementia      0                1     0.0859  0.1086       0.63      0.4291                  1.090
161   Dyslipidemia  0                1     0.0452  0.0488       0.86      0.3543                  1.046
162   Gender        F                1     0.1286  0.1138       1.28      0.2584                  1.137
163   HIV           0                1    -1.4303  0.1024     195.28     <.0001                   0.239
164   Hemiplegia    0                1     0.0426  0.1260       0.11      0.7354                  1.044
165   Hepatitis     0                1     0.1364  0.0842       2.62      0.1054                  1.146
166   IMP_BMI                        1    -0.0200  0.0191       1.10      0.2940     -0.0279       0.980
167   IMP_Race  American Indian-Alaskan  1    -0.3870  0.4524     0.73      0.3923                  0.679
168   IMP_Race  Asian-Pacific Islander   1     0.4082  0.3012     1.84      0.1753                  1.504
169   IMP_Race  Black                1    -0.0589  0.1931       0.09      0.7604                  0.943
170   RA            0                1     0.1331  0.1585       0.71      0.4010                  1.142
171
```

Step 27: Review Odds Ratio Results. Odds can range from 0 to infinity and is the ratio of success to failure. If the odds of success are 4 (or 4 to 1), the odds of failure would be the inverse (or 1% to 4% or 25%). The odds ratio can be calculated from the odds of each of the groups and compares the odds of an event that occurs divided by the odds that the event does not occur. Odds ratio interpretation varies based on whether the target variable is interval or nominal. Interval (or continuous) variables that have an odds ratio over 1 explain that the target variable is more likely to occur as the input variables increase. For target variables that are nominal or categorical, an odds ratio greater than 1 explains that the category is more likely to occur. An Odds Ratio for Level 0 versus Level 1 is 4 indicates that the event of 1 is 4 times greater that the odds for level 0 (Minitab, 2016; IDRE, 2017).

Figure 6.17: Odds Ratio

			Point
Effect			Estimate
Admit_Age			1.102
Alcoholism	0 vs 1		1.203
Anemia	0 vs 1		1.087
COPD	0 vs 1		0.938
Cancer	0 vs 1		0.424
Dementia	0 vs 1		1.187
Dyslipidemia	0 vs 1		1.095
Gender	F vs M		1.293
HIV	0 vs 1		0.057
Hemiplegia	0 vs 1		1.089
Hepatitis	0 vs 1		1.314
IMP_BMI			0.980
IMP_Race	American Indian-Alaskan vs White		0.654
IMP_Race	Asian-Pacific Islander vs White		1.448
IMP_Race	Black vs White		0.908
RA	0 vs 1		1.305

Output — Odds Ratio Estimates (lines 173–193)

Step 28: Review Event Classification Table Results. The Event Classification Table shows the overall model performance. When training the model, we received 455 False Negatives, 9107 True Negatives, 12 False Positives, and 25 True Positives. To calculate the error rate, we add the False Negatives and False Positives and divide by the total number: $(455 + 12) / (455 + 9107 + 12 + 25) = 467/9599 = 4.87\%$ error rate. We can do the same for the Validate data set: $(296 + 10) / (296 + 6071 + 10 + 24) = 306/6401 = 4.78\%$ error rate. Both of these values indicate that our model performs well for determining the status of HAAKI. If the training error rate was much lower than the validation error rate, this can indicate that we overfit the model. In this application, our model performed well when presented with a new set of validation data.

Figure 6.18: Event Classification Table

```
264
265     Event Classification Table
266
267     Data Role=TRAIN Target=HAAKI Target Label=HAAKI
268
269       False        True         False        True
270     Negative     Negative     Positive     Positive
271
272       455         9107          12           25
273
274
275     Data Role=VALIDATE Target=HAAKI Target Label=HAAKI
276
277       False        True         False        True
278     Negative     Negative     Positive     Positive
279
280       296         6071          10           24
```

Step 29: Review Fit Statistics Results. Alternatively, instead of manually calculating the error rate from the Event Classification Table, the Fit Statistics output window can be used for various error rate comparisons. Notice the last entry on Misclassification Rate, which matches the manual calculation above at 4.87% for Train and 4.80% for Validation.

Figure 6.19: Fit Statistics

Fit Statistics	Statistics Label	Train	Validation
AIC	Akaike's Information Criterion	3493.423	.
ASE	Average Squared Error	0.044019	0.044344
AVERR	Average Error Function	0.180197	0.183863
DFE	Degrees of Freedom for Error	9582	.
DFM	Model Degrees of Freedom	17	.
DFT	Total Degrees of Freedom	9599	.
DIV	Divisor for ASE	19198	12802
ERR	Error Function	3459.423	2353.812
FPE	Final Prediction Error	0.044175	.
MAX	Maximum Absolute Error	0.990296	0.99104
MSE	Mean Square Error	0.044097	0.044344
NOBS	Sum of Frequencies	9599	6401
_NW_	Number of Estimate Weights	17	.
RASE	Root Average Sum of Squares	0.209807	0.210581
RFPE	Root Final Prediction Error	0.210178	.
RMSE	Root Mean Squared Error	0.209993	0.210581
SBC	Schwarz's Bayesian Criterion	3615.303	.
SSE	Sum of Squared Errors	845.0724	567.6978
SUMW	Sum of Case Weights Times Freq	19198	12802
MISC	Misclassification Rate	0.048651	0.047805

Step 30: In the Logistic Regression properties, the best model is selected based on the lowest Validation Misclassification rate in a Stepwise selection. Verify these match with your property settings and re-run the logistic regression node. In other words, when the logistic regression adds one variable at a time and tests the validation misclassification rate after each variable is added, the lowest or best error rate model is used.

Figure 6.20: Logistic Stepwise Property

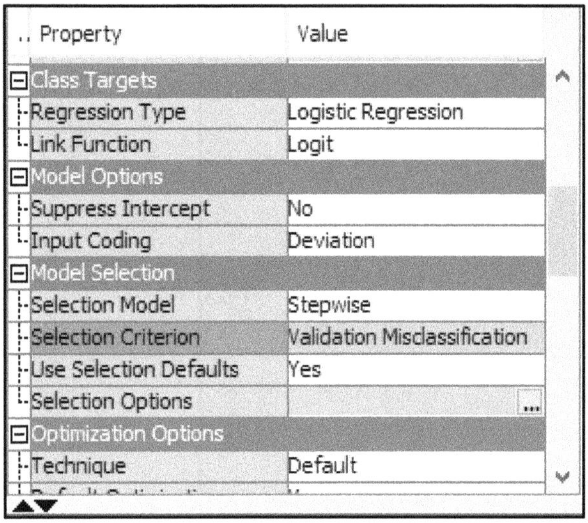

.. Property	Value
⊟ Class Targets	
┆─Regression Type	Logistic Regression
└─Link Function	Logit
⊟ Model Options	
┆─Suppress Intercept	No
┆─Input Coding	Deviation
⊟ Model Selection	
┆─Selection Model	Stepwise
┆─Selection Criterion	Validation Misclassification
┆─Use Selection Defaults	Yes
└─Selection Options	...
⊟ Optimization Options	
┆─Technique	Default

Step 31: Review Logistic Stepwise Results. In the Logistic Regression results, the best model selected is in step 2 with two variables of Admit_Age and HIV. The results indicate that, with these two variables, a low validation misclassification rate is achieved. The model is sometimes called the parsimonious model, or, put another way, is the simplest-best model. In other words, all things considered, we want to choose the simplest and best performing model instead of an overly complex model. The parsimonious model is chosen due to a number of reasons, including diminishing improvements in validation error rates, cost to collect additional data variables, and performance when running the model in a live operational setting. The stepwise selection guides us in choosing the simplest-best model or the parsimonious model. (You might also find humor in the irony that the parsimonious term name is not so simple!)

Figure 6.21: Logistic Stepwise Results

```
Output

                         Analysis of Maximum Likelihood Estimates

                                      Standard      Wald              Standardized
      Parameter          DF  Estimate   Error    Chi-Square  Pr > ChiSq   Estimate   Exp(Est)

      Intercept           1   -8.1410   0.6324     165.70      <.0001                 0.000
      Admit_Age           1    0.0963   0.00904    113.62      <.0001       0.3085    1.101
      Cancer     0        1   -0.4280   0.0491      76.05      <.0001                 0.652
      HIV        0        1   -1.4125   0.1017     192.80      <.0001                 0.244


                              Odds Ratio Estimates

                                                Point
      Effect                                  Estimate

      Admit_Age                                1.101
      Cancer       0 vs 1                      0.425
      HIV          0 vs 1                      0.059


      NOTE: No (additional) effects met the 0.05 significance level for entry into the model.


                              Summary of Stepwise Selection

                                                              Validation
                  Effect          Number     Score     Wald   Misclassification
      Step       Entered    DF     In     Chi-Square  Chi-Square  Pr > ChiSq    Rate

        1        HIV         1      1      321.6329               <.0001       0.0500
        2        Admit_Age   1      2      121.7590               <.0001       0.0472
        3        Cancer      1      3       79.7974               <.0001       0.0473


      The selected model, based on the misclassification rate for the validation data, is the model trained in Step 2. It consists of the following effects:

      Intercept  Admit_Age  HIV
```

Model Summary

In summary, the logistic regression can contain 1 or more inputs and 1 target. The inputs can be continuous or categorical, whereas the target will be binary (0/1 or yes/no). To evaluate the logistic regression, we can use error, lift, p-value, odds ratio, and misclassification rate.

- Model: Regression node
- Logistic Regression: 1+ input and 1 target variable
- Input: Interval, Binary, or Nominal
- Target: Binary or Nominal
- Evaluation: Error, Lift, p-value, Odds Ratio, Misclassification Rate

Experiential Learning Application: Immunizations

Executive Summary

According to the World Health Organization (WHO), immunizations help stop 2-3 million annual deaths. Polio, which is an infectious viral disease, that can cause paralysis, was once considered to be eradicated. It is now returning to at-risk populations such as children. It is estimated that nearly 19 million infants have missing vaccinations. In some cases, reports about vaccination results, such as links to other diseases, have led to distrust, along with personal, religious, or moral objections and have caused children not be vaccinated. In Ukraine, nearly 50% of children have not received all immunizations. In California, an outbreak of measles in 2014 resulted in a legislative bill to require all children to be vaccinated, with opposition to overturn the requirement. Structural or cultural issues can also create issues due to lack of resources or competing priorities. WHO is working on a Global Vaccine Action Plan (GVAP) with a goal of 90% national vaccination coverage by 2020. Early results show that the progress is behind schedule due to public acceptance even when reports or previously eradicated diseases arise (Nesson, 2016).

Past studies have attempted to review factors associated with vaccination. These factors reviewed include education, age, socio-economic status, religion, health usage, urban and rural areas, and media exposure. In a study in Zimbabwe, three sets of factors were reviewed to determine the immunization status of children 12-23 months. Predisposing factors such as gender, birth order, mother age, marital status, media exposure, religion, place of delivery, and natal care were reviewed. Enabling factors included education, wealth, rural-urban, region, and employment status. Region of the country was also found to have a significant effect on vaccination, with some regions being five times more likely to be vaccinated. The education level of the mother was also significant. Mothers with secondary education were two times more likely to have their children fully vaccinated. High wealth or income mothers were two times more likely to receive full vaccination. Birth order, place of delivery, and care during pregnancy also were significant factors in vaccination (Munkungwa, 2015; Wiley, 2017).

The CDC also releases annual vaccination schedules for children and those under 18 years of age, along with separate vaccination schedules for adults. The immunizations are provided each year based on approved recommendations from the Advisory Committee on Immunization Practices (ACIP). The yearly CDC schedule includes a number of immunizations and updates, including recommended ages and frequency. Immunizations include influenza, measles mumps and rubella (MMR), hepatitis A and B, tetanus diphtheria and acellular pertussis (Tdap), pneumococcal, meningococcal, polio, and others. The CDC also maintains a Behavioral Risk Factor Surveillance System (BRFSS), which includes prevalence of immunization data within the U.S. (CDC, 2017b; CDC, 2017c, CDC, 2017d).

A sample has been provided for you to get started, the Immunization data set includes 16 variables and 1817 records. Help your management team answer the following questions, follow the SEMMA process and provide recommendations.

Question: Which factors best determine whether a patient will receive a flu shot immunization?

Data Set File: 6_EL2_Immunizations.xlsx

Variables:

- Year
- State, CA
- FluShotVaccine, (1=True, 0=False)
- Gender, (Female, Male)
- Age Group, in years (65-74, 75+)
- Race/Ethnicity, (Black non-Hispanic, Hispanic, Multiracial non-Hispanic, White non-Hispanic, Other non-Hispanic)
- Education_Attained (College graduate, Some post-H.S., H.S. or G.E.D., Less than H.S.)
- Immunization, (0/1, No/Yes)

Follow the SEMMA process for your experiential learning application. A template has been provided below that can be reused across future projects.

Figure 6.22: SEMMA Process

Title	Immunizations
Introduction	Provide a summary of the business problem or opportunity and the key objective(s) or goal(s).
	Create a new SAS Enterprise Miner project.
	Create a new Diagram.
Sample	Data (sources for exploration and model insights)
	Identify the variables data types, the input and target variable during exploration.
	Add a FILE IMPORT
	Provide a results overview following file import:
	Input / Target Variables
	Generate a DATA PARTITION
Exploration	Provide a results overview following data exploration
	Add a STAT EXPLORE
	Add a GRAPH EXPLORE
	Add a MULTIPLOT
	Summary statistics (average, standard deviation, min, max, and so on.)
	Descriptive Statistics
	Missing Data
	Outliers
Modify	Provide a results overview following modification
	Add an IMPUTE
Model	Discovery (prototype and test analytical models)
	Apply a stepwise regression model and provide a results overview following modeling.
	Add a REGRESSION
	Model description
	Analytics steps
	Regression results (p-value, odds ratio)
	Model results (Classification Matrix, Lift, Error, Misclassification Rate)
	Correlation Matrix (which variables measure the same thing)
	Selection Model
Assess and Reflection	Provide overall recommendations to business
	Model advantages / disadvantages
	Performance evaluation
	Model recommendation
	Summary analytics recommendations
	Summary informatics recommendations
	Summary business recommendations
	Summary clinical recommendations
	Deployment (operationalization plan: timeline, resources, scope, phases, project plan)
	Value (return on investment, healthcare outcomes)

Learning Journal Reflection

Review, reflect, and retrieve the following key chapter topics only from memory and add them to your learning journal. For each topic, list a one sentence description/definition. Connect these ideas to something you might already know from your experience, other coursework, or a current event. This follows our three-phase learning approach of 1) Capture, 2) Communicate, and 3) Connect. After completing, verify your results against your learning journal and update as needed..

Key Ideas – Capture	Key Terms – Communicate	Key Areas - Connect
Provider Anamatics		
Provider Inpatient/Outpatient Setting		
EHR/EMR		
EHR Implementation Process		
EHR Implementation Success Factors		
Open vs Closed Source		
Cloud vs On-Premise		
Logistic Regression		
Odds Ratio		
Event Classification Table		
Stepwise Regression		
Parsimonious		
Hospital Acquired Conditions Application		
Immunizations Application		

Chapter 7: Modeling Payer Data

Chapter Summary

The purpose of this chapter is to develop data modeling skills using SAS Enterprise Miner, and with respect to the Model capabilities within the SEMMA process. The chapter explores modeling with the decision tree model. This chapter also includes experiential learning application exercises on patient mortality indicators and self-reported general health. The focus of this chapter is detailed in Figure 7.1.

Figure 7.1: Chapter Focus – Model

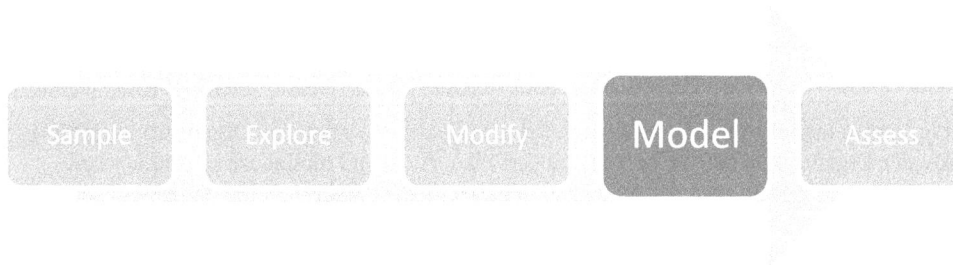

Chapter Learning Goals

- Describe the Model process steps
- Understand payer-level data sources
- Develop data modeling skills

- Apply SAS Enterprise Miner model data functions
- Master the decision tree model

Payers

Payers include insurance agencies and third-party payment processors. Typically, payers are separated into categories of commercial payers, governmental payers, and third-party payers. The most commonly known type of payer includes commercial payers such as UnitedHealth Group and Blue Cross Blue Shield. Government payers include Medicare and Medicaid, which will be covered further in chapter 8. Third-party payers include self-insured health plans by an employer and other insurance providers such as health costs covered through care insurance or workers' compensation (PHDSC, 2007). In some cases, providers are setting up independent health plans due to the shift in value-based healthcare to maximize returns. Hospitals could partner with a physician-owned health plan in order to share the cost savings from the value-based care. Hospitals and providers continue to work in hand with insurers to identify high-cost or high-risk patients and to improve coordination of care (Livingston, 2016).

Government insurances were originally designed out of a need, due to a lack of coverage available in commercial markets. Medicare was enacted for people over 65, since those over 65 were three times more likely to use medical services, and the costs were unaffordable for both patients and insurers. Medicare is a federal program in the U.S. and is paid for through payroll taxes. By pooling resources, it enables the protection of individuals in the event of a high-cost healthcare requirement. In addition, there are no exclusions to the program based on age beyond the minimum age of 65, health status, or income. By contrast, commercial insurance aims to avoid risk in order to ensure that a profit is made and the company remains in business. The for-profit aspect allows commercial insurers to exclude those of high risk or high cost, and create barriers to payment of all claims (Archer and Marmor, 2012). TRICARE is another government-based U.S. healthcare program for uniformed service members and families and covers general healthcare, prescriptions, and dental plans. The program is managed through the Defense Health Agency and seeks to provide a world-class healthcare system for its over 9.4 million participants. TRICARE offers plans that meet the ACA requirement to maintain minimum coverage (Tricare, 2017). Another related program is the Civilian Health and Medical Program of Veterans Affairs (CHAMPVA), and covers the majority of health expenses. To be eligible for CHAMPVA, members cannot be eligible for TRICARE, and the program has requirements for those with disabilities. CHAMPVA is managed through the Veterans Health Administration and has over 1,700 locations with over 8.7 million participants. Veterans programs also meet the ACA requirements, and veterans might choose among plans including from the VA and TRICARE (VFW, 2017; VA.gov, 2017).

Commercial insurance models typically have a model of a Health Maintenance Organization (HMO), Preferred Provider Organization (PPO), or Self-Funded Plan. An HMO model uses a primary care physician (PCP) to act as a main point of contact for a patient's care and refers the patients to specialists or other plans of care as appropriate. The PPO model permits patients to see different providers directly and, as a result, usually carry a higher premium. Both HMO and PPO plans typically have an assigned network of providers with which they contract and negotiate rates. Both HMO and PPO plans also typically pay providers per service rendered or also known as a fee-for-service model (MedicalBillingandCoding, 2017). A self-insured health plan, also known as a self-funded plan, is where the employer pays the financial cost for the healthcare for their employees. With an HMO or PPO plan, a monthly fixed premium is paid for each member, and the insurer pays the financial cost of healthcare. Self-funded plans might represent up to half of all commercial plans. Employers choose this type of plan to allow more customization of providers and coverage, maintain control, and reduce regulations and taxes. Disadvantages include financial risk that can be unpredictable. The self-funded type of plan can be challenging for small businesses or those with a high-cost population. Typically, an employer will contract with an existing insurer to administer the self-funded plan, giving a similar network and coverage options with the main difference of financial risk. Over 90% of employers with 5,000 of more workers are self-funded or partially self-funded (SIIA, 2015; Kaiser Family Foundation, 2016).

In the U.S., the largest commercial health insurers collect over $700 billion in annual premiums, and in 2017 the average annual family premiums were $18,764 (NCSL, 2017). Top insurers in the U.S. include UnitedHealth Group, Kaiser Foundation, WellPoint, Aetna, Humana, Cigna, Highmark, and Blue Cross Blue Shield (BCBS) within various state organizations such as BCBS of California (Heilbronn, 2017; NCSL, 2017). There have been previous attempts between health insurers to pursue mergers and acquisitions within the industry. Anthem offered $48 billion to acquire Cigna, and Aetna sought to acquire Humana for $34 billion. In both cases, federal judges blocked the acquisitions after U.S. Justice Department officials believed the combinations would lead to increased premiums due to reduced competition. Insurance companies argued that these would help them negotiate better prices from pharmaceutical companies and hospitals for their customers. The companies are still considering options for appeals, pending potential changes in federal administration, and pursuing various litigation with regard to termination fees and damages due to the failed mergers. With potential changes also planned to the Affordable Care Act and related U.S. healthcare legislation, many insurers are in a wait-and-see mode, holding on to their cash stockpiles and determination options for their investment (Tracer et al., 2017; Murphy, 2017). Other healthcare organizations are moving forward, with CVS proposing a nearly $70 $1billion merger with Aetna (Ramsey, 2018). Cigna acquired Express Scripts, a pharmacy benefit and healthcare management company for $67 billion. Walmart is reportedly reviewing their own options for acquiring Humana health insurance (Pearson, 2018). Walgreens Boots Alliance and wholesale drug distributor AmerisourceBergen have met to discuss a potential $25 billion deal. Other competitors including Amazon, Kroger, and Albertsons have been exploring varying strategies including acquisitions, mergers or alliances with healthcare payers and intermediaries (Hirsch and Sherman, 2018).

Payer Anamatics

Health anamatics has great potential to transform payer cost efficiency and coverage. Although analytics and informatics have been used by payers for many years for actuarial purposes, usage in other areas of payer operations has varied. Payers are currently focusing on patients, providers, and customers to improve savings. Payers also review financial and operational measures such as forecasting, operations, and fraud monitoring. On the patient side, payers have attempted to identify patients that will have future high costs, in order to increase interventions through preventative care. To assist providers, payers have focused on pay-for-performance programs and fraudulent billing. Similar to the providers, most payers are at stage 2 – localized analytics, of the five levels of analytics capability. Although localized analytic capabilities exist, a more complete organizational strategy is still in development, and organizational data warehouse and common analytics toolsets are not prevalent. Payers such as UnitedHealthcare through the acquisition of Ingenix as a subsidiary have developed more advanced analytics capabilities. Payers are also seen as being in an ideal position to use health anamatics given the large amounts of claims and transaction data available to them (Cheek, 2014)

Payer Data

Payer Systems and Sources of Data

Payers also generate and collect large amounts of data on patients, providers, and outside sources. Common payer systems include claims, population health management, financial and billing. As part of population health management, insurance sponsored or run personal health records and screenings also collect and store information about patients. Following the screening, wellness systems are offered through insurance carriers such as Florida Blue to promote healthier lives through eHealth education and incentives. Insurers also have an incentive to offering these programs as part of their plans aimed at reducing incurred healthcare costs. Insurers also collect personal data such as medical records and health history when reviewing coverage applications and claims. For example, some insurers might even access social media data to use while reviewing a claim. Although several payer systems are developed in-house, software vendors of provider systems also develop payer systems. Epic, a company known for their EHR software, has a product called Tapestry for managing health insurance. Features include enrollment verification, member portals, care management to improve health outcomes, customer relationship management modules, utilization management, and claims adjudication, processing and billing (Epic, 2017).

Another well-known company for claims processing is TriZetto, which permits electronic claims processing. Trizetto has software edits to reduce errors and improve payments. It creates secondary billing for claims with additional insurance payers, sends electronic bill remittances, and converts image files to HIPPA compliant EDI formats. Trizetto has built-in business intelligence and analytics for reports and tracking claims data for improving decision-making. In 2014, Cognizant announced a $2.7 billion acquisition of privately held Trizetto. The combined company includes 350 health payers and 180 million covered lives within the U.S. (Cognizant, 2014; Trizetto, 2017). McKesson also offers supporting software for payers such as ClaimsXten, an auditing software to improve accuracy of payments, increase auto-adjudication, convert between ICD 9 and 10 codes, reduce administrative costs, such as through identification of waste and abuse items, and local coverage decisions (McKesson, 2017).

Payer Systems Process

Health claim processing varies by vendor and plan, with advantages for each method. Overall, a common system framework exists despite the individual process differences. The first layer includes the data storage, contained within a database management system. The database stores the claims data, provider data, and member data. The system should contain a set of reporting tools for making business decisions, producing regulatory reporting, and providing customers with information. A system should contain a processing engine for developing rules to adjudicate or process claims, including automatic adjudication. The system should also provide a method for customization and modifications of benefits and rules, which is due to the increasing complexity of health claims processing (TM Floyd, 2006).

A typical process would begin first with the completion of a medical treatment. The medical treatment would then be followed by a claim submission to the insurer. The claim form might be mailed on paper or sent electronically. The claim form can be scanned and read or manually entered. The health plan will then review the claim to determine whether payment should be made. The payment determination and amount is called adjudication. During adjudication, the health plan will also check the patient benefits and eligibility for services. The provider will also be verified for processing and payment terms. Further checks for other insurance coverage will be made and various quality checks conducted, such as a duplicate claims check. Health plans might also reject the claim or deny the claim without payment. Once adjudication is complete and payment determinations made, an Explanation of Benefits (EOB) will be generated and sent to the insured, which is typically the patient (TM Floyd, 2006).

Claim Forms

Two primary paper claim forms are used for billing payers: the CMS-1500 previously known as an HCFA (Healthcare Financing Administration) form for outpatient claims, and a CMS-1450 previously known as a UB-92 (uniform billing) form for inpatient claims. Electronic claims use Electronic Data Interchange (EDI) transactions, which are standardized through HIPAA. The HIPAA 837 is the equivalent EDI transaction for the CMS-1500 and CMS-1450 paper forms (CMS, 2014; CMS, 2016b). EDI transactions for claims were modeled after the paper forms with additional capabilities. Within the forms and EDI transactions there are a number of common insurance billing terms. Table 7.1 includes a brief summary of key terms.

Table 7.1: Claims Terminology

Term(s)	Definition/Example
Guarantor, Health Plan, Payer	This is the financially responsible party for the claim, such as Aetna.
Subscriber, insured party, enrollee, member, beneficiary	This is the patient that represents the claim. Historically, this might have included the parent as the subscriber. However, most health plans now bill using the patient information only.
Member number, policy number, insurance ID	This is the unique identifier for the patient. Historically, this might have been a social security number. However, most health plans now assign each patient a unique identifier that does not identify the individual patient.
Group number	This is the unique identifier of the coverage group. Typically, this number is the same for all employees of a given employer and identifies the coverage and benefits.
Adjudication	This is the process of reviewing and paying the claim by the insurer. The claim can be automatically adjudicated or require a human review to determine coverage and payment.
Explanation of benefits (EOB), remittance advice	The explanation of benefits is provided to the patient following insurer adjudication. This form explains the charges from the provider and what was covered or paid by the insurer after adjudication.
Billed amount	This is the original amount billed by the provider for the service(s) rendered, and can include one or more services, such as an office visit, medical supplies, and so on.
Allowed amount	This is the maximum amount permitted for the service(s) as per the contract between the insurer and provider.
Contractual Adjustment	This is the difference between the billed amount and the allowed amount per the contract between the insurer and provider.

Term(s)	Definition/Example
Coordination of benefits, crossover or piggyback claims	This can occur if a patient has more than one coverage or priority coverage. For example, a claim can be first paid by Workman's Compensation as the primary payer if the employee was injured at work. Then any remaining amounts can be covered by the employee's commercial insurance as the secondary payer.
Copay, coinsurance, deductible, out of pocket amount	These are the amounts the patient is responsible for, even if the claim is paid by the insurer. These help offset the monthly premium costs and are referred to as shared patient responsibilities.

To learn more the claims process and claim forms, we will review claim billing and payment processing examples using the paper-based forms.

Experiential Learning Activity - Claim Forms Billing

Claim Forms Billing

Description: CMS-1500 is the most commonly used claim form, and is used for professional claims such as physician office visits. Navigate to http://CMS.gov, search for "CMS-1500," and open the CMS-1500 form. For this activity, you will practice completing the paper-based forms.

The form is completed through a series of numbered boxes, 1-33. Some boxes have subparts such as 1a. The form completion order is like reading a page, left to right, top to bottom. Complete the form using sample data for yourself as the patient. Some boxes will require a code lookup. In box 21, you will need to enter the ICD-10 diagnosis code. In box 21 d., you will need to enter the CPT/HCPCS code. There is a more detailed CMS-1500 form instruction available for field by field assistance.

CMS-1500 Form:

https://www.cms.gov/Medicare/CMS-Forms/CMS-Forms/Downloads/CMS1500.pdf

CMS-1500 Form Instructions:

http://www.cms.gov/Regulations-and-Guidance/Guidance/Manuals/Downloads/clm104c26.pdf

ICD-10 Lookup:

https://www.cms.gov/Medicare/Coding/ICD10/2018-ICD-10-PCS-and-GEMs.html

http://www.icd10data.com

HCPCS Fee Lookup:

http://www.cms.gov/apps/physician-fee-schedule/overview.aspx

Experiential Learning Activity: Claims Adjudication Processing

Claims Adjudication Processing

Description: For this experiential learning activity, you will take the role of a claims adjudicator. The claim is adjudicated following the payer receipt of the paper claim form from the provider. In order to adjudicate the claim properly, coverage information has been provided. Calculate the amounts following the review.

Eligibility: Span of insurance coverage based on service date.
Premium: Monthly payment for insurance.
Copay: Fixed amount due at time of each service regardless of deductible/OOP, can vary by type or location.
Deductible: Direct amount of payment before insurance begins.
Coinsurance: Percentage of allowed amount payment after deductible is met.
Out-of-Pocket Maximum: The highest cumulative payment in a calendar year, generally deductible, co-pay and co-insurance all count toward your out-of-pocket maximum.
Contracted Payment: Amount your insurance has agreed to pay the provider of care.

Benefit Coverage Details:
Eligibility Dates: 1/1/2018 – 12/31/2018
Plan: Florida BlueCare Everyday Health
Premium: $325 Per Month
Allowed Amount - Contracted Payment Rate: 80% of Billed
Copay: $0 Preventative (e.g. immunization) / $20 Primary Care Provider / $35 Specialist / $75 ER
Deductible: $500 Per Person / $1,600 Per Family
Coinsurance: 10% of the Allowed Amount (after deductible is met)
Out-of-Pocket Maximum: $2,500 Per Person / $5,000 Per Family

2018 Beneficiary Records (Totals YTD):
Member 11111111:
- Deductible Individual / Family: $0 / $0
- OOP Individual / Family: $0 / $0

Member 22222222:
- Deductible Individual / Family: $500 / $1600
- OOP Individual / Family: $2500 / $5000

Member 33333333:
- Deductible Individual / Family: $100 / $100
- OOP Individual / Family: $20 / $20

Claims Adjudication Processing

Claim # 1000

Member	11111111
Service Date	3/1/2018
Service/Procedure	Office Visit
Billed Amount	$100
Allowed Amount	
Not Covered Amount	
Copay Amount	
Deductible Amount	
Coinsurance Amount	
Patient Responsible	
Insurance Responsible	

Claim # 1001

Member	22222222
Service Date	12/5/2017
Service/Procedure	Office X-Ray
Billed Amount	$150
Allowed Amount	
Not Covered Amount	
Copay Amount	
Deductible Amount	
Coinsurance Amount	
Patient Responsible	
Insurance Responsible	

Claim # 1002

Member	33333333
Service Date	5/15/2018
Service/Procedure	Specialist Visit
Billed Amount	$250
Allowed Amount	
Not Covered Amount	
Copay Amount	
Deductible Amount	

Claims Adjudication Processing	
Coinsurance Amount	
Patient Responsible	
Insurance Responsible	

Electronic Data Interchange

Now that you are familiar with the paper-based forms for claims billing, we will cover the alternative electronic method known as EDI, which captures a similar set of data as the paper form and includes the ability to send additional information that is not included in the paper forms. Electronic Data Interchange (EDI) is the computer-to-computer exchange of information that uses international standards. Large retailers such as Wal-Mart, the automotive industry, and the healthcare industry all use EDI. EDI uses computerized technology to exchange data and improve processing efficiencies, delivery times, reliability, and quality over existing methods. EDI allows for standardized and efficient transmission of data between organizations. EDI is included as part of the Health Insurance Portability and Accountability Act (HIPAA) standards to facilitate administrative cost savings and efficiencies. HIPAA required the Secretary of Department of Health and Human Services to adopt standards to support the electronic exchange of administrative and financial healthcare transactions primarily between healthcare providers and plans. Transaction standards and specifications were adopted by the Secretary to enable health information to be exchanged electronically. Implementation guides for each standard have been produced at the time of adoption, and consistent usage of the standards (including loops, segments, and data elements) across all guides is mandatory to support the Secretary's commitment to standardization (Woodside, 2013).

The typical healthcare data process flow involves setting the standard transaction set in batch mode through a file transfer protocol (FTP) or other similar transport method over a Value-Added Network (VAN). The process typically results in a transmission/receipt occurrence once per day. The alternative would be a real-time transmission/receipt resulting in multiple transmission/receipts per day and would use HTTP or a similar protocol as the transport method. The EDI x12 standard is then converted to XML through a variety of third-party applications or custom-built software. The XML data is then stored in a database, typically as a character large object (CLOB). Existing applications that need to interface with the various EDI transactional data (such as billing systems, claims systems, membership systems, authorization systems, and financial systems), typically cannot read EDI or XML. This results in a secondary conversion to a fixed file format readable by the source system. The data is then stored within the source system for use, resulting in data redundancy within internal systems. For EDI transactions responses (such as a 271, 277, 278, and 835), the process repeats. The source system produces a file in a fixed file format. The file is then converted to XML, which is then converted to the EDI x12 standard (Woodside, 2013).

A clearinghouse, which is an intermediary between providers and health plans, might be used. The typical role of a clearinghouse is to receive EDI transactions from the provider and payer, convert them to the appropriate format, and send them on to the appropriate party. A clearinghouse might take non-HIPAA formatted data and translate it to the standard HIPAA EDI format. A clearinghouse might also run quality or edit checks and analytics on the transactions. Typically, a per transaction fee is assessed by the clearinghouse. A clearinghouse is permitted to transmit PHI, which is considered one of the covered entities under HIPAA. Change Healthcare is one of the largest clearinghouses in the U.S. with over 2,100 payer connections, 5,500 hospitals, and 800,000 physicians. In the latest fiscal year, Change Healthcare processed over 12 billion healthcare transactions and $2 trillion in claims (Change Healthcare, 2017).

In an effort to reduce the costs of healthcare, which in the U.S. has averaged double the inflation rate per year since the 1970s, EDI standards were created as part of the 1996 HIPAA act. The U.S. is not alone in these efforts. China has also implemented measures to promote EDI, including policies and infrastructure investment. Problems confronting healthcare organizations include increasing costs and inefficiencies in resources. Hospitals began using EDI to communicate with other hospitals, suppliers, insurance companies, and banks. The relatively limited EDI presence is explained by high EDI start-up costs as compared with labor, unfamiliar new relationship-making, and technical infrastructure and complexity. A New Jersey state study, the HINT project, estimated the cost savings from the application of computerized systems. Their findings included estimates that 17% of costs are related to processing, and a minimal reduction in those costs would amount to several billion dollars across the industry. Most payers have already put significant investment into computer technology and can further tap into EDI. One of the most detailed and comprehensive analysis for EDI standards was created by Workgroup for Electronic Data Interchange (WEDI). A large number of estimates were provided and included pilot projects. WEDI mentioned that although estimated savings might not result in hard-dollar savings, it will allow for efficiency to be improved and resources to be re-allocated to improve quality, care, and service. Additional studies list benefits, which include near-term reduction of paperwork and a long-term potential to use information technology to improve quality and cost effectiveness of healthcare. System data standards integrated across parties will allow for improved accuracy, reliability, and data usage (Woodside, 2007b).

As part of the HIPAA legislation, a set of approved EDI transactions was developed to simplify processing and reduce costs. The transactions were developed in compliance with ANSI standards and some documentation includes the ANSI prefix. EDI is popular across many different industries (such as finance and manufacturing) with different transaction types and was applied similarly to healthcare. The standard set includes:

Table 7.2: EDI Transactions

Category	Transaction	Description
Authorization	278	Referral certification and authorization. This is used to request prior authorization for a service and provider referral to ensure payment. Precertification or preauthorization is the prior approval by the payer of a certain action to be taken by the provider during treatment. The claim might be denied if authorization is not requested prior to the service. EDI reduces time spent by the payer contacting one another or the provider. Additional time is reduced by documenting and/or entering data received manually. Assuming 30% referrals, admissions and emergency room visits require review/approval; the payer savings from using the EDI 278 transaction is $0.81 to $1.23 per transaction, with provider savings of $0.65 to $0.98 per transaction.
Claims	837	Claims or equivalent encounters and coordination of benefits. The 837 is used for billing the claim, which is similar to the paper form CMS-1450 and CMS-1500. The 837 might have an associated sub-designation, such as 837-I or 837-P, which is a designated institutional or professional claim to match with the same billing of CMS-1450 (institutional claims) and CMS-1500 (professional claims). Claims transactions are simplified through EDI. Information can be entered and transmitted electronically from the provider to the payer. Claim information can be re-sent easily. Re-sent information includes claims corrections and adjustments. The estimate of payer savings ranges from $0.50 to $1.50, minus a transaction cost of $0.17. The provider cost per transaction for physician claims varies from $0.51 to $1.96, with hospital claims from $0.11 to $1.07. Coordination of benefits transactions enables electronic transmission on a single claim. The cost savings potential for payers is $0.22 per transaction. The savings for providers is $0.95 to $1.16 per transaction, based on savings by not identifying, copying, and re-submitting remittances from one payer to another.

Category	Transaction	Description
Claims – Additional Information	275	Patient Information in Support of a Health Claim or Encounter. The 275 is used for attaching electronic information such as clinical information, lab reports, emergency department, rehabilitative services, ambulance services, and medications. This is for supplemental information not included on the 837 EDI transaction.
Claims - Status	276-277	Claim status inquiry (276) and response (277). The 276-277 is a paired transaction to check on the payment status of a claim. The 276 is sent by the provider to the insurer, and the insurer returns a 277 with the current claim status. In the past, providers might have called on the telephone and waited on hold while the status of the claim was checked. Now they can have real-time updates as needed. Claims status transactions typically are received by mail or phone. It is estimated that public and private healthcare payers receive over 60 million claim status inquiries per year, and EDI is estimated to save payers $1.06 to $2.72 net per inquiry, and save providers $3.56 to $3.88 per inquiry.
Claims - Response	835	Remittance and payment advice, which is known as an Electronic Remittance Advice (ERA). The 835 is used to provide the explanation of benefits and payment and describe how the claim was adjudicated and provides details about the claim payments. Payment and remittance transactions include transfer of funds typically by check, and the explanation of the benefit payments from the payer. Potential savings include electronic remittance and electronic funds transfer transaction. The savings result from the elimination of postage and handling. The manual costs of processing a remittance and payment range from $0.45 to $1.00, and the costs of processing under EDI are $0.11 to $0.35. The net savings range between $0.10 to $0.89. Approximately $73,432 can be saved per year per hospital, and $1,918 in savings per year per physician practice.

Category	Transaction	Description
Membership - Eligibility	270-271	Eligibility benefit inquiry (270) and response (271). The 270-271 is a paired transaction to check on the eligibility and benefits of the patient. The 270 is sent by the provider to the insurer, and the insurer returns a 271 with the current claim status. In the past, providers might have had to call by telephone and wait on hold to have the status of eligibility checked. Now they can have real-time updates as needed. A patient might not know their co-pay amount, and a provider can easily verify the amount. Eligibility transactions allow confirmation of an individual's eligibility for healthcare services payment by a third party, as well as a determination of benefit coverage including patient liabilities. An estimation of 150 million transactions occur each year, primarily by telephone. The savings estimated for payers is of $0.50-1.00 per inquiry. The savings estimated for providers is $1.10 to $2.09 per inquiry.
Membership - Payment	820	Health plan premium payments. The 820 is used to make monthly payments for the insurance enrollment, which is typically the set of employees.
Membership - Enrollment	834	Enrollment and unenrollment in a health plan. The 834 is to add to or remove from monthly membership in the insurance, which typically occurs as employees are hired and leave each month.
Pharmacy	NCPDP 5.1	Retail drug claims, coordination of drug benefits, and eligibility inquiry. NCPDP is used for billing pharmacy services such as at Walgreens. Electronic prescribing (e-prescribing) savings are estimated at $27 billion per year in the U.S. Savings are due to reduced errors, improved efficiency, and easier access to payer drug formularies or approved drugs when prescribing. Also, savings result from the ability to substitute lower cost generic drugs or formulary options when available (Porterfield et al., 2014).

EDI Structure

To review the EDI transaction structure, first picture a blank text file. Then to complete a transaction, within the text file, you might have different data elements such as the patient name or the amount billed. If everyone completed the text file, in whatever way they thought best, you would wind up with many different variations of the text file. EDI creates a set of standards and exact positions within the text file to place your data elements such as patient name and amount billed. With this solution, everyone sends and receives their text files the same way and they can be easily translated using EDI.

To begin with basic EDI terminology, there are a series of loops within an EDI text file. Think of these as headings when writing a paper: you have your title, introduction, analysis, and conclusion. Similarly, with EDI loops you have transaction file title information, submission information, patient information, claim information, and individual service information. The loops come into play because you can repeat information at each of the levels. A submitter might submit 100 claims even though their information is needed only once. Likewise, a patient might have multiple services and their information is needed only once. Table 7.3 includes the common loops. Each loop is assigned a standard alphanumeric number, beginning with the 1000A loop for submitter, and continuing through 2400 for services. The table of loops and segments are contained within an 837-P claim transaction. For simplicity, the key loops, segments, and elements are included. Other loops and segments might include default or standard information about each file. Each EDI transaction such as the 837, 835 or 270, has a slightly different set of loops and segments, though following a similar structure.

Table 7.3: EDI Loops and Segments

Loop	Name	Segment	Elements
1000A	Submitter Name	NM1	NM101-NM109
1000A	Submitter Contact Info	PER	PER01-PER09
1000B	Receiver Name	NM1	NM101-NM109
2010AA	Billing Provider Name	NM1	NM101-NM109
2010AA	Billing Provider Address	N3	N301
2010AA	Billing Provider City/State/Zip	N4	N401-N403
2010AA	Billing Provider ID	REF	REF01-REF02
2000B	Subscriber Info	SBR	SBR01-SBR-09
2010BA	Subscriber Name	NM1	NM101-NM109
2010BA	Subscriber Address	N3	N301
2010BA	Subscriber City/State/Zip	N4	N401-N403
2010BA	Subscriber Demographic Info	DMG	DMG01-DMG03
20101BB	Payer Name	NM1	NM101-NM109
2300	Claim Info	CLM	CLM01-CLM09
2300	Claim ID	REF	REF01-REF02
2300	Health Diagnosis	HI	HI01-HI02
2400	Service Line	LX	LX01
2300	Professional Line	SV1	SV101-SV109

Loop	Name	Segment	Elements
2400	Service Date	DTP	DTP01-DTP03

Within each loop is a series of segments, and each segment has a name and designation. Within the 1000A submitter loop, there is an NM1 segment for name, and a PER segment for contact info. Next, within each segment there are a series of elements, which are numbered with the segment plus 01, 02, 03, and so on. Some segments might have more or fewer elements. The NM1 segment has nine elements, numbered from NM101, NM102, NM103, NM104, NM105, NM106, NM107, NM108, and NM109. In many files, a comma is used to separate values. However, a comma might be used in someone's name such as Name, Jr. Using the comma would create an issue when separating a file into elements. Instead, a unique delimiter is used to separate the fields using an asterisk. The special character of tilde ~ is used to end a row or segment. A segment might therefore look as follows, with each element listed below:

NM1*NM101*NM102*NM103*NM104*NM105*NM106*NM107*NM108*NM109~

Below is another example of a completed segment with a Florida hospital included as the provider.

NM1*85*2*FLORIDA HOSPITAL*****XX*1033239991~
N3*3565 S. MAGNOLIA AVE.~
N4*ORLANDO*FL*32806~

Note that some elements contain special codes to designate the following field. In position NM103, there is a '2', which according to the EDI standard designates an organization, and a code of '1' indicates a person such as an individual provider. Note also that there are continued asterisks '*' in sequence, which indicate e no values. In a segment, there are also required and optional elements. The 'XX' in NM108 indicates that the next value is an NPI number such as 103323991. The segment positions are counted by using the asterisks '*' as the delimiters. Start from the left and count the segment and element such as NM101 = '85', NM012 = '2', ... NM109 = '1033239991. The last part would be the tilde '~' to indicate that the segment is complete. The next segment that follows the billing name would be the address, and city, state, and zip. Each segment has a designation. In this case, N3 is used for the address, and N4 is used for the city, state, and zip.

Experiential Learning Activity: EDI Translation

EDI Translation

Description: For each of the EDI transactions, translate the information and location of the information by connecting your knowledge of EDI and claim processing.

2010AA: NM1*85*1*CARE*SAM****XX*1234567890~

Loop Name:
Provider Name:

EDI Translation

Provider ID Qualifier Value:
Provider ID Quality Location: NM1___
Provider ID Value:
Provider ID Location: NM1___

2010BA:
NM1*IL*1*SMITH*JANE****MI*222334444~
DMG*D8*19431022*F~

Loop Name:
Patient Name:
Patient ID:
Patient DOB:
Patient Gender:
DMG02 Designation:
DMG03 Designation:

What is the NPI number in the EDI line:
NM1*85*2*MAYO CLINIC*****XX*1922074434~

What position is the Provider Name located in the EDI line:
NM1*85*2*NEMOURS*****XX*1234567890~

Build the EDI segments for the following provider:
Name: JOHN HOPKINS EMERGENCY MEDICAL SYSTEM
NPI: 1619903622
Address: 5755 CEDAR LN, COLUMBIA, MD 21044-2912

Now that you are familiar with the payer system, data, and claim processing, a variety of payer data is captured that can now be used for analysis. We'll continue with our SEMMA modeling process using data available through payer systems, and review the decision tree model.

SEMMA: Model

Model Process Step Overview

During the modeling process step, the data mining model is applied to the data. During the partitioning phase, data is segmented into training and validation data sets. The training data set is used to fit the model, and the validation data set is used to validate the model on a new set of data to demonstrate the reliability of the model. Based on the results, the model can then be tuned to optimal performance.

There are many different models that can be selected during this step. The decision to choose each model is based on earlier exploration of data and knowledge of each model. We will continue this chapter with the model of decision tree.

Model Tab Enterprise Miner Node Descriptions

Model Application Examples

Decision trees can be applied to a variety of areas within the healthcare setting. Researchers in Taiwan examined ICD-9-CM codes within claims data to identify cases of coronary artery bypass graft infections from a sample of 1,017 surgeries. The overall goal of the researchers was to accurately predict infection sites in an effort to improve quality. Their decision tree model performed well in terms of true positive predictive performance. A set of regression models was also run to compare performance. However, researchers noted limitations in the regression model's ability to handle the highly dimensional data, and the decision tree was able to more easily classify highly dimensional data. The first branch (or split) in the decision tree was the length of stay variable (Yu et al., 2014).

In another study, MVP Health Care, with over 750,000 members in the eastern U.S., implemented a set of decision trees for prior claim authorization. MVP Health Care estimated $2 million in savings associated with the improved prior authorizations. Typical prior authorizations previously took 2-3 days' turnaround, cost $75 each, and required a telephone call. The decision tree made the results available through a web-based interface. The interface walked the provider staff through a few short questions, with the results determined based on the individual patient benefits and medical information. Most medical policies and technology assessments are not standardized and are often out-of-date. The medical policies and guidelines can be updated dynamically and can be linked to electronic health records to improve information transparency among stakeholders. The policies can be standardized for systematic communication and centralized for more timely updates (Moeller, 2009).

DecisionTree Node

The decision tree node is used for the decision tree model and is associated with the **Model** tab in SAS Enterprise Miner.

Figure 7.2: DecisionTree Node

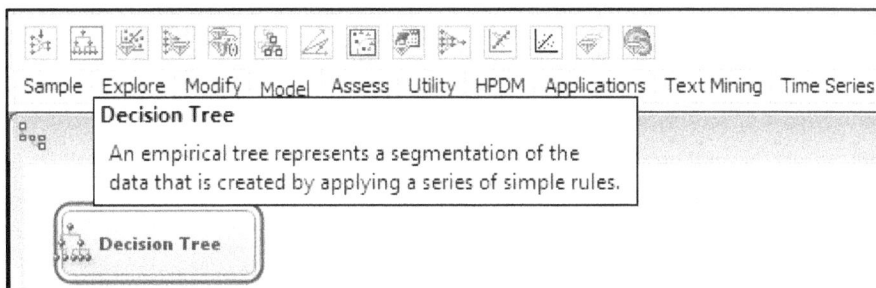

Decision Tree Model Description

Decision Trees are a flexible model capable of handling various input and target data types, along with missing and non-standardized data. Decision trees can handle binary, continuous, or nominal inputs and output variables, whereas most models have more specific variable type requirements. Decision trees also do not require the statistical assumptions that must be met with models such as multiple linear regression. As a result, decision trees are one of the most popular and widely used techniques because they can also be easily communicated. Decision trees are modeled after actual trees, although in contrast with living trees, decision trees are often depicted top to bottom or left to right. Just like real trees, decision trees are grown starting with the primary node or trunk of the tree and following a series of branches, segments (or splits) based on the variables in the data set. The decision tree is fully grown following a series of splits (or branches) to the terminal nodes or leaves of the tree (Klimberg and McCullough, 2013).

Model Assumptions and Data Preparation

Due to the overall flexibility, decision trees carry fewer model assumptions and requirements than earlier models such as regression. However, decision trees, like regression, might suffer from overfitting the model, because the model is perfected for training data and has become unable to model new data. To avoid overfitting, we should again select the parsimonious model or simplest-best model. An advantage of the decision tree is that limited data preparation is required as compared with other models. Decision trees handle missing data and outliers to a greater extent and are less affected than other methods. Although data quality is always important, often time constraints impact model selection.

Partitioning Requirements

For decision trees, we typically create two data sets: 1. Training and 2. Validation. In some cases, three data sets might be used: 1. Training, 2. Validation and 3. Test. When using the third data set for the decision tree process, train the tree (or grow the tree to its full potential), and then validate (or prune) the tree to remove extemporaneous or invaluable branches and paths to simplify or improve the shape of the tree. Then, test the tree using the pruned model from validation.

Model Results Evaluation

To evaluate the results of our decision tree, we can use several items including errors, lift, misclassification rate, and English rules.

Errors

Errors are calculated from the predicted value less the actual value, which is also known as the residuals. Common measures of errors are sum of squared errors (SSE) and root mean square error (RMSE). The measures are calculated by taking the square or square root of the errors.

Lift

Lift is a measurement between a random or baseline model against the analytical model. A higher lift or outperformance of the random selection is best.

Misclassification Rate

For models with nominal or binary targets, the percentage of total records is misclassified as false positive or false negative.

English Rules

For decision trees, a tree model is available that can produce a set of IF…THEN conditions, which are also known as English rules, and they assist with the interpretability of the model. The rules are often used in decision support systems to model human decision-making. If you visit your physician, based on their experience, they will instinctively walk through a set of IF…THEN rules to make a diagnosis. For example, if you have a fever and cough, and stuffy nose, then you are diagnosed with influenza (or the flu).

Now that we have covered the decision tree model, let's continue with an experiential learning application to connect your knowledge of decision trees with a health application on patient mortality indicators.

Experiential Learning Application: Patient Mortality Indicators

Many quality improvement approaches to improve quality of care are based on manual activities without a direct link to the data within the healthcare information system, which can influence quality of treatment and cost of care. Payer and provider systems can supply patient outcome information and clinical pathways to support patient care. Data mining through decision trees allows for knowledge discovery from large sets of data that can be used to identify patterns or rules (Woodside, 2010b).

A decision tree can be used to determine how inpatient mortality rates compare to overall proportions, and which segments to focus on. In one study, a set of 8,405 patients were used for indicators of inpatient mortality as part of decision tree analysis to determine inpatient mortality factors. Factors and indicators included gender, discharge location (such as surgery department), age group, and disease class. The results found that for patients discharged from Internal medicine departments, mortality was nearly three times more likely than for other discharge locations. Patients with length of stay (LOS) over 16 days, resulted in a six times higher mortality rate. The variable significance included LOS and discharge department, followed by age group. Although logistic regression could be used, the output would be missing the segment characteristics that would be useful (Chae et al., 2003).

Previous studies have examined factors (such as gender, discharge department, age group, and disease class) to determine a relationship with mortality. For this experiential learning application, we want to verify which of these factors might have a relationship with mortality. To start our process, we first want to identify our input and target variables. In this application, our inputs (x) are gender, discharge location, age group, and disease class. Our target (y) is patient mortality. With a decision tree, one advantage is that the model can handle varying input and target variable types. In this application, our inputs are nominal, and our target variable is binary.

Data Set File: 7_EL1_Patient_Mortality.xlsx

Variables:

- ID, unique identifier
- Gender, (Female, Male)
- Discharge Department, (Internal Medicine, Surgery)
- Age, (Under 20, 21-40, 41-60, 61 or older)
- Disease Class, (Circulatory, Congenital, Eye and Ear, Gastrointestinal, Miscellaneous, Muscle, Neoplasm, Pulmonary, Urinary)
- Length of Stay, (1-5 Days, LOS 17-341 Days, LOS 6-16 Days)
- Inpatient Mortality, (1=True, 0=False)

Step 1: Sign in to SAS Solutions OnDemand.

Step 2. Open SAS Enterprise Miner (Click SAS Enterprise Miner link).

Step 3. Create a New Enterprise Miner Project (Click **New Project**).

Step 4: Use the default SAS Server, and click **Next**.

Step 5: Add Project Name *PatientMortalityIndicators*, and click **Next**.

Step 6: SAS will automatically select your user folder directory (If using the desktop version, choose your folder directory), and click **Next**.

Step 7: Create a new diagram *PatientMortalityIndicators* (Right-click **Diagram**).

Step 8: Add a File Import node (Click the **Sample** tab, and drag the node onto the diagram workspace).

Step 9: Click the File Import node, and review the property panel on the bottom left of the screen.

Step 10: Click **Import File** and navigate to the *7_EL_1_Patient_Mortality.xlsx* Excel File.

Step 11: Click **Preview** to ensure that the data set was selected successfully, and click **OK**.

Step 12: Right-click the File Import node, and click **Edit Variables**.

Step 13: Set Inpatient_Mortality to the Target variable role, set ID to the ID role, and all other variables to the Input role. Set the remaining variables according to their nominal, interval, or binary levels. To review an individual variable and to verify its role and level assignment, click the variable name and click **Explore**. You have finished setting all variables, click **OK**.

Figure 7.3: Edit Variables

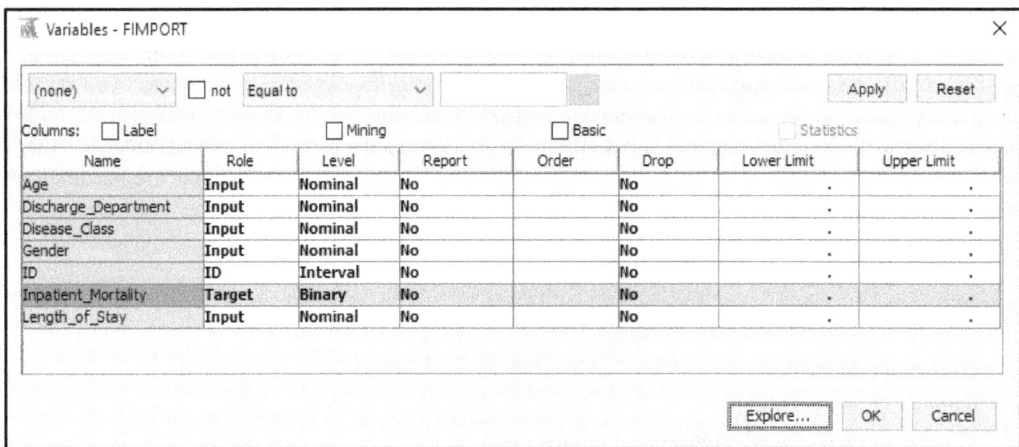

Name	Role	Level	Report	Order	Drop	Lower Limit	Upper Limit
Age	Input	Nominal	No		No	.	.
Discharge_Department	Input	Nominal	No		No	.	.
Disease_Class	Input	Nominal	No		No	.	.
Gender	Input	Nominal	No		No	.	.
ID	ID	Interval	No		No	.	.
Inpatient_Mortality	Target	Binary	No		No	.	.
Length_of_Stay	Input	Nominal	No		No	.	.

Step 14: Add a Stat Explore node, Graph Explore node, and MultiPlot node (click the **Explore** tab, and drag the nodes onto the diagram workspace). Set the Graph Explore Property Sample Size to **Max**.

Figure 7.4: StatExplore and Graph Explore Nodes

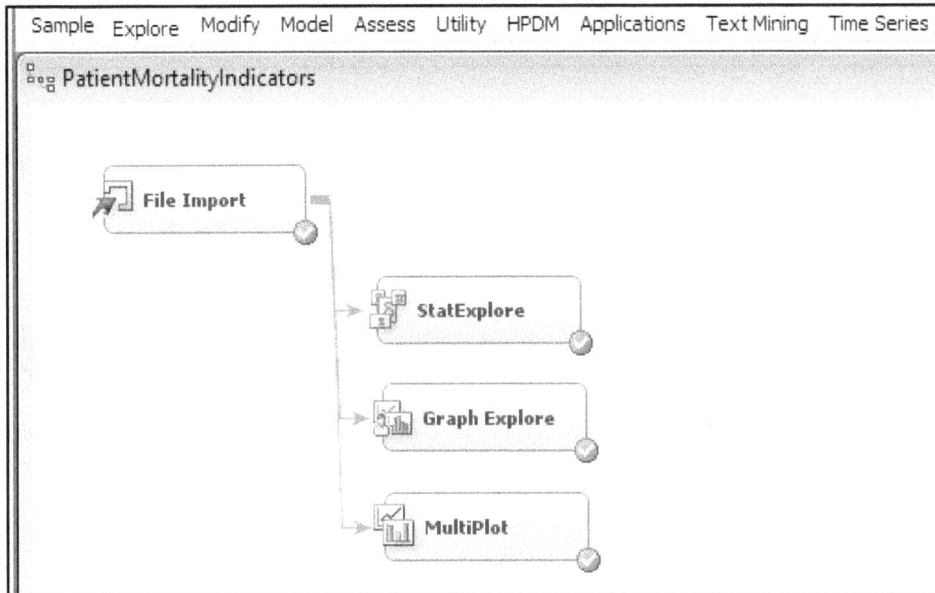

Step 15: Review the results. From the Stat Explore descriptive statistics results, we identify a good data quality result, which is verified through 0 missing records across variables. The breakout of the target variable Inpatient_Mortality is shown with 8235 records with a value of 0 (or False), and 170 records with 1 (or True).

Figure 7.5: StatExplore Results

```
Output
37    Data Role=TRAIN
38
39                                      Number
40    Data                              of                                    Mode                        Mode2
41    Role      Variable Name    Role   Levels  Missing  Mode              Percentage  Mode2           Percentage
42
43    TRAIN     Age              INPUT    4       0      41-60                31.22    Under 20            25.02
44    TRAIN     Discharge_Department INPUT 2      0      Internal Medicine    72.71    Surgery             27.29
45    TRAIN     Disease_Class    INPUT    9       0      Neoplasm             28.76    Miscellaneous       22.39
46    TRAIN     Gender           INPUT    2       0      Male                 52.96    Female              47.04
47    TRAIN     Length_of_Stay   INPUT    3       0      1-5 Days             52.91    LOS 6-16 Days       31.95
48    TRAIN     Inpatient_Mortality TARGET 2      0      0                    97.98    1                    2.02
49
50
51
52    Distribution of Class Target and Segment Variables
53    (maximum 500 observations printed)
54
55    Data Role=TRAIN
56
57    Data                               Frequency
58    Role      Variable Name    Role   Level   Count    Percent
59
60    TRAIN     Inpatient_Mortality TARGET  0    8235    97.9774
61    TRAIN     Inpatient_Mortality TARGET  1     170     2.0226
```

Step 16: Review the results. From the Stat Explore Variable Worth results, Length_of_Stay has the greatest variable worth with regard to our target variable of Inpatient_Mortality

Figure 7.6: StatExplore Results Variable Worth

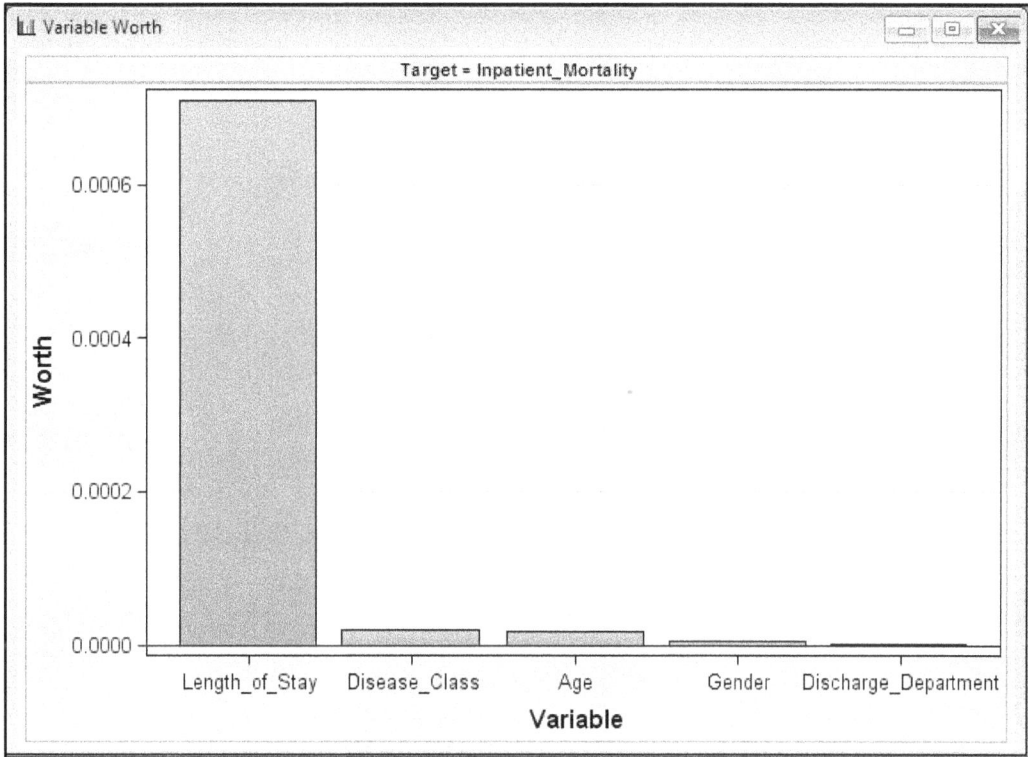

Step 17: Review the results. From the Graph Explore results, we also see the breakdown of the Inpatient_Mortality.

Figure 7.7: Graph Explore Results

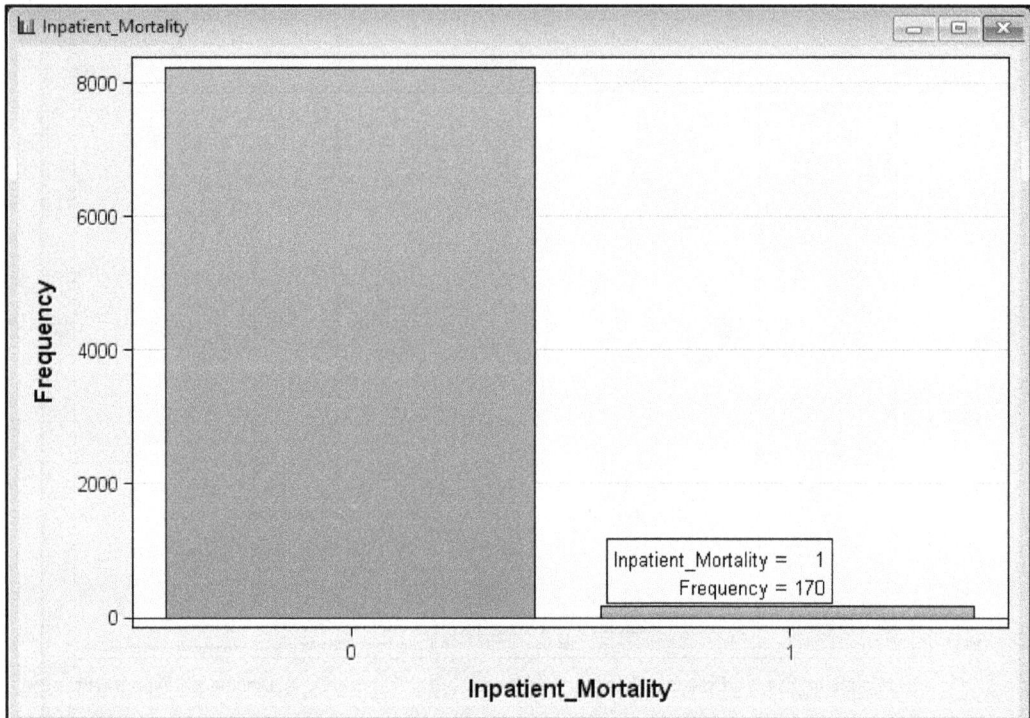

Step 18: Review the results. From the MultiPlot results, review each of the variables. For example, Age by Inpatient_Mortality shows a distribution across all age groups with both 0 and 1 frequencies.

Figure 7.8: MultiPlot Results

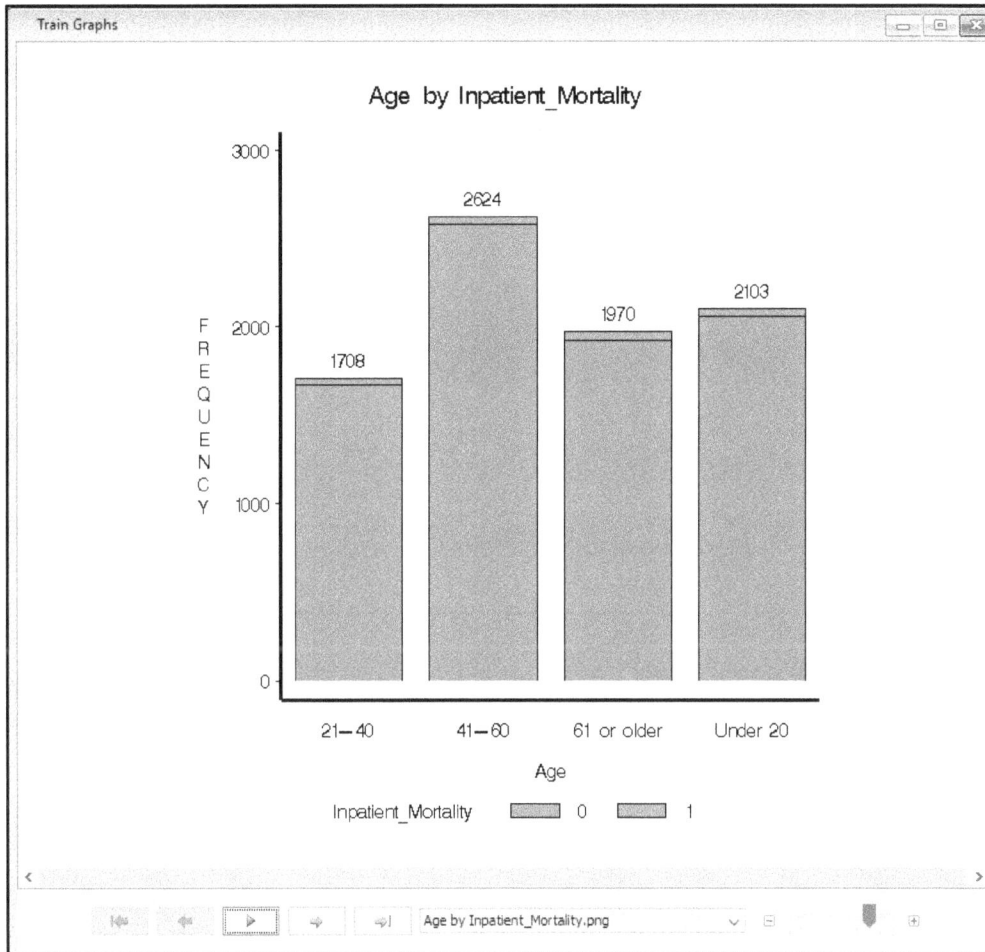

Step 19: From our Stat Explore, Graph Explore, and MultiPlot results, we see that the inpatient mortality is a rare event in terms of occurring only 2% in our data set. As a result, we will include a sampling node to conduct a rare event sampling to improve the final results. From the **Sample** tab, add a Sample node to the diagram and connect to File Import.

Figure 7.9: Add Sample Node

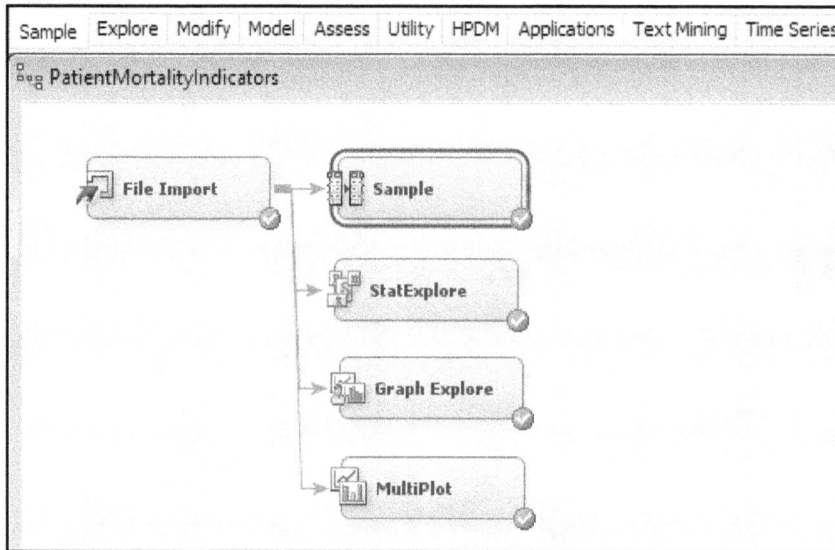

Step 20: Click the Sample node, and click the properties section. Then click **Variables** and set the Inpatient_Mortality to the Stratification Sample Role. The setting will allow us to select a sample based on the Inpatient_Mortality variable. Click **OK**.

Figure 7.10: Sample Properties Stratification

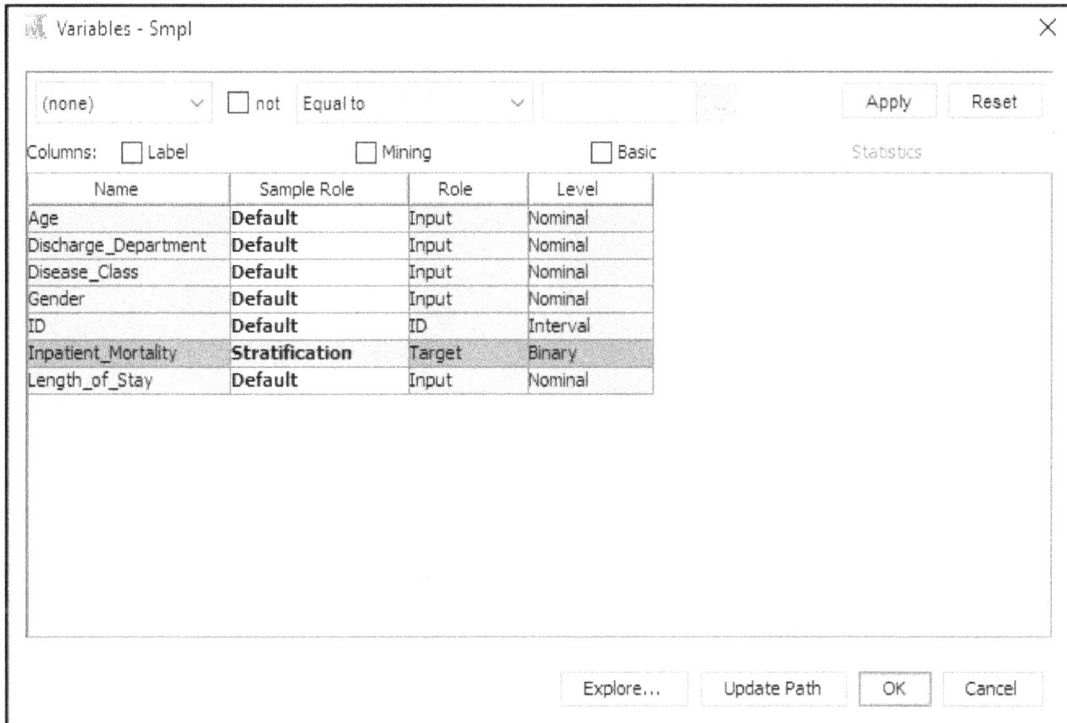

Name	Sample Role	Role	Level
Age	Default	Input	Nominal
Discharge_Department	Default	Input	Nominal
Disease_Class	Default	Input	Nominal
Gender	Default	Input	Nominal
ID	Default	ID	Interval
Inpatient_Mortality	Stratification	Target	Binary
Length_of_Stay	Default	Input	Nominal

Step 21: Click the Sample node properties, and set the Type to **Percentage** and set the Percentage to **100**. For the Stratified property, set the Criterion to **Equal**. The settings will select an equal sample of the Inpatient_Mortality rare event and the Inpatient_Mortality non-event. In other words, we will select an equal sample of both true and false cases of Inpatient_Mortality. If we selected the normal sample size, the results might be limited given the size of non-events, since all occurrences would favor a non-event scenario. Our goal is to find the factors leading to Inpatient_Mortality.

Figure 7.11: Sample Properties

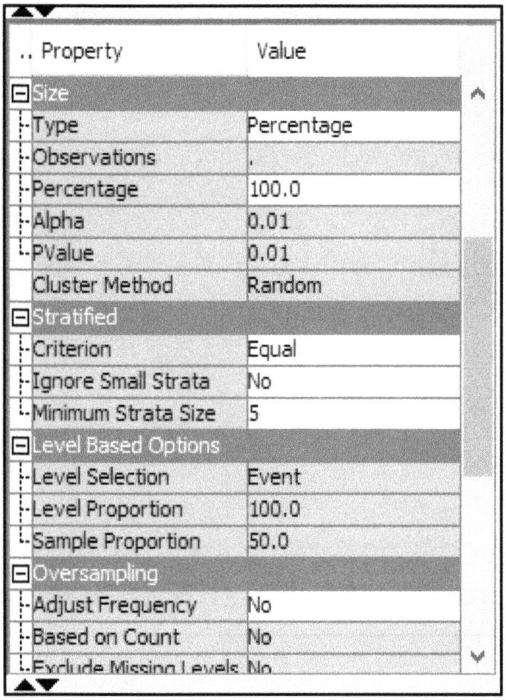

.. Property	Value
Size	
Type	Percentage
Observations	.
Percentage	100.0
Alpha	0.01
PValue	0.01
Cluster Method	Random
Stratified	
Criterion	Equal
Ignore Small Strata	No
Minimum Strata Size	5
Level Based Options	
Level Selection	Event
Level Proportion	100.0
Sample Proportion	50.0
Oversampling	
Adjust Frequency	No
Based on Count	No
Exclude Missing Levels	No

Step 22: Run the Sample node and view the results. From the output, the original data set is shown with Inpatient_Mortality = 1/True occurring 170 times for 2% of the total data set. After sampling, Inpatient_Mortality = 1 occurs the same amount as Inpatient_Mortality = 0 for an equal sample data set.

Figure 7.12: Sample Results

```
Output
46
47     Data=DATA
48
49                         Numeric    Formatted    Frequency
50         Variable         Value       Value        Count      Percent          Label
51
52     Inpatient_Mortality    0           0          8235       97.9774    Inpatient Mortality
53     Inpatient_Mortality    1           1           170        2.0226    Inpatient Mortality
54
55
56     Data=SAMPLE
57
58                         Numeric    Formatted    Frequency
59         Variable         Value       Value        Count      Percent          Label
60
61     Inpatient_Mortality    0           0           170          50      Inpatient Mortality
62     Inpatient_Mortality    1           1           170          50      Inpatient Mortality
```

Step 23: Add a Data Partition node (Click the **Sample** tab, and drag the node onto the diagram workspace). Set the Data Partition Property Data Set Allocations to **60.0** for Training, **40.0** for Validation, and **0.0** for Test. Run the Data Partition node.

Figure 7.13: Data Partition Node

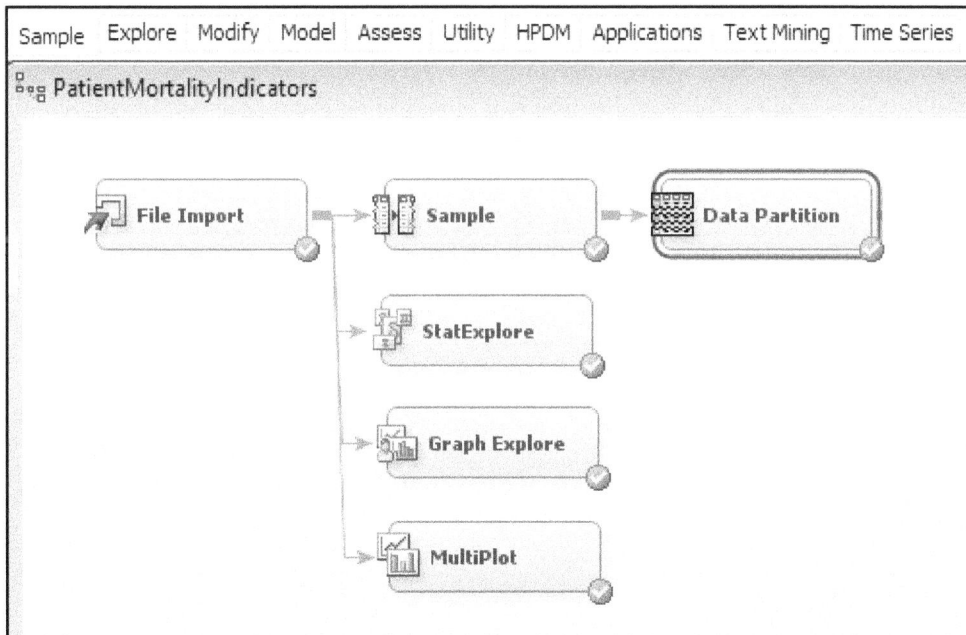

Step 24: Review the data Partition Results.

Figure 7.14: Data Partition Results

```
 Output

  24      Partition Summary
  25
  26                                            Number of
  27      Type                Data Set          Observations
  28
  29      DATA          EMWS1.Smpl_DATA              340
  30      TRAIN         EMWS1.Part_TRAIN             203
  31      VALIDATE      EMWS1.Part_VALIDATE          137
  32
```

Step 25: Add an Impute node (Click the **Modify** tab, and drag the node onto the diagram workspace). Verify that the Impute Property is set to **Count** for Class variables and **Mean** for Interval variables.

Figure 7.15: Impute Node

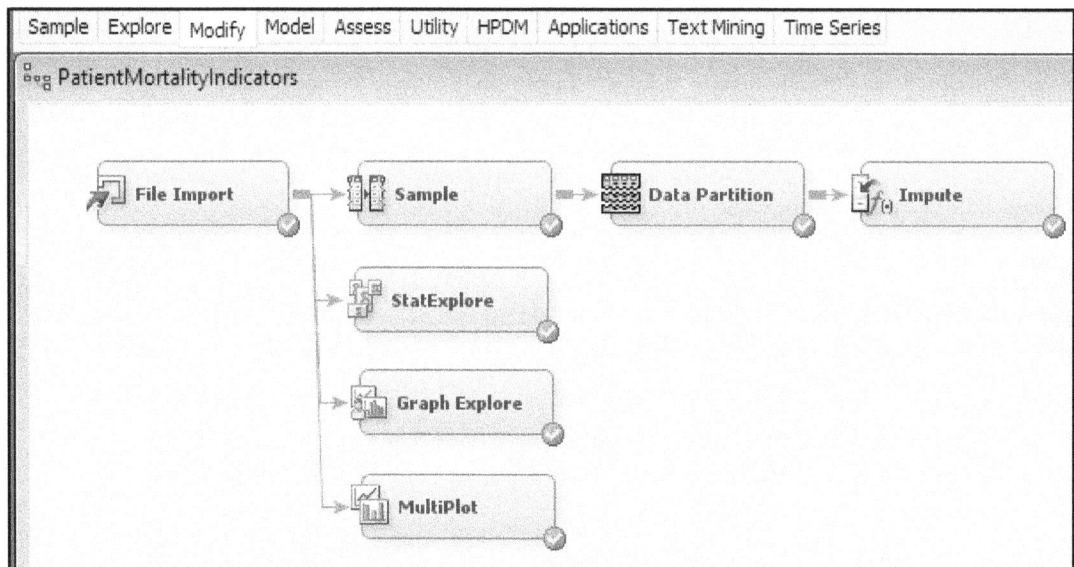

Step 26: Add a Decision Tree node (Click the **Model** tab, and drag the node onto the diagram workspace).

Figure 7.16: Decision Tree Node

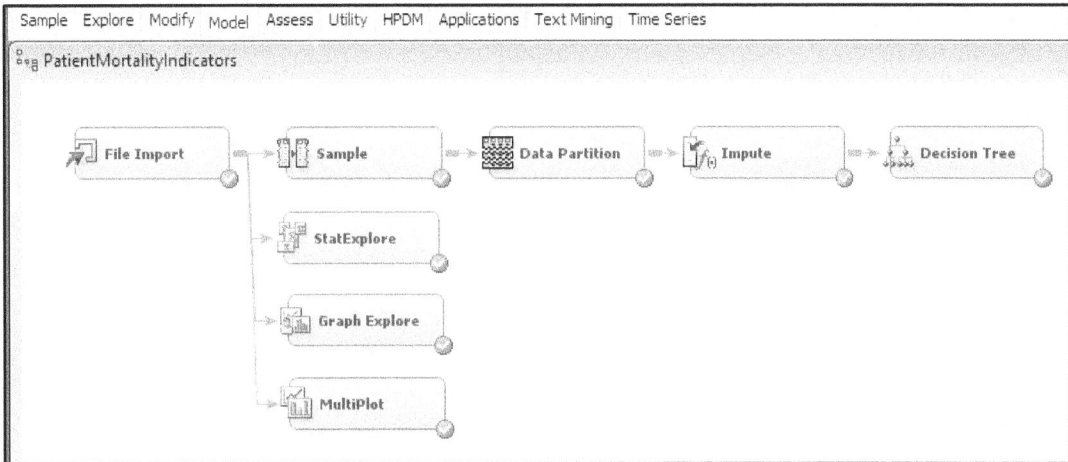

Step 27: Select the Decision Tree node. In the Tree Property, under Splitting Rule, set Minimum Categorical Size to **2**, and under Node, set Leaf Size to **1**. The settings allow a tree to grow with a 2-category split and a single leaf or a single record.

Figure 7.17 Decision Tree Node Properties

.. Property	Value
Splitting Rule	
Interval Target Criterion	ProbF
Nominal Target Criterion	ProbChisq
Ordinal Target Criterion	Entropy
Significance Level	0.2
Missing Values	Use in search
Use Input Once	No
Maximum Branch	2
Maximum Depth	6
Minimum Categorical Size	2
Node	
Leaf Size	1
Number of Rules	5
Number of Surrogate Rules	0
Split Size	.

Step 28: Select the Decision Tree node. Under the Subtree property, select **Largest**. The property setting will run the full tree to all its branches and leaves, or all splits and decision points.

Figure 7.18: Decision Tree Node Properties

Property	Value
Subtree	
Method	Largest
Number of Leaves	1
Assessment Measure	Decision
Assessment Fraction	0.25
Cross Validation	
Perform Cross Validation	No
Number of Subsets	10
Number of Repeats	1
Seed	12345
Observation Based Import	
Observation Based Import	No
Number Single Var Importa	5
P-Value Adjustment	
Bonferroni Adjustment	Yes

Step 29: Right-click the Decision Tree node, and click **Run**.

Step 30: Expand the Output Window Results and Review Model Results. The misclassification rate for the Training set is 31.5% and the Misclassification Rate for the Validation set is 35.0%.

Figure 7.19: Decision Tree Node Results

Fit Statistics

Statistics Label	Train	Validation
Sum of Frequencies	203	137
Misclassification Rate	0.315271	0.350365
Maximum Absolute Error	0.896552	0.896552
Sum of Squared Errors	84.24056	62.3695
Average Squared Error	0.207489	0.227626
Root Average Squared Error	0.45551	0.477102
Divisor for ASE	406	274
Total Degrees of Freedom	203	.

Step 31: Review the Model Results. The cumulative lift shows that the model outperforms a random model. At the top 10% of records (or depth), the train model outperforms a random model by nearly 1.8 times and the validation model by 1.5 times.

Figure 7.20: Decision Tree Node Results

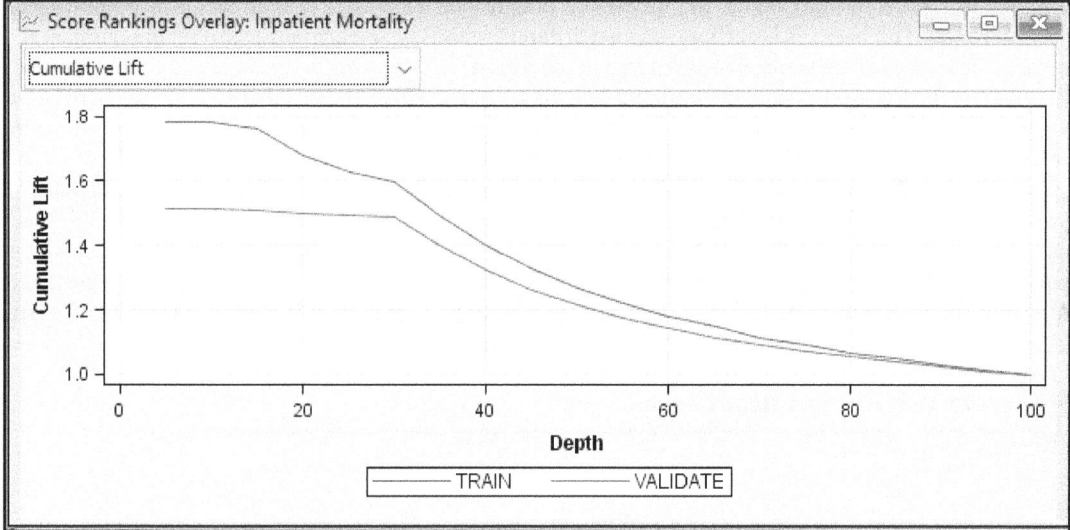

Step 32: Review the Model Results. The decision tree is presented as a visual model. Think of the decision tree as modeled after a living tree. At the top, we have the root or Node Id 1. The top is where the tree starts is like the trunk of a tree. Node Id 1 also gives a breakdown of our data with about 50% 0 and 50% 1 cases and a split between Train with 203 records and Validation with 137 records. The lines can be considered branches, and the remaining nodes are leaves. Therefore, we build (or grow) our tree starting with the root or Node Id 1 and branching out all the way to the final leaves Node Id 5. A final leaf is also known as a terminal node. The lines (or branches) are also different widths or thicknesses based on the number of records. Try to visualize an upside-down tree in the Figure 7.21 Starting again from the root Node Id 1, the first split in the tree occurs with the variable Length_of_Stay. This indicates that Length_of_Stay has a high variable importance in our model. If we follow the split to the left, we have LOS 17-341 Days, this means that all records with LOS 17-341 Days follow the left side of the split. The right side of the Length_of_Stay split contains all records with LOS 1-5 Days, 6-16 Days, or missing values. Looking closer at Node Id 2, you can see that the record breakdown is also given. For Validation, 26.19% are 0 cases and 73.81% are 1 cases. This indicates that using a LOS of 17-341 Days split, we can identify 80% of the 1 (or true) cases for patient mortality. We can further follow the tree to the next split and branches for Gender. Follow the tree to Gender of Male and Node Id 5. For the Validation breakdown, we find 75% are 1 cases and 25% are 0 cases, with true cases slightly higher for males than females.

Figure 7.21: Decision Tree Node Results

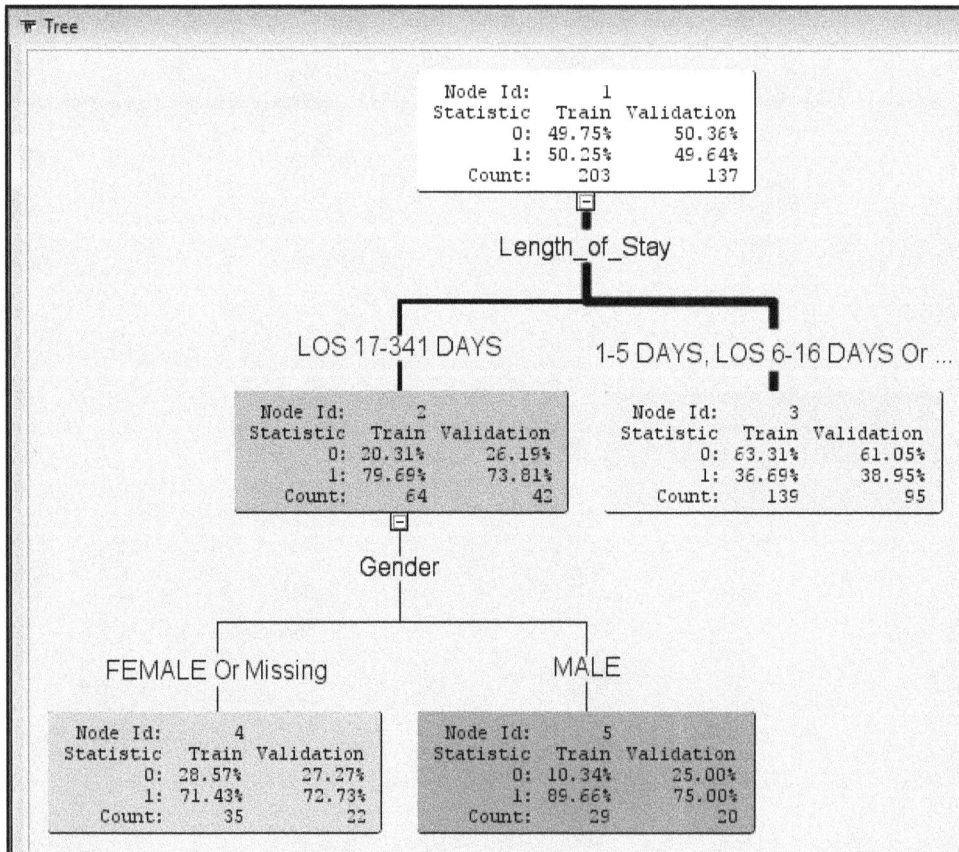

Step 33: Review the Model Results. We can further develop programmatic IF…THEN rules to describe the branches, which are also known as English Rules. The rules can be valuable for automating decision-making in a system such as a decision support or EHR system that used by medical professionals for assessing risk of morality and developing an appropriate plan of care. Click Node Id 5 and right-click and select **Tools ▶ Display** English rule.

Figure 7.22: Decision Tree Display English Rule

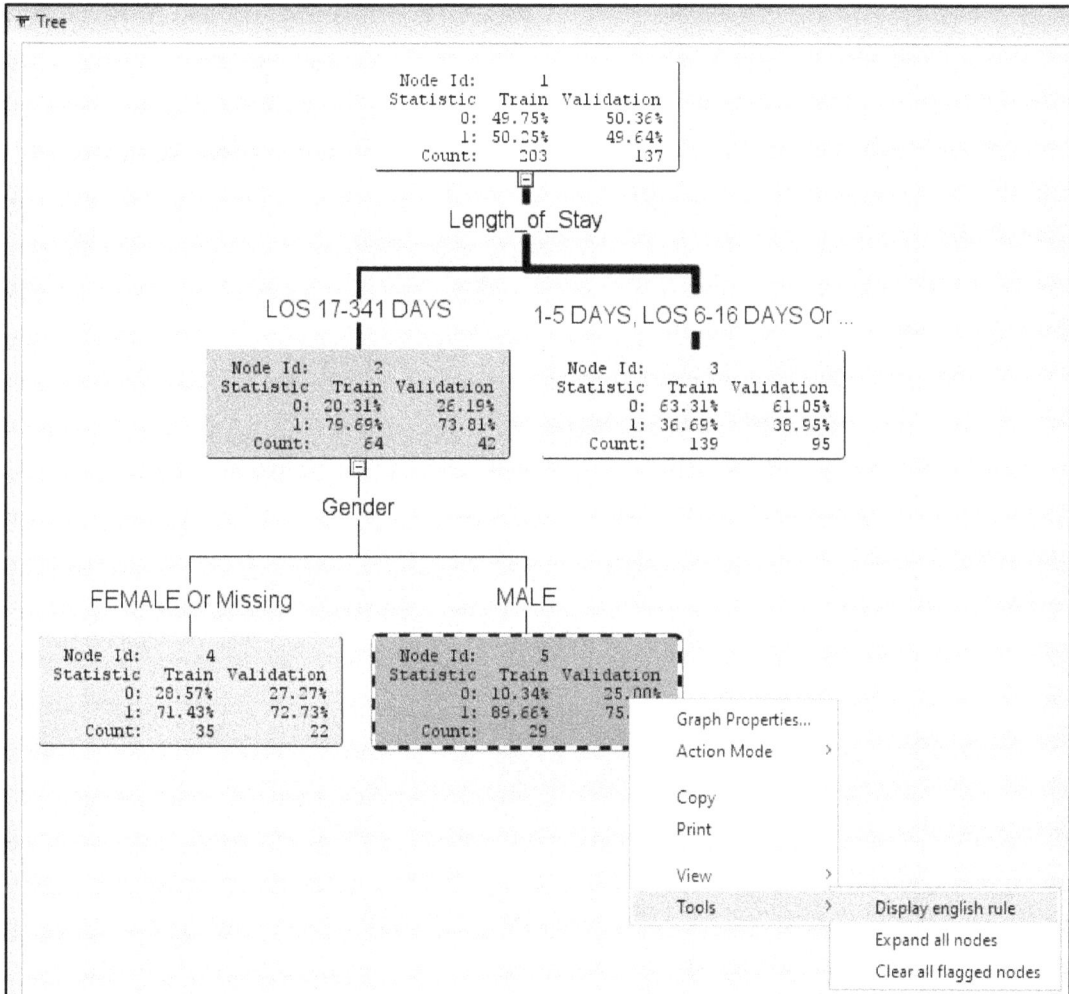

Step 34: Review the Model Results. The results will display the English Rule following the tree structure, Where Length_of_Stay (LOS) is **17-341** AND Gender is **MALE**. The rules are easily communicated to other clinical and non-clinical individuals. For example, if the length of stay is between 17 and 341 days and the patient is male, nearly 90% of our training records and 75% of our validation records indicate patient mortality.

Figure 7.23: Display English Rule

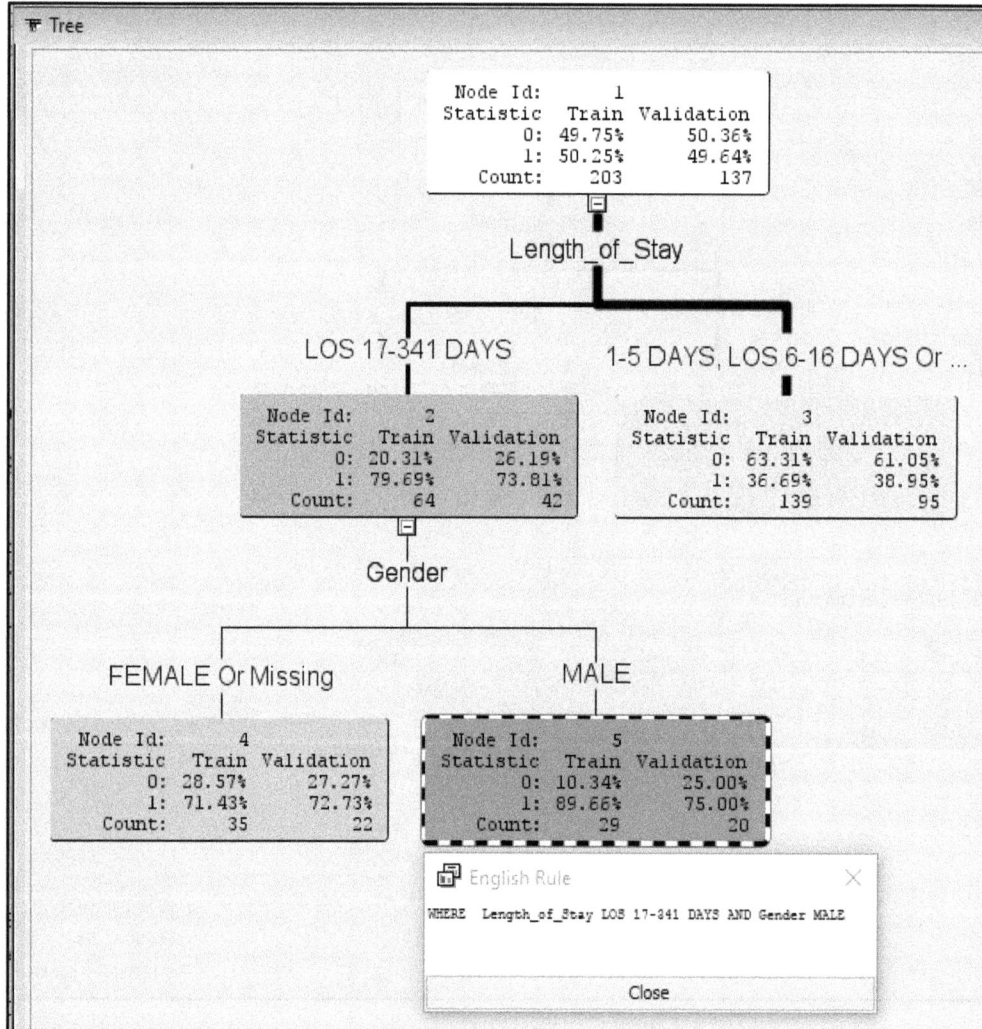

Step 35: Add a Regression node. From the previous chapter, we can also include a Regression model for our data set. The node can be connected in a way that is similar to the Decision Tree node. Run the Regression node.

Figure 7.24: Regression Node

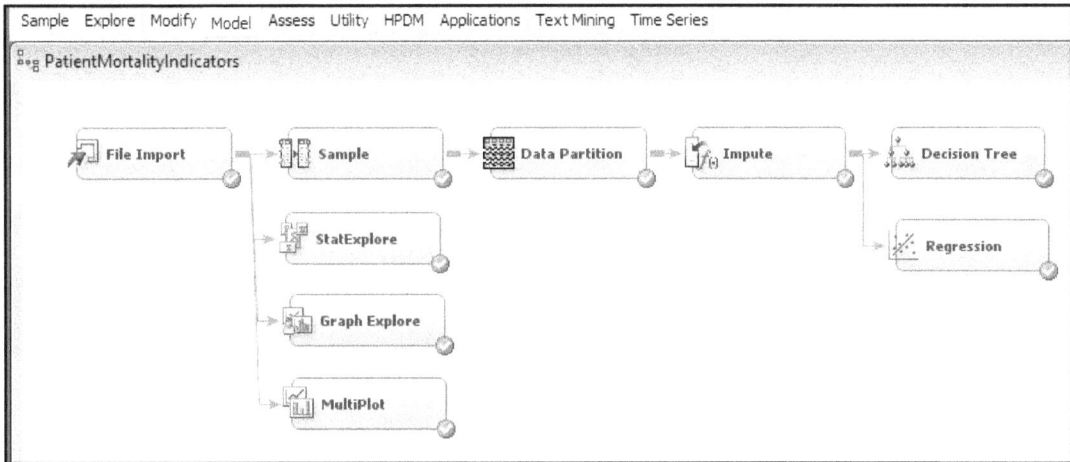

Step 36: Review the Model Results. The misclassification rate for the Training set is 29.1%, and the misclassification rate for the Validation set is 38.7%. The results perform similarly to our Decision Tree model, with 31.5% and 35% misclassification rate, respectively.

Figure 7.25: Regression Node Fit Statistics

Statistics Label	Train	Validation
Akaike's Information Criterion	260.2229	.
Average Squared Error	0.191212	0.244899
Average Error Function	0.562125	0.697723
Degrees of Freedom for Error	187	.
Model Degrees of Freedom	16	.
Total Degrees of Freedom	203	.
Divisor for ASE	406	274
Error Function	228.2229	191.1761
Final Prediction Error	0.223933	.
Maximum Absolute Error	0.878243	0.930906
Mean Square Error	0.207573	0.244899
Sum of Frequencies	203	137
Number of Estimate Weights	16	.
Root Average Sum of Squares	0.437278	0.494873
Root Final Prediction Error	0.473216	.
Root Mean Squared Error	0.455601	0.494873
Schwarz's Bayesian Criterion	313.2342	.
Sum of Squared Errors	77.63219	67.1024
Sum of Case Weights Times Freq	406	274
Misclassification Rate	0.29064	0.386861

Step 37: Review the Model Results. Adding the Regression node also provides additional results such as Odds Ratio. From the Analysis of Maximum Likelihood Estimates output, the Length_of_Stay is a significant variable with Pr > ChiSq, p-value less than 0.01. From the odds ratios, the length of stay, 17-341 days, carries a 6.752 times the odds of mortality than a length of stay 6-16 days.

Figure 7.26: Regression Node Output

			Analysis of Maximum Likelihood Estimates				
Parameter		DF	Estimate	Standard Error	Wald Chi-Square	Pr > ChiSq	Exp(Est)
Intercept		1	0.1234	0.2284	0.29	0.5890	1.131
Age	21-40	1	-0.5902	0.2989	3.90	0.0483	0.554
Age	41-60	1	-0.2092	0.2891	0.52	0.4692	0.811
Age	61 or older	1	0.1752	0.2921	0.36	0.5487	1.191
Discharge_Department	Internal Medicine	1	-0.1337	0.1955	0.47	0.4939	0.875
Disease_Class	Circulatory	1	-0.4136	0.5356	0.60	0.4400	0.661
Disease_Class	Congenital	1	-1.1430	0.7537	2.30	0.1294	0.319
Disease_Class	Eye and Ear	1	-0.5042	0.5956	0.72	0.3973	0.604
Disease_Class	Gastrointestinal	1	0.3806	0.7446	0.26	0.6092	1.463
Disease_Class	Miscellaneous	1	0.5120	0.3786	1.83	0.1763	1.669
Disease_Class	Muscle	1	0.9644	0.6053	2.54	0.1111	2.623
Disease_Class	Neoplasm	1	0.0536	0.3308	0.03	0.8712	1.055
Disease_Class	Pulmonary	1	-0.4396	0.5382	0.67	0.4140	0.644
Gender	Female	1	-0.2189	0.1677	1.70	0.1916	0.803
Length_of_Stay	1-5 Days	1	-1.0193	0.2272	20.13	<.0001	0.361
Length_of_Stay	LOS 17-341 Days	1	1.4645	0.2716	29.07	<.0001	4.326

		Odds Ratio Estimates
Effect		Point Estimate
Age	21-40 vs Under 20	0.297
Age	41-60 vs Under 20	0.435
Age	61 or older vs Under 20	0.638
Discharge_Department	Internal Medicine vs Surgery	0.765
Disease_Class	Circulatory vs Urinary	0.367
Disease_Class	Congenital vs Urinary	0.177
Disease_Class	Eye and Ear vs Urinary	0.335
Disease_Class	Gastrointestinal vs Urinary	0.811
Disease_Class	Miscellaneous vs Urinary	0.925
Disease_Class	Muscle vs Urinary	1.455
Disease_Class	Neoplasm vs Urinary	0.585
Disease_Class	Pulmonary vs Urinary	0.357
Gender	Female vs Male	0.645
Length_of_Stay	1-5 Days vs LOS 6-16 Days	0.563
Length_of_Stay	LOS 17-341 Days vs LOS 6-16 Days	6.752

Model Summary

In summary, the decision tree will take the form of 1 or more inputs and 1 target variable. The inputs can be interval, binary, or nominal, and the target can also be interval, binary, or nominal. To evaluate the decision tree, we can use error, lift, English rules, and misclassification rate for a binary, nominal, or categorical target variable.

- Model: Decision Tree node
- Decision Tree: 1+ input and 1 target variable
- Input: Interval, Binary, or Nominal
- Target: Interval, Binary, or Nominal
- Evaluation: Error, Lift, English Rules, Misclassification Rate

Experiential Learning Application: Self-Reported General Health

Executive Summary

One study found that people often evaluate their overall health and wellness based on their lived health rather than their experience of biological health. The self-reported general health (SRGH), contains the levels of very good, good, fair, bad, and very bad. SRGH is one of the most commonly used measures of health in population health and clinical health surveys and is used to compare populations. Nearly 2,000 scientific studies have been conducted using SRGH (or general survey questions) about how you would rate your health. The question is also used internationally and is included as part of the European Organization of Research and Treatment of Cancer (EORTC) Quality of Life Questionnaire. SRGH has been used as an input variable for predicting health outcomes, such as mortality, or it is used as a target variable based on inputs such as diagnosis and symptoms (Bostan et al., 2014).

For the experiential learning application that you have been provided, a sample of 27,446 records with SRGH is the target variable, and setting type, gender, age group, education, number of health conditions, biological health score, and lived health score are the input variables. Help your management team answer the following question.

Question: Which factors might influence SRGH?

Data Set File: 7_EL2_SRGH.xlsx

Variables:

- Population Type: 17,739 Community-Dwelling, 9,707 Institutionalized Population
- SRGH: Very Good, Good, Fair, Bad, very bad
- Gender (male, female)
- Age groups (<=65,>65)
- Education (no school, primary-school-incomplete, primary-school-complete, secondary school first step, secondary school finished, professional school medium, professional school superior, University)
- Number of health conditions (0, 1-2, > 2). The health conditions were: Spinal cord injury, Parkinson's, Lateral sclerosis, Multiple sclerosis, Agenesis/Amputation, Laryngectomy, Arthritis, Rheumatoid arthritis or Ankylosing spondylitis, Muscular dystrophy, Spina bifida/hydrocephaly,

Myocardial infarction or Ischaemic cardiopathy, Cerebrovascular accidents, Down's Syndrome, Autism and other disorders associated with autism, Cerebral paralysis, Acquired brain damage, Senile Dementia of the Alzheimer Type, Other types of dementia, Schizophrenia, Depression, Bipolar disorder, Pigmentary retinosis, Myopia magna, Senile macular degeneration, Diabetic retinopathy, Glaucoma, Cataract, HIV/AIDS, Rare illnesses, Cancer (only for community dwelling population).

- Biological Health Score 0 (best biological health) to 100 (worst biological health)
- Lived Health Score 0 (best lived health) to 100 (worst lived health)

Follow the SEMMA process for your experiential learning application and provide recommendations. A template has been provided below that can be reused across future projects.

Title	Self-Reported General Health
Introduction	Provide a summary of the business problem or opportunity and the key objective(s) or goal(s). Create a new SAS Enterprise Miner project. Create a new Diagram.
Sample	Data (sources for exploration and model insights) Identify the variables data types, the input and target variable during exploration. Add a FILE IMPORT Provide a results overview following file import: Input / Target Variables Generate a DATA PARTITION
Exploration	Provide a results overview following data exploration Add a STAT EXPLORE Add a GRAPH EXPLORE Add a MULTIPLOT Summary statistics (average, standard deviation, min, max, and so on.) Descriptive Statistics Missing Data Outliers
Modify	Provide a results overview following modification Add an IMPUTE

Title	Self-Reported General Health
Model	Discovery (prototype and test analytical models)
	Apply a decision tree model and provide a results overview following modeling.
	Add a DECISION TREE
	Model description
	Analytics steps
	Decision Tree results (tree model, English rules)
	Model results (Lift, Error, Misclassification Rate)
	Selection Model
Assess and Reflection	Provide overall recommendations to business
	Model advantages / disadvantages
	Performance evaluation
	Model recommendation
	Summary analytics recommendations
	Summary informatics recommendations
	Summary business recommendations
	Summary clinical recommendations
	Deployment (operationalization plan: timeline, resources, scope, phases, project plan)
	Value (return on investment, healthcare outcomes)

Learning Journal Reflection

Review, reflect, and retrieve the following key chapter topics only from memory and add them to your learning journal. For each topic, list a one sentence description/definition. Connect these ideas to something you might already know from your experience, other coursework, or a current event. This follows our three-phase learning approach of 1) Capture, 2) Communicate, and 3) Connect. After completing, verify your results against your learning journal and update as needed..

Key Ideas – Capture	Key Terms – Communicate	Key Areas - Connect
Payer Anamatics		
Payers		
Claims System and Process		
Claims Forms		
EDI		
Claims Adjudication		

Key Ideas – Capture	Key Terms – Communicate	Key Areas - Connect
Decision Tree		
Inpatient Mortality Application		
Self-Reported General Health		

Chapter 8: Modeling Government Data

Chapter Summary

The purpose of this chapter is to develop data modeling skills using SAS Enterprise Miner with respect to the Model capabilities within the SEMMA process. This chapter explores modeling with the neural network model. This chapter also includes experiential learning application exercises on fraud detection and hospital readmissions. The focus of this chapter is shown in Figure 8.1.

Figure 8.1: Chapter Focus – Model

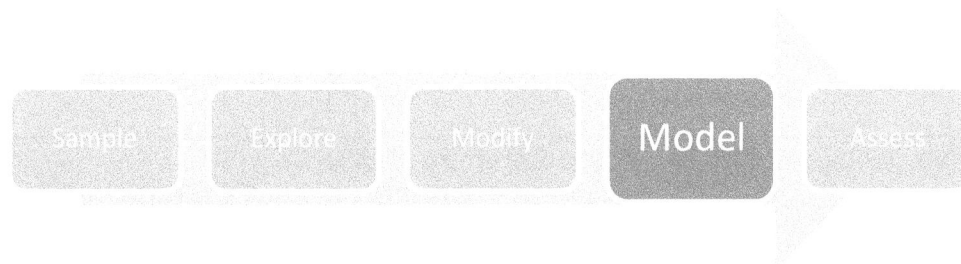

Chapter Learning Goals

- Describe the Model process steps
- Understand government-level data sources
- Develop data modeling skills
- Apply SAS Enterprise Miner model data functions
- Master the neural network model

Government Agencies

The U.S. government has established a number of organizational agencies involved in various aspects of healthcare. Key government agencies include the Department of Health and Human Services (HHS), Centers for Disease Control and Prevention (CDC), Centers for Medicare and Medicaid Services (CMS), Food and Drug Administration (FDA), National Institutes of Health (NIH), Drug Enforcement Agency (DEA), and Office of Civil Rights (OCR) (HHS, 2017a). To give an idea of the size of the organizational scope in 2017, CMS had an annual budget of $1 trillion. The FDA had an annual budget of $5 billion, the CDC had an annual budget of $12 billion, and the NIH had an annual budget of $33 billion (HHS, 2017b).

Department of Health and Human Services (HHS)

The Department of Health and Human Services (HHS) is charged with enhancing the health and well-being of Americans by providing for effective health and human services and by fostering sound and sustained advances in the sciences underlying medicine, public health, and social services. To support this mission, the budget provides $82.8 billion in discretionary funding for HHS as well as new mandatory investments to expand mental health services, opioid abuse treatment availability, and research and development (HHS, 2017c). HHS is an umbrella organization with eleven operating divisions and many agencies providing services and protecting Americans through improving health and well-being. Some of the larger and well-known agencies include the Centers for Disease Control and Prevention (CDC), Centers for Medicare and Medicaid Services (CMS), Food and Drug Administration (FDA), and National Institutes of Health (NIH) (HHS, 2017a). Similar to HHS in the U.S., the National Health Service (NHS) in the United Kingdom (U.K.) is responsible for establishing priorities to improve health care and well-being. NHS has a budget of over £100 billion and aims to ensure efficient and effective care is being provided (NHS, 2017).

Centers for Disease Control and Prevention (CDC)

The Centers for Disease Control (CDC) is responsible for protecting Americans from health, safety, and security threats both nationally and internationally. Disease can include chronic conditions, curable conditions or preventable conditions, and deliberate attacks. CDC provides health information and supports scientific research in order to protect against and respond to threats. CDC has an annual budget of over $12 billion, working in over 60 countries with over 13,000 staff members (CDC, 2017e; CDC, 2017f).

Centers for Medicare and Medicaid Services (CMS)

The Centers for Medicare and Medicaid Services (CMS) covers over 100 million individuals with an overall goal of low-cost, high-quality care. President Lyndon B. Johnson signed a bill on July 30, 1965 that initiated Medicare and Medicaid and included two primary components: 1) Part A for hospital insurance, and 2) Part B for medical insurance. In 1972, Medicare was expanded to include additional coverage and benefits for the disabled (including those with end-stage renal disease and those who are over 65 of age) and prescription drug coverage. Medicaid was also expanded its services to account for income, pregnancy, disabilities, and long-term care. In 2003, The Medicare Prescription Drug Improvement and Modernization Act (MMA) was a major update to Medicare, including a Part C for private health plan coverage and Part D for optional prescription drug benefits. In 1997, the Children's Health Insurance Program (CHIP) was added for uninsured children. In 2010, the Affordable Care Act (ACA) established the Health Insurance Marketplace, which is a central location for patients to enroll in health insurance plans (CMS, 2017c; CMS, 2017d).

Food and Drug Administration (FDA)

The Food and Drug Administration (FDA) is responsible for ensuring safety, efficacy, and security of drugs (both prescription and non-prescription). It also manages biologics (such as vaccines and medical devices that range from simple items such as tongue depressors to more complex items such as pacemakers), foods (including supplements and additives), cosmetics (including makeup and perfume), veterinary products (including pet foods), items that emit radiation (such as microwave ovens), and tobacco regulation (including cigarettes) (FDA, 2017a; FDA, 2017b).

National Institutes of Health (NIH)

The National Institutes of Health (NIH) is responsible for discovering and applying knowledge to improve health and longevity and to decrease illness and disabilities. The mission is accomplished through innovative research and scientific discoveries. Over 80% of the $32 billion NIH budget is distributed through nearly 50,000 competitive grants to 300,000 researchers at over 2,500 institutions (NIH, 2017a; NIH, 2017b).

Drug Enforcement Agency (DEA)

The Drug Enforcement Administration (DEA) was established to enforce controlled substances in the U.S. and support non-enforcement programs to reduce the availability of controlled substances nationally and internationally. The DEA was included in an announcement by the U.S. Justice Department on the seizure of the largest international online marketplace. The announcement was part of a coordinated law enforcement effort by the U.S., Thailand, the Netherlands, Lithuania, Canada, the U.K., and Europe. The online marketplace seized by the DEA was a dark web location for illegal drugs and other criminal activities. At the time of the shutdown, there were over 250,000 illegal drug and toxic chemical listings (DEA, 2017a; DEA, 2017b).

Office for Civil Rights (OCR)

The Office of Civil Rights (OCR) protects nondiscrimination and privacy rights through the Health Insurance Portability and Accountability Privacy Rule (HIPAA) and the Patient Protection and Affordable Care Act (ACA). The nondiscrimination provision of the Affordable Care Act (ACA) prohibits discrimination on the basis of race, color, national origin, sex, age, or disability in certain health programs or activities. The OCR assists with nondiscrimination and privacy rights through education of health and social workers and of communities on privacy and confidentiality laws. In addition, OCR is responsible for investigating complaints about civil rights, privacy, confidentiality, and patient safety in order to identify violations of the law and take appropriate corrective actions (OCR, 2017).

Government Health Anamatics

Governments around the world are in a position to have a tremendous impact on healthcare. Nevertheless, government adoption of health anamatics has proved challenging. In the U.K., a project was started in 2002 to upgrade the National Health Service (NHS) information systems, including electronic health records and integrated systems. The project wound up being one of the costliest failed projects in world history. The project was ultimately abandoned yet still cost taxpayers an estimated £10 billion (or $13-$20 billion USD, depending on the exchange rate. The project was plagued with technical challenges, changing requirements, management problems, contractual issues, and disputes with suppliers (UK Politics, 2013; Syal, 2013). In an interview conducted by the World Health Organization (WHO) with Desmond Tutu, chairman of the Global eHealth Ambassadors Program (GeHAP), the importance of governments in fueling technology and e-health adoption was discussed. Governments establish the guidelines and regulations for healthcare and have a great influence on the industry in setting the right policies and strategies, including building an environment for e-health adoption through government funding and health-related training. Hamadoun Touré, secretary-general of the International Telecommunications Union, cites technical issues, scalability, and politics as major barriers to successful completion. He underscored the necessity for cooperation between government and all stakeholders (WHO, 2012).

In the U.S., the governmental share of healthcare spending is expected to grow from 64.3% in 2014 to 67.1% by 2024. The share includes programs such as Medicare, Medicaid, public employee health insurance coverage, and tax savings. The U.S. government's share is one of the highest global rates, even exceeding countries known for universal healthcare programs. The statistic counters most individuals' beliefs that in the U.S. the majority of healthcare is privately funded and operated (Himmelstein and Woolhander, 2016; Almberg, 2016). Government healthcare cost growth at 5.9% per year is expected to exceed private healthcare cost growth at 5.4% per year through 2025. The faster increase for governmental spending is projected due to Medicare growth from the baby boomer generation along with government subsidies for health insurance enrollments for low-income applicants (CMS, 2017e).

Government Regulations

In the U.S., key legislative regulations have occurred over several decades and have had a significant impact on various healthcare entities. The growth started in 1996 with HIPAA and continued through ARRA in 2009, ACA in 2010, and proposed AHCA and BCRA legislation in 2017. A more detailed view of each follows.

Figure 8.2: Key Government Regulations Timeline

| 1996 HIPAA | 2009 ARRA | 2010 ACA | 2017 AHCA/BCRA |

Government Regulations

- 1996 - Health Insurance Portability and Accountability Act (HIPAA)
- 2009 - American Recovery and Reinvestment Act (ARRA)
  - Health Information Technology for Economic and Clinical Health Act (HITECH)
- 2010 - Affordable Care Act (ACA)
- 2017 - American Health Care Act (AHCA)
- 2017 - Better Care Reconciliation Act (BCRA)

1996 – Health Insurance Portability and Accountability Act (HIPAA)

The Health Insurance Portability and Accountability Act (or HIPAA) was passed in 1996 and was a Congressional measure to reform healthcare. The overall objectives of the act were to ensure that individuals could move their health insurance between jobs (the portability) and ensure security and privacy of patient data with standardization of data. HIPAA further defined a set of rules for implementation including transactions, code sets, and identifiers for data standardization, privacy, and security (UCMC, 2010).

HIPAA Transactions and Codesets Rule

Within healthcare, The Department of Health and Human Services (DHHS) estimated that as many as 400 different data formats were being used to exchange data between healthcare entities in addition to the use of paper or other manual methods being used. In an effort to improve the process, the transactions and code-sets rule described a set of standard Electronic Data Interchange (EDI) transactions, which is covered in Chapter 7. The rule included standards for sending and receiving healthcare claims, healthcare enrollment, healthcare eligibility, healthcare payment, healthcare status, healthcare referrals and authorizations, coordination of benefits, and reporting of injury. Later, other transactions included paperwork attachments (UCMC, 2010).

Security and Privacy Rules

The privacy rule's main component is the area of Protected Health Information (PHI). PHI includes items such as name, address, birthdate, Social Security number, and biometric information. In addition, patient permission is required to use or disclosure information. Civil and criminal penalties were included for those who violate or disobey the rule, and the penalties range up to $1,500,000 in fines and/or ten years in prison (HHS, 2016). The security rule develops standards and safeguards to protect electronic protected health information (ePHI) (UCMC, 2010). The security and privacy rules are further discussed in the context of health care administration in Chapter 9.

2009 - American Recovery and Reinvestment Act (ARRA)

Despite spending billions on U.S. health information technology over the last several decades, there are still issues relating to costs, errors, efficiency, and coordination of care, which reflect the limited saturation of health information technology within healthcare systems. Inefficient paper-based systems, inaccessible medical information during care, limited patient access to health information, misinterpreted handwriting, and unavailable best treatment options affect the current healthcare systems (Woodside, 2007a). In 2004, the Bush Administration called for the next ten years to the be decade of health information technology, and, by 2014, EHRs would be universally adopted. An Office of the National Coordinator of Health Information Technology (ONCHIT) was established to lead the effort and was supported by the Health Information Technology and Clinical Health Act (HITECH) in 2009 (Burke and Will, 2013).

2010 – Affordable Care Act (ACA)

The Patient Protection and Affordable Care Act, or commonly known as the ACA (or Obamacare), was passed in 2010 by the U.S. Senate and contained over 900 pages of legislation with varying implementation dates. The goal was to provide access to quality and affordable healthcare while reducing costs. The Congressional Budget Office (CBO) estimated that the ACA would provide over 94% of Americans with coverage within a $900 billion budget limit that President Obama established, which would reduce healthcare costs and the deficit over the next decade. The ACA contains nine major sections (or titles) for healthcare reform, including:

- Quality and affordable healthcare for all Americans (such as health insurance exchanges)
- Role of public programs (such as Medicaid)
- Improving quality and efficiency (such as through value-based and pay for performance programs)
- Prevention of chronic disease and public health (such as healthy communities)
- Healthcare workforce (such as support through education, and training)
- Transparency and program integrity (such as public information reporting and research)
- Improvements in access to innovative medical therapies (such as biologics or biosimilar products and drug discounts)
- Class act (such as community living assistance support and long-term care insurance)
- Revenue provisions (such as taxes on high cost employer health coverages and voluntary procedures)
- Strengthening quality and affordable healthcare (such as incentives and value-based programs)

As the main component of the ACA, the first title is quality and affordable healthcare for all Americans. The section includes a number of key components including: Elimination of discriminatory practices such as excluding coverage due to pre-existing conditions, provided coverage for everyone, provided tax credits to ensure affordability, eliminated limits on benefits, provided coverage for immunizations and preventive services, increased coverage for dependents through age 26, developed similar coverage documents to more easily compare insurances, created limits on non-medical and administrative expenses by insurance companies, and created shared responsibility. As a method to create shared responsibility, if coverage is not maintained, a penalty will be due for most individuals. The penalty will increment over time and is approximately $700 per adult. In 2017, due to potential changes in legislation, the Internal Revenue Service (IRS) indicated that tax returns would still be processed even if the ACA requirement is not met (US Senate, 2010; NCSL, 2011; Goldstein, 2017). The tax position was later clarified by the IRS in early 2018. As a result of the Tax Cuts and Jobs Act, the IRS would require taxpayers to report coverage, have an exemption, or pay the penalty (IRS, 2018). The ongoing legislative landscape of 2017 and 2018 takes us to our next section on AHCA.

2017 – American Health Care Act (ACHA)

The American Health Care Act (AHCA) of 2017 was intended to replace the ACA. The ACA was commonly referred to as Obamacare as advocated by President Obama, and the AHCA was commonly referred to as Trumpcare (or Ryancare) as the legislation was advocated by President Trump and Speaker of the House Ryan. The naming is a result of the U.S. President in office at the time of the legislation and as the major driving force behind the legislation as a result of campaign promises to reform healthcare. The main components of AHCA include a repeal of the ACA and Health Care and Education Reconciliation Act of 2010, a tax deduction for health insurance, and an update to Health Savings Accounts (HSAs), among other changes. The bill was introduced to the House of Representatives on January 4, 2017. Before a formal vote was made, the bill was withdrawn, in part based on some opposition to provisions within the bill (Roe, 2017). Due to uncertainty in government payments through the ACA health insurance exchange system due to planned AHCA legislature, Anthem announced plans to exit state markets in 2018 such as Ohio. Anthem was the only insurance company that sold health insurance exchange produces in all 88 Ohio counties and the only insurer in 20 counties (Humer, 2017). After changes to the bill, including protections for pre-existing conditions, the bill was passed by the House on May 4, 2017 in a 217-213 vote (Soffen, et al., 2017).

2017 – Better Care Reconciliation Act (BCRA)

Although the AHCA was the U.S. House of Representatives version, the Better Care Reconciliation Act (BCRA) was the U.S. Senate draft version of a new healthcare bill. Key differences between BCRA and ACHA included a longer schedule to reduce Medicaid and retention of some ACA components such as premium subsidies and taxes. BCRA, like AHCA, would not completely repeal ACA, and rules such as dependent coverage through age 26, exclusions on lifetime limits, Medicare provisions, and prohibition of discrimination would remain intact (Kendrick, et al., 2017). In July 2017, there were several Congressional discussions on bill amendments, including an abridged "skinny" repeal of ACA without a replacement, which is also known as the Healthcare Freedom Act (HCFA). Despite these modifications, the bill failed to pass and temporarily ended the passage of a major comprehensive health care bill (Caldwell and Hillyard, 2017). A later version of BCRA, named the Graham-Cassidy health care bill, was also shelved following lack of Congressional support. The Graham-Cassidy bill sponsors, Senator Lindsey Graham of South Carolina and Bill Cassidy of Louisiana, have indicated they will continue to pursue the repeal and replacement of the ACA (Mukherjee, 2017).

Experiential Learning Activity: Government Data Sharing

Government Data Sharing

Description: In the U.S., the http://healthcare.gov website administered by the federal CMS agency is intended to allow consumers to purchase subsidized health insurance. The website contains connections with outside vendors and advertisers that might be able to identify you on the site through information shared such as age, income, ZIP code, and health factors including smoking and pregnancy. Although no information breaches or security issues have been found, and vendors are prohibited from using information for company purposes, some consultants and advocates raise privacy and security concerns about data that is retained about individuals (Alonso-Zaldivar and Gillum, 2015). According to the terms of the website, CMS might share your information with contractors that perform functions for the healthcare.gov marketplace, government agencies to combat fraud and abuse, insurance companies, Internal Revenue Service (IRS), health plans, consumer credit agencies, homeland security, social security, and employers listed on your application to verify eligibility (CMS, 2017f).

Several governments entities have also developed health information exchanges (HIEs) where protected health information (PHI) is shared. In North Carolina, several health systems have agreed to supply data to the state-sponsored HIE known as NC HealthConnex containing nearly 4 million patients from 100 data providers (Snipes, 2017). A health information exchange (HIE) is intended to allow a standardized method to exchange electronic health information across organizations. The HIE can operate in the context of a regional health information organization (RHIO) typically at a state level, or a national health information network (NHIN) at a national level. The HIE can be used to exchange information, improve health outcomes for individuals, assist with medical provider decision making, monitor and improve quality, evaluate policy and allow accreditation. A few states have developed HIEs through several barriers exists including policies, financial, legal, and technical considerations (Hebda and Czar, 2013).

Government Data Sharing

To discuss this government information sharing in the context of a code of ethics, we turn to AHIMA. The American Health Information Management Association (AHIMA) is a worldwide association of over 103,000 health information management (HIM) professionals worldwide. The AHIMA was founded in 1928 with an aim of improving health record quality and management of health data according to world class practices and standards (AHIMA, 2017a). AHIMA has developed a set of ethical obligations for HIM professionals, which include safeguarding, protecting, maintaining, and ensuring access and integrity of health information. As consumers are becoming increasingly aware and concerned about personal health information security and privacy, there are often questions on how personal health information is used, disclosed, collected, shared, handled, accessed, and retained. Although compliance regulations exist at a federal, state, and employer level, ethical considerations must also be considered. The ethical considerations are a component of professional responsibility and extend beyond an employer. The following set of ethical principles was developed as part of the AHIMA core values (AHIMA, 2011).

A health information management professional will do the following:
- Advocate, uphold, and defend the individual's right to privacy and the doctrine of confidentiality in the use and disclosure of information.
- Put service and the health and welfare of persons before self-interest and conduct oneself in the practice of the profession so as to bring honor to oneself, their peers, and to the health information management profession.
- Preserve, protect, and secure personal health information in any form or medium and hold in the highest regards health information and other information of a confidential nature obtained in an official capacity, considering the applicable statutes and regulations.
- Refuse to participate in or conceal unethical practices or procedures and report such practices.
- Advance health information management knowledge and practice through continuing education, research, publications, and presentations.
- Recruit and mentor students, peers and colleagues to develop and strengthen professional workforce.
- Represent the profession to the public in a positive manner.
- Perform honorably health information management association responsibilities, either appointed or elected, and preserve the confidentiality of any privileged information made known in any official capacity.
- State truthfully and accurately one's credentials, professional education, and experiences.
- Facilitate interdisciplinary collaboration in situations supporting health information practice.
- Respect the inherent dignity and worth of every person.

Does the government data sharing comply with the AHIMA code of ethics? Include the specific code components that you believe comply/do not comply.

> **Government Data Sharing**
>
>
>
> Would you recommend any additions, changes or revisions to the code of ethics?

Government Billing and Payments

Government insurance billing works similar to the process for commercial insurance billing, typically through electronic data interchange (EDI), though can also occur through paper-based claims. Given the share of government payments in the overall healthcare marketplace, fraud and abuse have been a critical area of focus, as healthcare fraud and abuse costs both public and private sectors billions of dollars. In the U.S., the costs to public and private programs are estimated as high as 10% of annual spending or $100 billion per year. Many health systems rely on human experts for manual review, which is often expensive and ineffective. Data mining can reduce costs and identify previously unknown patterns and trends (Yang 2006; Liou et al., 2008). Increasingly, healthcare entities are using data mining tools to identify fraudulent behaviors. Data mining methods (including classification trees, neural networks, and regression) have been applied to healthcare. The Utah Bureau of Medicaid Fraud, Australian Health Insurance Commission, and Texas Medicaid Fraud and Abuse Detection mined data to identify fraud and abuse, saving and recovering millions of dollars. Although most fraud and abuse cases are associated with diagnosis and services, some studies used provider name, ID, demographics, claim patient, procedure, charge, bill date, and payment deductible, copayments, insurance, and payment dates to detect fraud (Viaenea 2005; Liou et al., 2008).

In Chapter 3, we discussed the capabilities of statistical sampling, which can also be applied to test for and identify fraud. In a recent case, industry groups have attempted to prevent statistical sampling from being used to prove liability for cases beyond the initial case. Opponents argue that preventing statistical sampling would allow larger scale fraud to continue. In another case. there was an allegation that a senior center and hospice facility in South Carolina conducted services that were not medically necessary. The whistleblowers identified only one facility. However, the case plaintiffs proposed the use of statistical sampling to apply the findings to all facilities. With this method, a random sample of services would be taken, and the percentage found to be fraudulent would be calculated across all claims. A lower court ruled that statistical sampling could not be used, with the 4th U.S. Circuit Court of Appeals declining to address the question of statistical sampling (Lauer et al., 2017). Other lower court rulings on statistical sampling have been mixed in the past. Based on the False Claims Act, damages of up to $11,000 per claim can be assessed, which could have a major impact on the tens of thousands of claims against a single facility or across all facilities (Schencker, 2017).

In 2017, The U.S. Justice Department added to a previous lawsuit that claims that the UnitedHealth Group, Inc. Medicare Advantage program overbilled hundreds of millions to billions of dollars by adding conditions (or more severe levels of conditions) to patients. This is a claim that UnitedHealth vowed to fight as untrue. The news caused shares to decline over 4% (Court, 2017b). The U.S. Justice Department also released information that an investigation into four health insurers had occurred due to a claim that Medicare was being defrauded by means of billing for treatments that patients did not receive or did not need (Raymond, 2017). In July 2017, the U.S. Justice Department announced that charges had been filed for over 400 individuals (including 56 physicians) for healthcare fraud totaling $1.3 billion in fraudulent charges. Officials indicated that this was one of the largest targeted healthcare fraud actions in history (Wilber, 2017).

Experiential Learning Activity: Billing Issues and Fraud and Abuse

Billing Issues and Fraud and Abuse

Description: In healthcare, an estimated 3-10% of all costs are attributed to fraud and abuse, and detection remains an important issue as healthcare is highly susceptible to fraud and abuse. Fraud can be classified as unauthorized benefits along with abuse, which often might not be against any laws or might occur without deliberate intent. Fraud and abuse might occur by patients, providers, and payers. Examples of patient fraud include obtaining multiple prescriptions for the same medication from different physicians, identity theft of another patient's health information, or using one patient's insurance for another patient's services. Examples of provider fraud and abuse include billing for services not provided, upcoming or including a more expensive service than what was performed, unbundling or charging separately for a single service, and receiving payment or kickbacks from other medical providers. Examples of insurer or payer fraud and abuse include falsifying reimbursements and collecting payment without providing services. General abuse examples include providing below acceptable levels of standard care, providing unnecessary tests or services, insufficient documentation, and charging additional items for insured patients versus uninsured patients (Rashidian et al., 2012).

Methods to address fraud generally fall into three categories: prevention, detection, and response. Prevention strategies attempt to stop fraud beforehand. Prevention might be done through influencing cultural norms and compliance systems and processes. Studies in the U.S. showed that between 10-50% of physicians reported willingness to manipulate reimbursement rules in order for patients to receive necessary care. Patients were also willing to allow deception if it was to their benefit. Detection strategies attempt to locate fraud as soon as possible and create a notification for investigation. In a study in Taiwan, logistic regression, neural networks, and decision trees were used to detect fraud and abuse of diabetic services. In a related model, the ability to detect fraud was approximately 65% accurate. The final category of response strategies uses legal or administrative options to address the fraud and abuse such as fines and penalties. In a related study, the more lenient policies in the U.S. between 1997-2001 might have contributed to the increase in fraud and abuse over the same time period (Rashidian et al., 2012).

Billing Issues and Fraud and Abuse

List the 3 intervention categories for fraud and abuse:

1.

2.

3.

Based on the study, which of the 3 intervention categories is the least common, and which do you feel is most important?

From the study, findings state that up to 40% of physicians reported manipulating reimbursement rules to allow for necessary patient care. Do you feel this is legal and ethical?

Find an outside article and describe an example of fraud and abuse. From Table 1, which type of fraud and abuse example does this fit best (For example, Provider Fraud → Up-coding)?

Now that you are familiar with government agencies, regulations, and government billing and payments, we'll continue with our SEMMA modeling process using data available through government billing systems, and review the neural network model.

SEMMA: Model

Model Process Step Overview

During the modeling process step, the data mining model is applied to the data. During the partitioning phase, data is segmented into training and validation data sets. The training data set is used to fit the model, and the validation data set is used to validate the model on a new set of data to demonstrate the reliability of the model. Based on the results the model can then be tuned to optimal performance.

There are many different models that can be selected during this step. The decision to choose each model is based on earlier exploration of data and knowledge of each model. We will continue this chapter with the neural network model.

Model Tab Enterprise Miner Node Descriptions

Model Application Examples

Neural networks are a class of nonlinear regression and statistical methods and have been studied extensively. Neural networks can be applied to a variety of applications including speech recognition, fraud, signature validation, mammogram screenings, identification of food infections, measurement of nurse burnout, and detection of heart failures.

Researchers have used neural networks for a variety of epidemiological events (or the patterns of diseases). Researchers predicted morbidity (or death) from environmental contaminants such as air pollution, weather, and conditions such as asthma. Another study used inputs of temperature, humidity, precipitation, sunshine amount, and rat density as viral carriers to study fever in Shenyang, China in the 1980s through early 2000s. Some epidemiological studies on environmental factors and morbidity showed a prediction accuracy rate using neural networks near 60% (Song et al., 2017).

In a study using neural networks to study nurse burnout, input factors including job stressors, personality, and syndromes were used. Nursing is considered an occupation that is susceptible to stress and affects professionals globally. The occupation has experienced a number of changes over the last several decades including readmission rates, efficiency emphasis, and demands of patients with acute and chronic conditions. Individual nurse interventions can then be made to reduce the risk of burnout and improve the work environment (Ladstatter, et al., 2016).

Researchers have also used EHRs in conjunction with neural networks to detect disease. Using EHR data, medical professionals can detect early onset of heart failure. Heart failure intervention often occurs only after diagnosis and carries a high risk of mortality near 50% within five years of diagnosis. Earlier detection of heart failure can improve patient outcomes through earlier treatment and changes such as reduced salt intake and mild exercise. Inputs to the model included age, gender, visits, diagnosis codes, medication codes, and procedure codes. The results were also compared with a logistic regression model. The neural network generally outperformed the logistic regression model, although the neural network took longer in terms of prediction time, in which logistic regression might be useful for real-time requirements, whereas neural network might be used for batch processing (Choi et al., 2017).

Neural Network Node

The Neural Network node is used for running a neural network model and is associated with the **Model** tab in SAS Enterprise Miner.

Figure 8.3: Neural Network Node

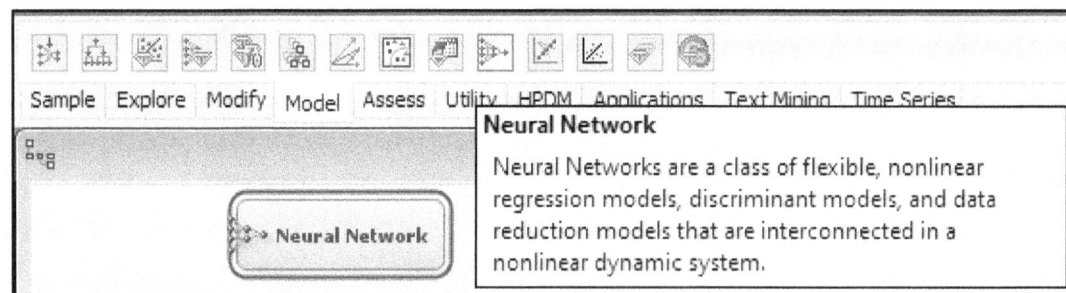

Neural Network Model Description

Neural networks are modeled after the human brain and were initially used for artificial intelligence to develop an ability to learn. In the brain, neurons accept inputs from other neurons, weight them, and produce outputs using a set of electrical signals. Similarly, in a neural network model, we have an input layer (or the variable inputs), a hidden layer where items are combined and weighted, and an output layer (or the target variable). Neural networks are sometimes referred to as a "black box," meaning that data is submitted as input to the neural network, and output is received. The inner workings of the algorithm are not easily presented (Klimberg and McCullough, 2013). The neural network model differs from a decision tree in which the algorithm creates easy to interpret IF...THEN rules (or regression) in which an easy to apply regression equation is created. Despite the interpretability drawback, neural networks often have one the lowest error rates and are a very effective model. Neural networks can also accept different types of input variables and target variables that are similar to a decision tree.

Model Assumptions

For neural networks, a parsimonious model and a model with uncorrelated variables are preferred due to the complexity and performance time required to complete the algorithm and return the solution. Similarly, although a decision tree (or regression) might be able to accommodate hundreds of variables, that type of setup would not be preferred due to performance and run times. Generally, a single hidden layer is used. As a starting point, the number of nodes used initially in the hidden layer can be a number that is between the number of inputs and outputs. If there are five input variables and one target variable, there might be 3three nodes in the hidden layer to begin with for the first model run. Although no formal rule exists, one strategy is to begin with a simple number of nodes (parsimonious model) and increase the number of nodes, while measuring performance run times against results as measured by error rates. Neural network models with uncorrelated variables are preferred, and neural networks might be affected by outlier or extreme values. Overfitting can also occur. This happens when we measure the model solely on the training sample data, and we attempt to perfect the model for the training data only. This might lead to cases where new data performs poorly, or we are unable to generalize the model beyond the initial data set.

Data Preparation

Unlike decision trees, neural networks are affected by outliers and missing data. In addition, non-normalized or standardized data might affect the model weighting. To ensure proper results, all variables should be standardized before running the model. Outliers and missing data should be corrected or removed from the data before standardization.

Partitioning Requirements

For neural networks, we partition our data into a training and validation data set. The model is formed based on our training data set, and is then run through our validation or holdout data set to confirm validity of the model. If the model performs well on the training data set though poorly on the validation data set, this can indicate that we have overfit the model. In other words, we tuned the model too well to the training data set, and now, it no longer performs well on a new or general data set. In this case, we might wish to remove variables to create a simpler, more general model.

Model Results Evaluation

To evaluate the results of our neural network, we can use several items including errors, lift, and misclassification rate. The p-value will be available for models with an interval target variable.

Errors

Errors are calculated from the predicted value less the actual value, which is also known as the residuals. Common measures of errors are sum of squared errors (SSE) and root mean square error (RMSE). The measures are calculated by taking the square or square root of the errors.

Lift

Lift is a measurement between a random or baseline model against the analytical model. A higher lift or outperformance of the random selection is best.

Misclassification Rate

For models with nominal or binary targets, misclassification rate is the percentage of total records misclassified as false positive or false negative.

Ensemble Node

The Ensemble node is used to combine different models into a single final model. A common approach is to combine multiple models such as neural networks or decision trees with different parameters applied to a training data set. The final model can be selected based on various criteria such as error rates or statistical tests. The Ensemble node is associated with the Model tab in SAS Enterprise Miner (SAS, 2018a).

Figure 8.4: Ensemble Node

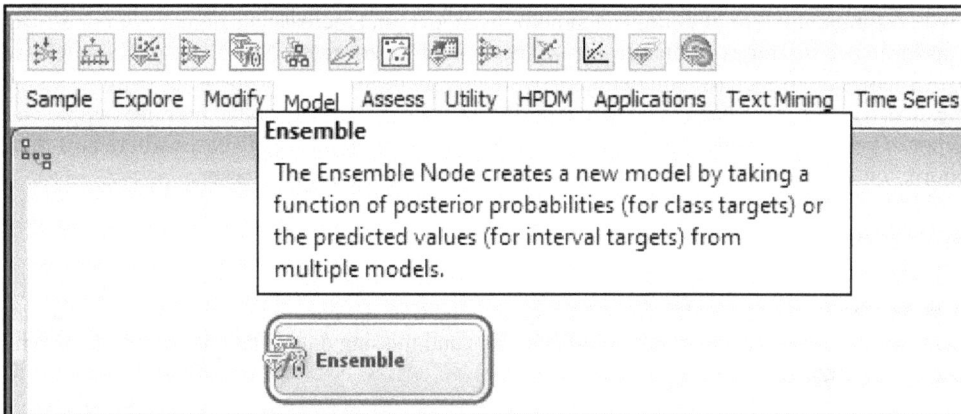

Ensemble Model Description

The Ensemble model combines several models into one model. The Ensemble model determines the target variables based on the average of the individual models. For interval targets, an average would be used. For categorical targets, a voting method where the most common class would be used. Each model can also be given different weights for the final prediction. Combining predictions from multiple models can improve accuracy. In neural networks, each model and algorithm can produce different outputs, based on the random weights and model parameters. Averaging these neural network models can improve results (SAS, 2018b).

Now that we have covered the neural network model often used in conjunction with an Ensemble model, let's continue with an experiential learning application to connect your knowledge of neural networks with a health application on fraud detection.

Experiential Learning Application: Fraud Detection

For our experiential learning application, let's consider the example where we want to study the detection of fraud (Liou et al., 2008). To start our process, we first want to review our file to identify our input and target variables. After reviewing, both our input variables are continuous interval variables and our target variable Fraudulent Hospital is binary.

Data Set File: 8_EL1_Fraud_Detection.xlsx

Variables:

- HospitalID
- AverageDaysOfDrugDispense
- AverageDrugCost
- AverageConsultationAndTreatmentFees
- AverageDiagnosisFees
- AverageDispensingServiceFees
- AverageMedicalExpenditures
- AverageAmountClaimed
- AverageDrugCostPerDay
- AverageMedicalExpenditurePerDay
- FraudulentHospital

Step 1: Sign in to SAS OnDemand for Academics.

Step 2. Open SAS Enterprise Miner Application (Click the SAS Enterprise Miner link).

Step 3. Create a New Enterprise Miner Project (Click **New Project**).

Step 4: Use the default SAS Server, and click **Next**.

Step 5: Add a project name, and click **Next**.

Figure 8.5 Create New Project

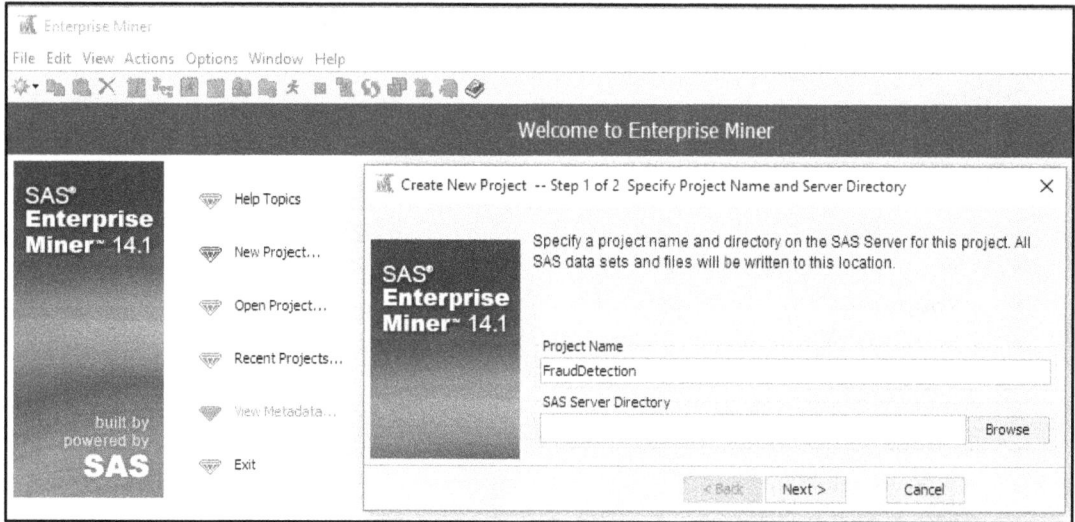

Step 6: SAS will automatically select your user folder directory (If you are using the desktop version, choose your folder directory), and click **Next**.

Step 7: Create a New Diagram (Right-click **D**iagram).

Figure 8.6 Create New Diagram and Add Name

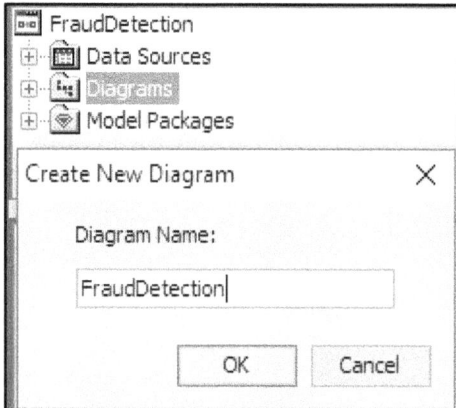

Step 8: Add a File Import node (Click the **Sample** tab, and drag the node onto the diagram workspace).

Step 9: Click the File Import node, and review the property panel on the bottom left of the screen.

Step 10: Click Import File and navigate to the *8_EL1_Fraud_Detection.xlsx* Excel file.

Figure 8.7 File Import

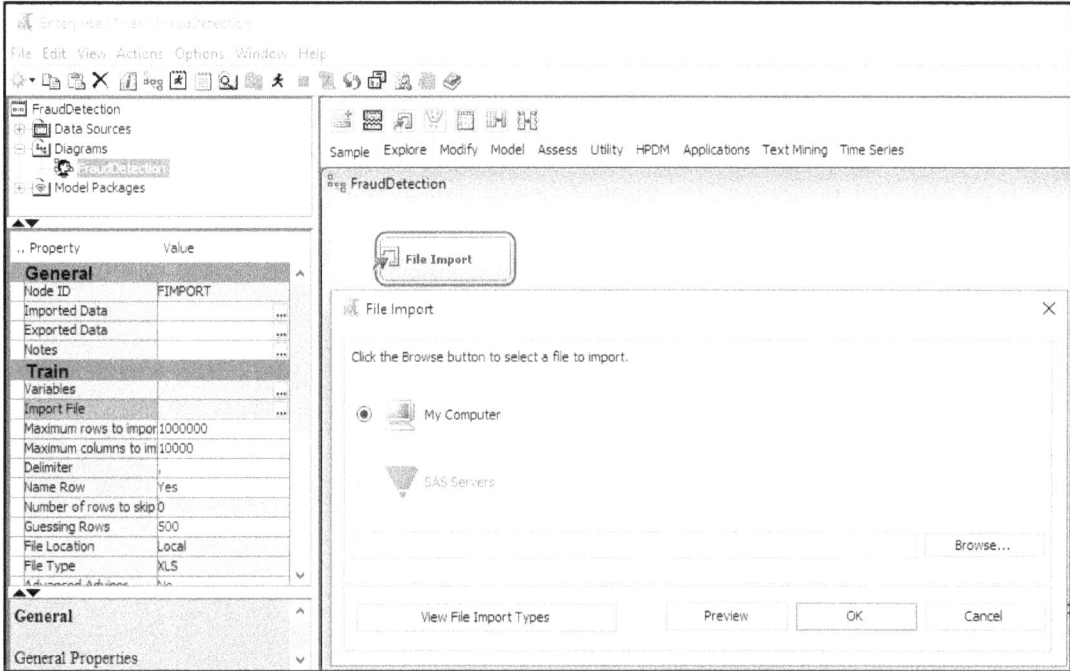

Step 11: Click **Preview** to ensure that the data set was selected successfully, and click **OK**.

Step 12: Right-click the File Import node and click Edit Variables.

Step 13: Set Fraudulent_Hospital to the Target variable role, set Hospital_ID to the ID role, and all other variables to the Input role. Explore and set the remaining variables according to their nominal, interval, or binary levels. To review an individual variable in order to verify its role and level assignment, click the variable name and click **Explore**. After you have finished setting all the variables, click **OK**.

Figure 8.8 Edit Variables

Name	Role	Level	Report	Order	Drop	Lower Limit	Upper Lin
AverageAmountClaimed	Input	Interval	No		No	.	
AverageConsultationAndTreatment	Input	Interval	No		No	.	
AverageDaysOfDrugDispense	Input	Interval	No		No	.	
AverageDiagnosisFees	Input	Interval	No		No	.	
AverageDispensingServiceFees	Input	Interval	No		No	.	
AverageDrugCost	Input	Interval	No		No	.	
AverageDrugCostPerDay	Input	Interval	No		No	.	
AverageMedicalExpenditurePerDay	Input	Interval	No		No	.	
AverageMedicalExpenditures	Input	Interval	No		No	.	
FraudulentHospital	Target	Binary	No		No	.	
HospitalID	ID	Interval	No		No	.	

Step 14: Add a Data Partition node (Click the **Sample** tab, and drag the node onto the diagram workspace). Set the Data Partition Property Data Set Allocations to **60.0** for Training, **40.0** for Validation, and **0.0** for Test. Review the data Partition Results.

Figure 8.9 Data Partition Node

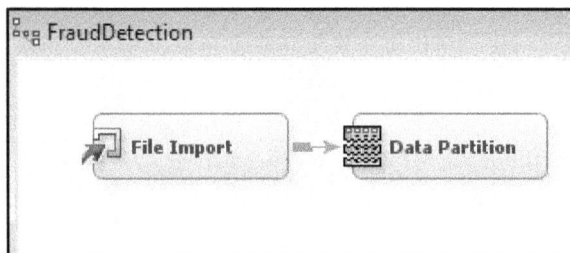

Step 15: Add a Stat Explore node, Graph Explore node, and MultiPlot node (Click the **Explore** tab, and drag the nodes onto the diagram workspace). Set the Graph Explore Property Sample Size to **Max**.

Figure 8.10 StatExplore, Graph Explore, and MultiPlot Nodes

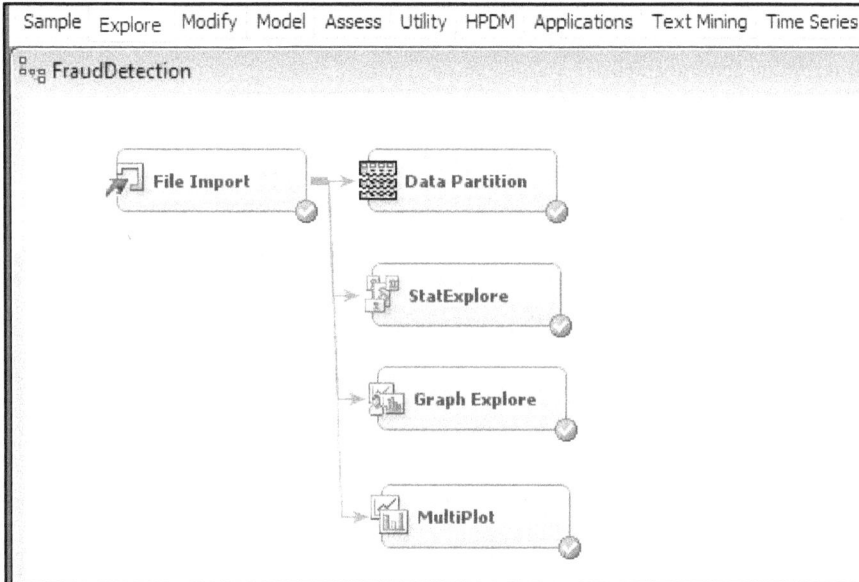

Step 16: Review the results. From the StatExplore results, we find that there is a total of 1267 records, with no missing data across variables, and initial indication of good data quality. The 1 (or True) cases for the FradulentHospital variable total 113 records (or 8.9%) of the data set. The skewness and kurtosis for all variables is within acceptable ranges of -2 to 2, indicating an absence of extreme outlier values. Since neural networks might be affected by missing or outlier values, we still want to add Impute and Transform nodes to our final model diagram as a best practice, since future unknown data sets might contain these issues and we want to reuse the same model.

Figure 8.11 StatExplore Results

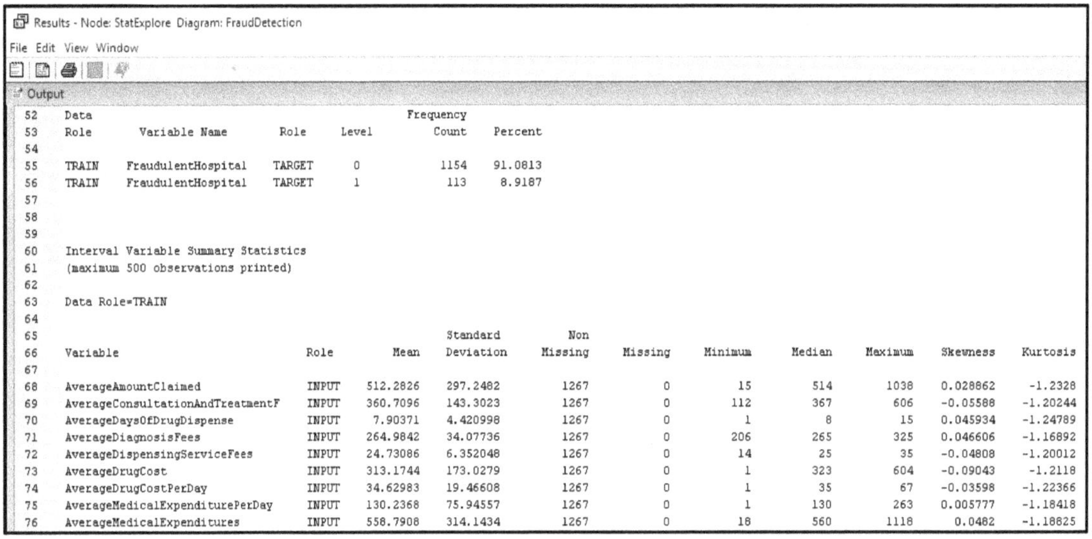

Results - Node: StatExplore Diagram: FraudDetection

File Edit View Window

Output

52	Data				Frequency	
53	Role	Variable Name	Role	Level	Count	Percent
54						
55	TRAIN	FraudulentHospital	TARGET	0	1154	91.0813
56	TRAIN	FraudulentHospital	TARGET	1	113	8.9187
57						
58						
59						
60	Interval Variable Summary Statistics					
61	(maximum 500 observations printed)					
62						
63	Data Role=TRAIN					

	Variable	Role	Mean	Standard Deviation	Non Missing	Missing	Minimum	Median	Maximum	Skewness	Kurtosis
68	AverageAmountClaimed	INPUT	512.2826	297.2482	1267	0	15	514	1038	0.028862	-1.2328
69	AverageConsultationAndTreatmentF	INPUT	360.7096	143.3023	1267	0	112	367	606	-0.05588	-1.20244
70	AverageDaysOfDrugDispense	INPUT	7.90371	4.420998	1267	0	1	8	15	0.045934	-1.24789
71	AverageDiagnosisFees	INPUT	264.9842	34.07736	1267	0	206	265	325	0.046606	-1.16892
72	AverageDispensingServiceFees	INPUT	24.73086	6.352048	1267	0	14	25	35	-0.04808	-1.20012
73	AverageDrugCost	INPUT	313.1744	173.0279	1267	0	1	323	604	-0.09043	-1.2118
74	AverageDrugCostPerDay	INPUT	34.62983	19.46608	1267	0	1	35	67	-0.03598	-1.22366
75	AverageMedicalExpenditurePerDay	INPUT	130.2368	75.94557	1267	0	1	130	263	0.005777	-1.18418
76	AverageMedicalExpenditures	INPUT	558.7908	314.1434	1267	0	18	560	1118	0.0482	-1.18825

Step 17: Review the results. From the StatExplore results, we find that the top three variables with a target of FraudlentHospital include AverageMedicalExpenditures, AverageAmountClaimed, and AverageMedicalExpenditurePerday. The variable worth can also help develop a parsimonious model (or the simplest-best model) by including only a few variables in the final data set and model, which would improve performance of the model with similar accuracy.

Figure 8.12 StatExplore Results

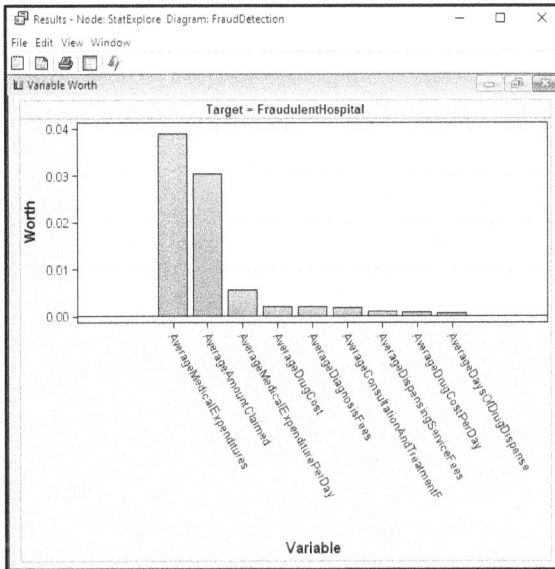

Step 18: Review the results. From the Graph Explore results, there are 113 cases with 1 (or True), and 1154 cases with 0 (or False).

Figure 8.13 Graph Explore Results

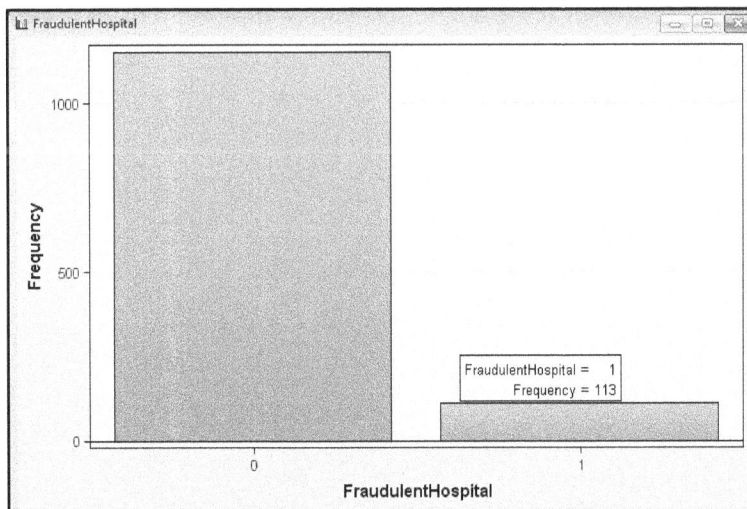

Step 19: Review the results. From the MultiPlot results, review the different variables. For AverageMedicalExpenditures by FradulentHosptial, we find a greater distribution of 1 (or True) cases as the AverageMedicalExpenditures amount increases.

Figure 8.14 MultiPlot Results

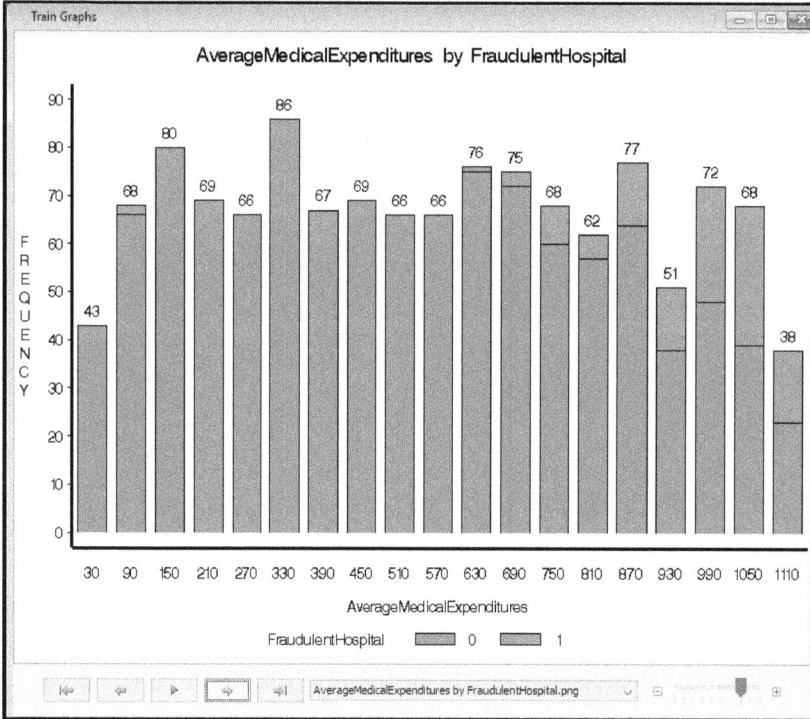

Step 20: Add an Impute node (Click the **Modify** tab, and drag the node onto the diagram workspace). Verify that the Impute Property is set to **Count** for Class variables and **Mean** for Interval variables. Although no missing data was identified within this data set, as a best practice, include the Impute node as future data sets used for this model might contain missing data.

Figure 8.15 Impute Node

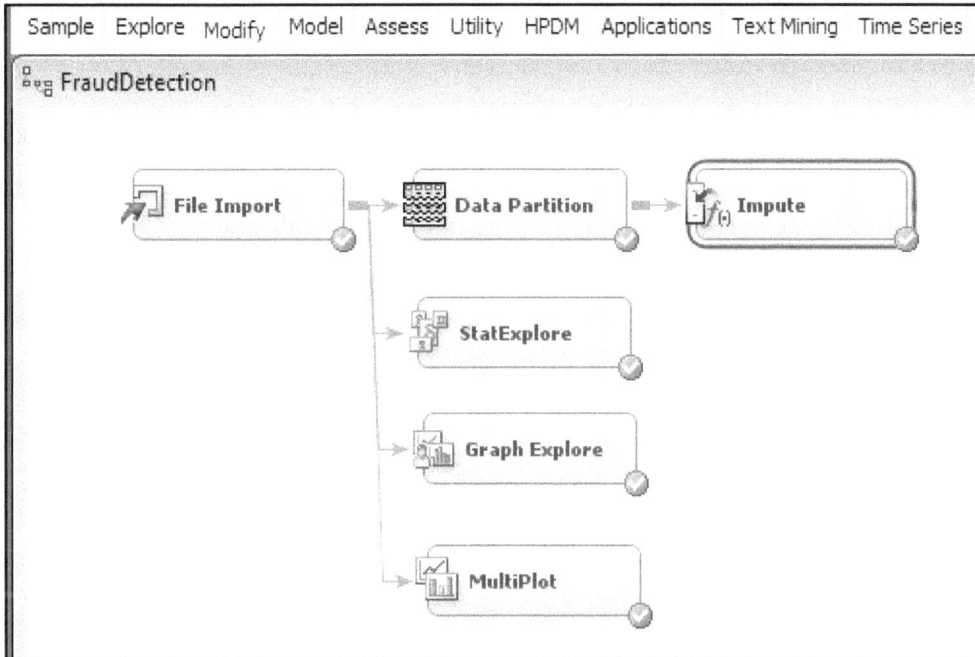

Step 21: Add a Transform Variables node (Click the **Modify** tab, and drag the node onto the diagram workspace). Right-click Transform Variables and click Edit Variables.

Figure 8.16 Transform Node

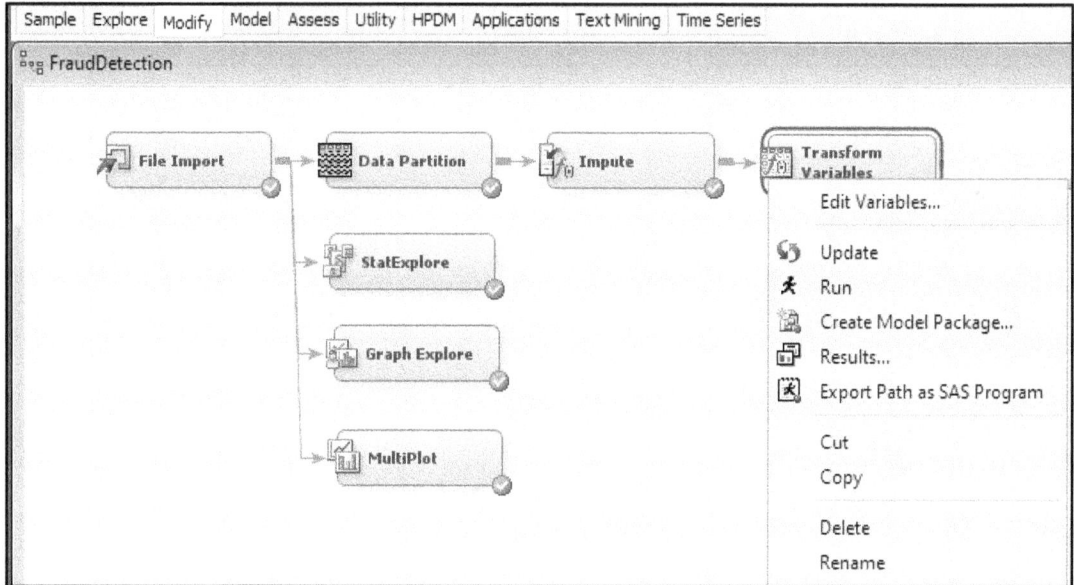

Step 22: Set the Transform Variables. Since neural networks are susceptible to outliers (or non-standardized variables), we want to standardize our data set using the Transform Variables node. Set all the Input Role records to Method **Standardize**. You can highlight multiple records by clicking the Name and holding the Shift key, and then change the Method all at once. By standardizing the data, we are using the same scale for all the variables, which is useful when applying the weights and training the neural network. For example, our variables contain both Days variables and Dollars variables. Days might range from 1-365, whereas Dollars might range from $1 - $1 million. Since these variables have varying ranges, we want to standardize all variables to be within a similar range and avoid one variable influencing the model results.

Figure 8.17 Transform Node Standardize

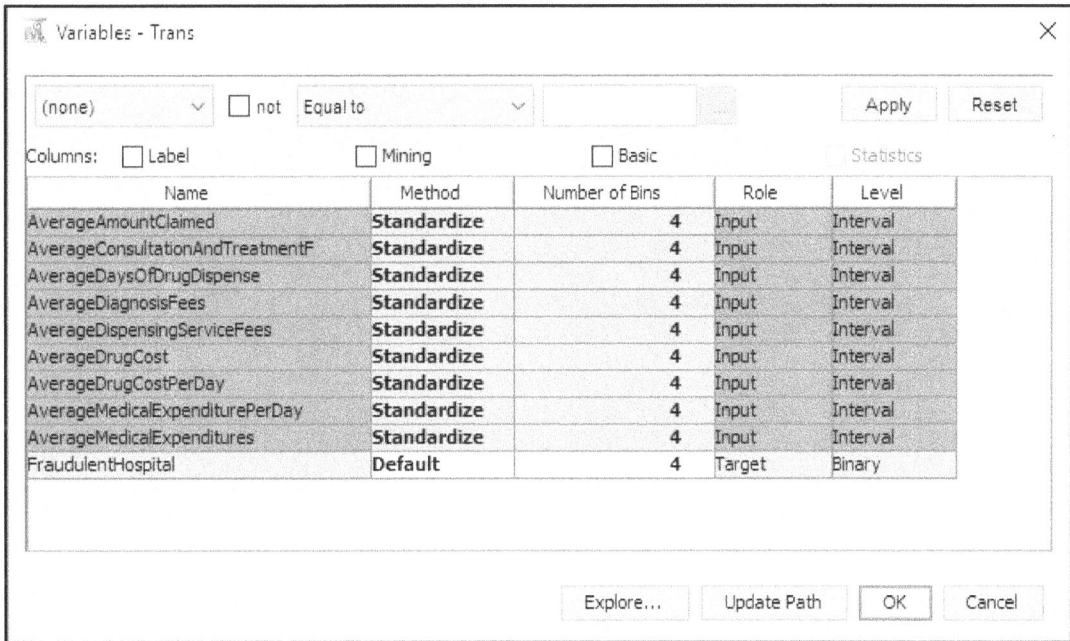

Name	Method	Number of Bins	Role	Level
AverageAmountClaimed	Standardize	4	Input	Interval
AverageConsultationAndTreatmentF	Standardize	4	Input	Interval
AverageDaysOfDrugDispense	Standardize	4	Input	Interval
AverageDiagnosisFees	Standardize	4	Input	Interval
AverageDispensingServiceFees	Standardize	4	Input	Interval
AverageDrugCost	Standardize	4	Input	Interval
AverageDrugCostPerDay	Standardize	4	Input	Interval
AverageMedicalExpenditurePerDay	Standardize	4	Input	Interval
AverageMedicalExpenditures	Standardize	4	Input	Interval
FraudulentHospital	Default	4	Target	Binary

Step 23: Review the Transform Variables Results. From the results, we now have a second set of variables created with a Method of Computed, and these variables are also prefixed with STD_for standardized. The range (or minimum and maximum) is significantly reduced between variables, and all variables now have a standard deviation of 1.

Figure 8.18 Transform Node Standardize Results

Method	Variable Name ▲	Minimum	Maximum	Mean	Standard Deviation
Original	AverageAmountClaimed	16	1029	501.2395	294.6897
Original	AverageConsultationAndTreatmentF	112	606	360.9711	143.1839
Original	AverageDaysOfDrugDispense	1	15	7.765789	4.383811
Original	AverageDiagnosisFees	206	325	265.1829	34.19655
Original	AverageDispensingServiceFees	14	35	24.60395	6.339515
Original	AverageDrugCost	2	604	310.8263	172.7746
Original	AverageDrugCostPerDay	1	67	34.72763	19.45985
Original	AverageMedicalExpenditurePerDay	1	263	131.4697	76.55462
Original	AverageMedicalExpenditures	18	1116	548.2947	309.1991
Computed	STD_AverageAmountClaimed	-1.64661	1.790902	1.43E-11	1
Computed	STD_AverageConsultationAndTreatm	-1.73882	1.711288	1.1E-11	1
Computed	STD_AverageDaysOfDrugDispense	-1.54336	1.65021	-3.6E-12	1
Computed	STD_AverageDiagnosisFees	-1.73067	1.749215	-9.2E-11	1
Computed	STD_AverageDispensingServiceFees	-1.67267	1.639882	6.64E-11	1
Computed	STD_AverageDrugCost	-1.78745	1.696856	-3E-12	1
Computed	STD_AverageDrugCostPerDay	-1.73319	1.658408	-2.7E-12	1
Computed	STD_AverageMedicalExpenditurePer	-1.70427	1.718123	2.75E-11	1
Computed	STD_AverageMedicalExpenditures	-1.71506	1.836051	6.81E-12	1

Step 24: Add a Neural Network node (Click the **Model** tool tab, and drag the node onto the diagram workspace).

Figure 8.19 Neural Network Node

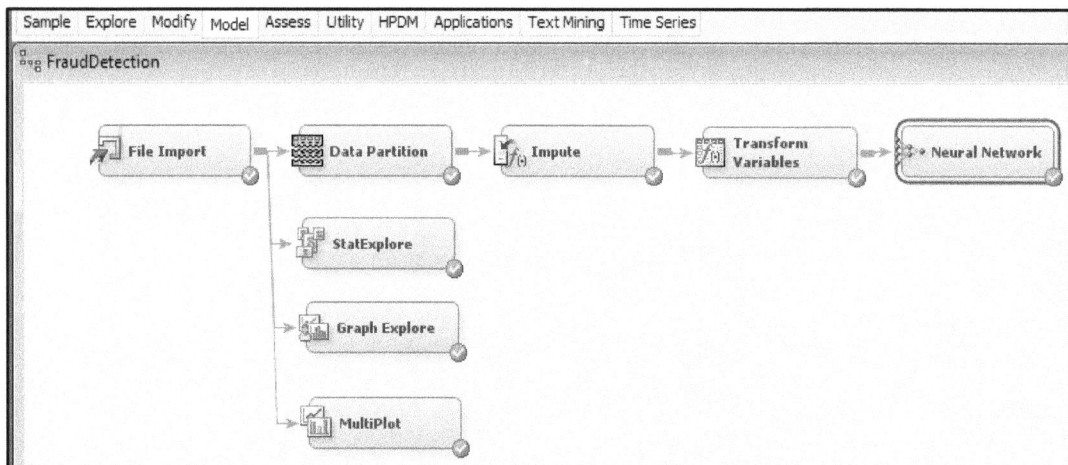

Step 25: Right-click the Neural Network node and click **Edit Variables**. Note that the standard deviation variables are prefixed with the STD_ and are used for the model by default. Click **OK**.

Figure 8.20 Neural Network Node Variables

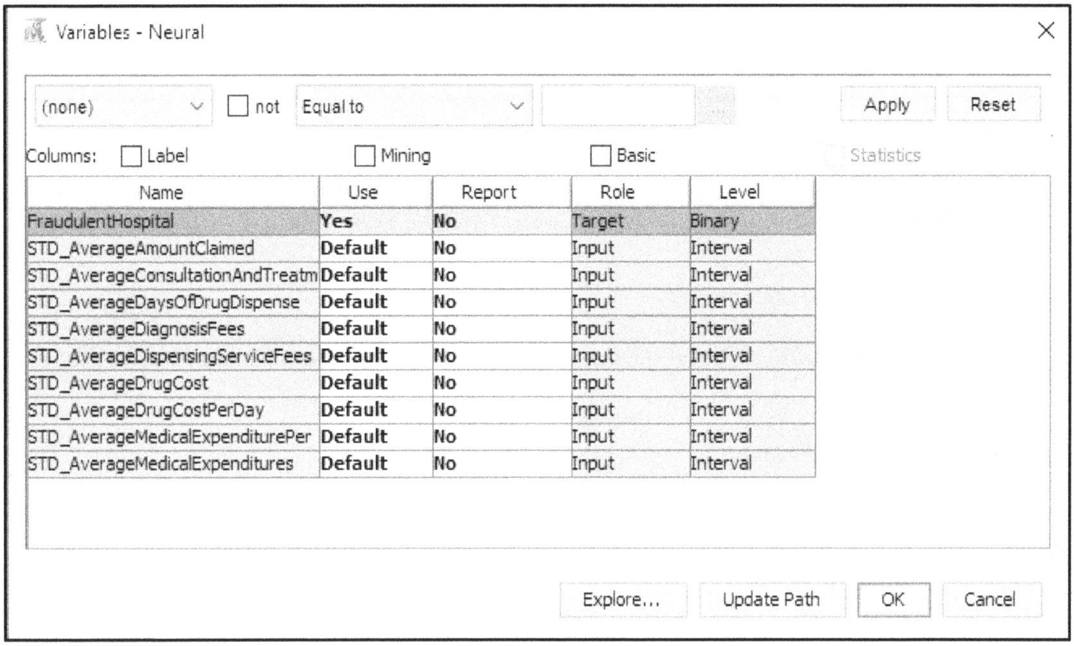

Step 26: Right-click the Neural Network node and click **Run**.

Step 27: Expand the Output Window Results.

Step 28: Review the Model Results. Since the target variable of FradulentHospital is binary (0/1), the results are presented within an Event Classification Table. The Event Classification Table shows the overall model performance. When validating the model, note that we received 5 False Negatives, 460 True Negatives, 2 False Positives, and 40 True Positives. To calculate the error rate, we add the False Negatives and False Positives and divide by the total number: $(5 + 2) / (5 + 460 + 2 + 40) = 7/507 = 1.38\%$ error rate.

Figure 8.21 Event Classification Table

```
 Results - Node: Neural Network  Diagram: FraudDetection

File  Edit  View  Window

 Output

298      Event Classification Table
299
300      Data Role=TRAIN Target=FraudulentHospital Target Label=FraudulentHospital
301
302       False        True         False        True
303      Negative     Negative     Positive     Positive
304
305         2           692            .           66
306
307
308      Data Role=VALIDATE Target=FraudulentHospital Target Label=FraudulentHospital
309
310       False        True         False        True
311      Negative     Negative     Positive     Positive
312
313         5           460            2           40
```

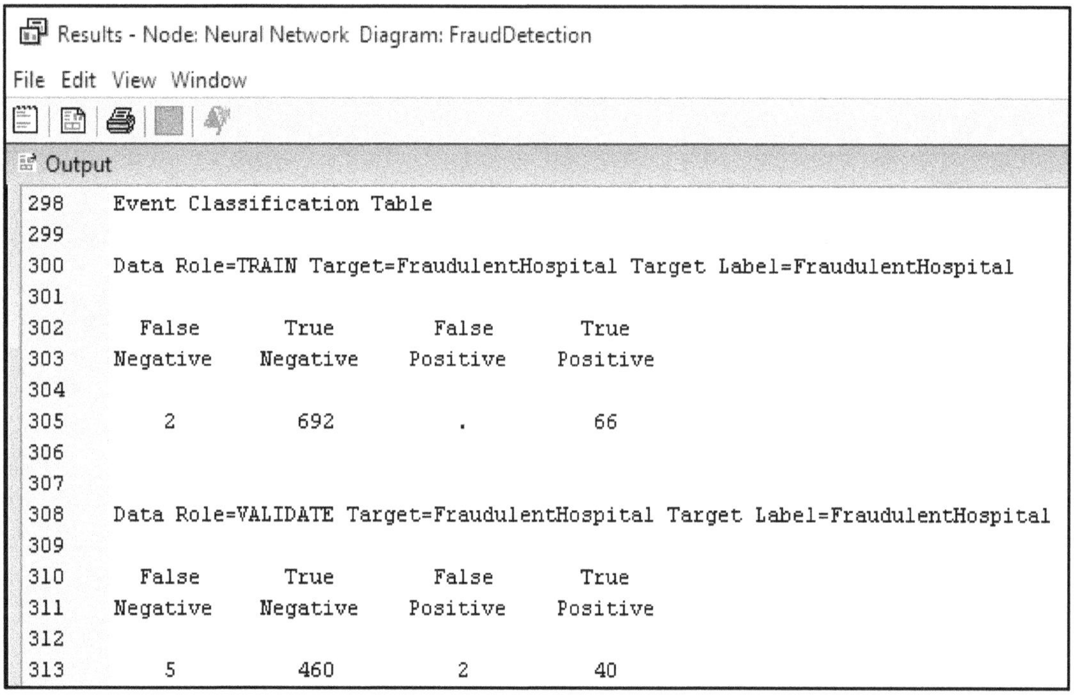

Step 29: Review the Fit Statistics Results. The Fit Statistics output window can be used for various error rate comparisons, including our misclassification rate from the event classification matrix. Notice that in the last entry on Misclassification Rate, 7 records were incorrectly classified, and the result matches the calculation above at 1.38% for Validation.

Figure 8.22 Fit Statistics

Target	Fit Statistics	Statistics Label	Train	Validation
FraudulentHospital	_DFT_	Total Degrees of Freedom	760	.
FraudulentHospital	_DFE_	Degrees of Freedom for Error	726	.
FraudulentHospital	_DFM_	Model Degrees of Freedom	34	.
FraudulentHospital	_NW_	Number of Estimated Weights	34	.
FraudulentHospital	_AIC_	Akaike's Information Criterion	80.76001	.
FraudulentHospital	_SBC_	Schwarz's Bayesian Criterion	238.2928	.
FraudulentHospital	_ASE_	Average Squared Error	0.002095	0.01375
FraudulentHospital	_MAX_	Maximum Absolute Error	0.989171	0.999992
FraudulentHospital	_DIV_	Divisor for ASE	1520	1014
FraudulentHospital	_NOBS_	Sum of Frequencies	760	507
FraudulentHospital	_RASE_	Root Average Squared Error	0.045774	0.11726
FraudulentHospital	_SSE_	Sum of Squared Errors	3.184785	13.94245
FraudulentHospital	_SUMW_	Sum of Case Weights Times ...	1520	1014
FraudulentHospital	_FPE_	Final Prediction Error	0.002292	.
FraudulentHospital	_MSE_	Mean Squared Error	0.002193	0.01375
FraudulentHospital	_RFPE_	Root Final Prediction Error	0.04787	.
FraudulentHospital	_RMSE_	Root Mean Squared Error	0.046834	0.11726
FraudulentHospital	_AVERR_	Average Error Function	0.008395	0.080978
FraudulentHospital	_ERR_	Error Function	12.76001	82.11205
FraudulentHospital	_MISC_	Misclassification Rate	0.002632	0.013807
FraudulentHospital	_WRONG_	Number of Wrong Classificati...	2	7

Step 30: Review the Model Results. The cumulative lift shows that the model outperforms a random model. At the top 10% of records (or depth), the train model outperforms a random model by nearly ten times, and at the top 20% of records, the train model outperforms a random model by nearly five times.

Figure 8.23 Cumulative Lift

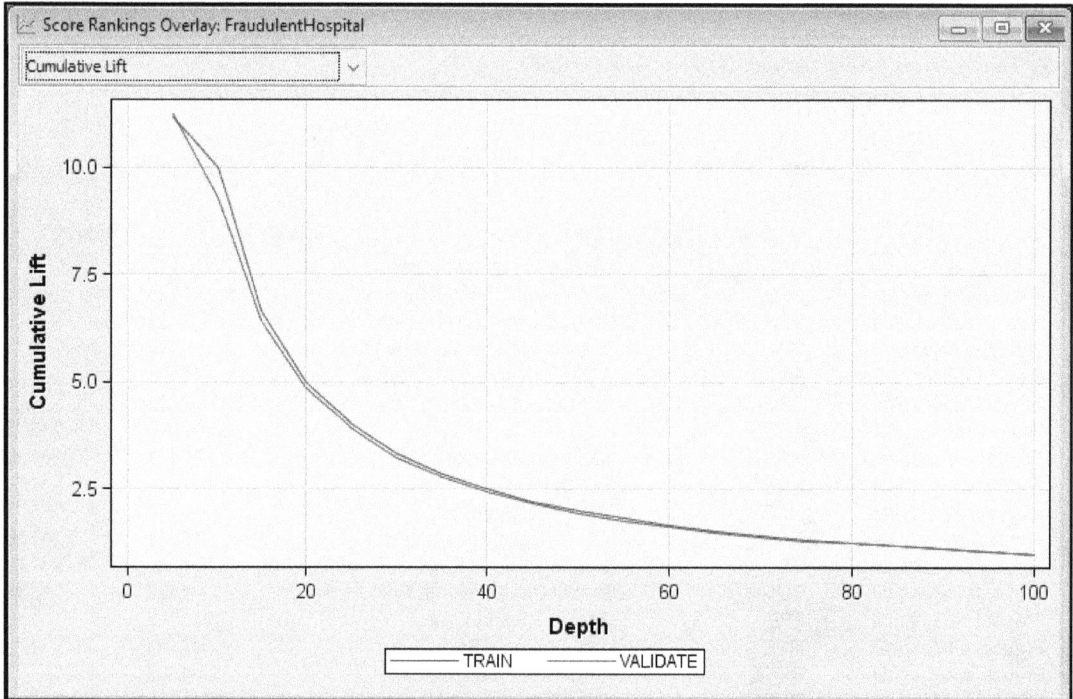

Step 31: Add a Decision Tree node. From the previous chapter, we can also include a Decision Tree model for our data set. The Decision Tree can be connected directly to the Impute node since the model is not affected by non-standardized variables, and the Decision Tree splits allows easier interpretation based on the actual values. Run the Decision Tree node.

Figure 8.24 Decision Tree Node

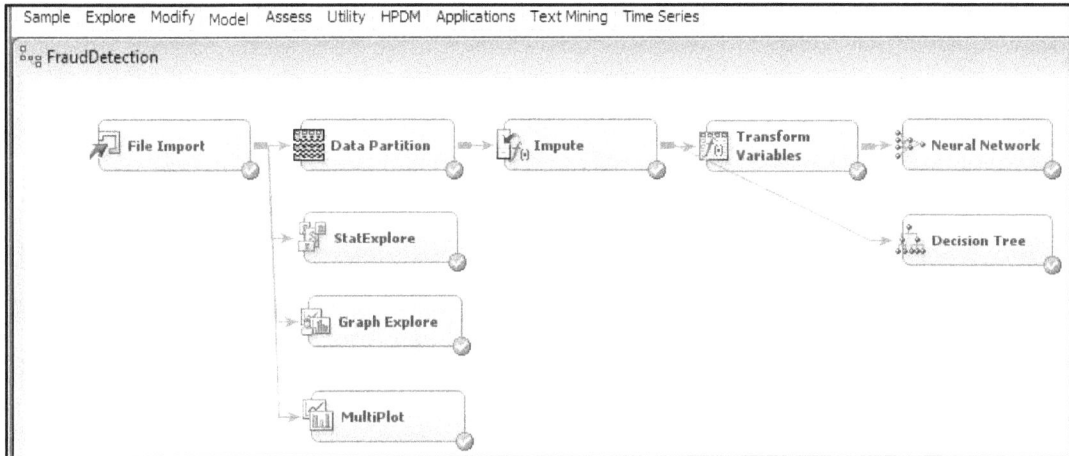

Step 32: Review the Model Results. The Decision Tree main branch (or first split) is Average Medical Expenditures, with a split value of 856.5, splitting to the left (or right) branches. The next branch split is Average Amount Claimed, and following the left branch Average Medical Expenditures and Average Diagnosis Fees to complete the final leaf at Node Id 19. The output could also be used to develop the set of English Rules to program potential fraud detection based on the model splits and parameters.

Figure 8.25 Decision Tree Results

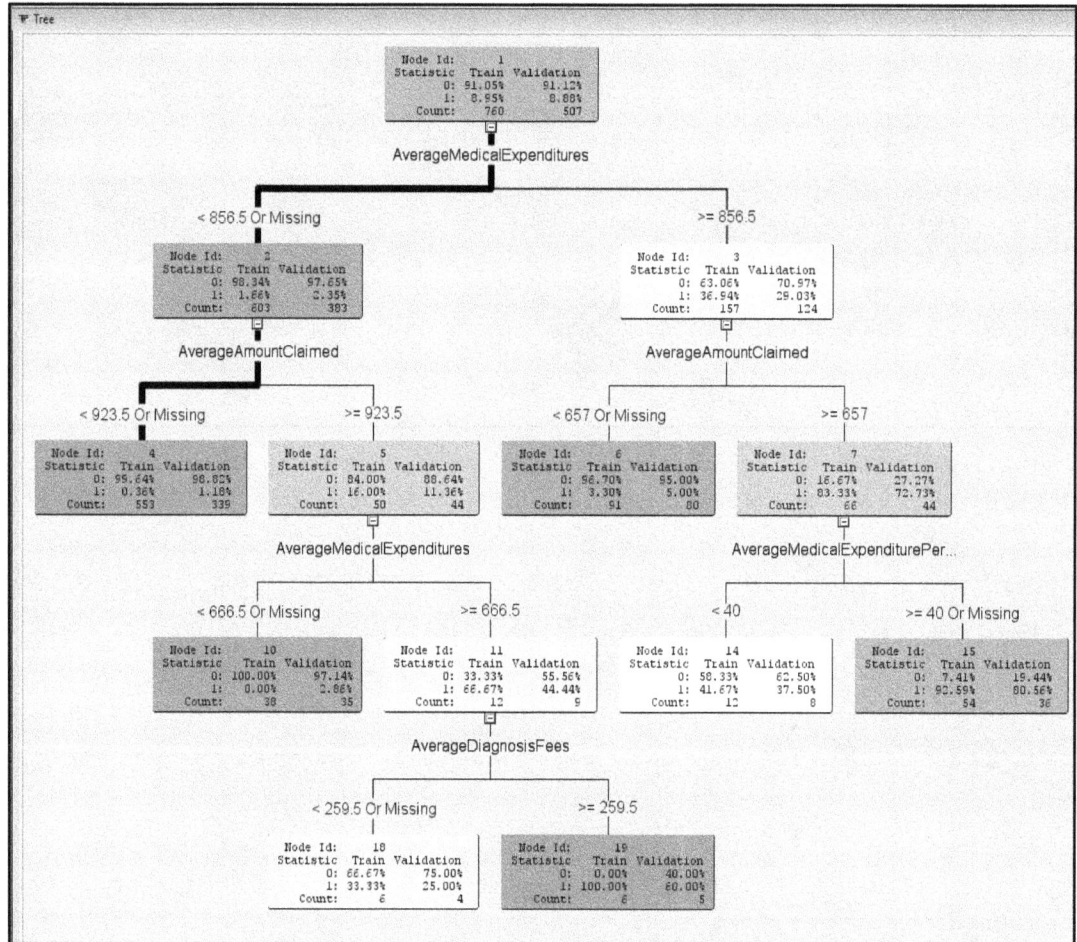

Step 33: Review the Model Results. From the fit statistics, the decision tree has a higher misclassification rate for the validation set at 4.44% versus 1.38% for the neural network. There are often tradeoffs between the models in terms of performance, interpretability, and error rates.

Figure 8.26 Decision Tree Results Fit Statistics

Statistics Label	Train	Validation
Sum of Frequencies	760	507
Misclassification Rate	0.021053	0.043393
Maximum Absolute Error	0.996383	1
Sum of Squared Errors	25.69514	39.25174
Average Squared Error	0.016905	0.03871
Root Average Squared Error	0.130018	0.196748
Divisor for ASE	1520	1014
Total Degrees of Freedom	760	.

Step 34: Add three additional Neural Network nodes for a total of four. Rename the four Neural Network Nodes to Neural Network 3, Neural Network 10, Neural Network 25, and Neural Network 50. The names reflect the number of hidden nodes for each. Adjust the hidden nodes by clicking on each node, and then click **Network** on the property panel. Update the hidden units to 3, 10, 25, and 50, respectively, to match the naming for each node.

Figure 8.27 Neural Network Nodes - Hidden Units

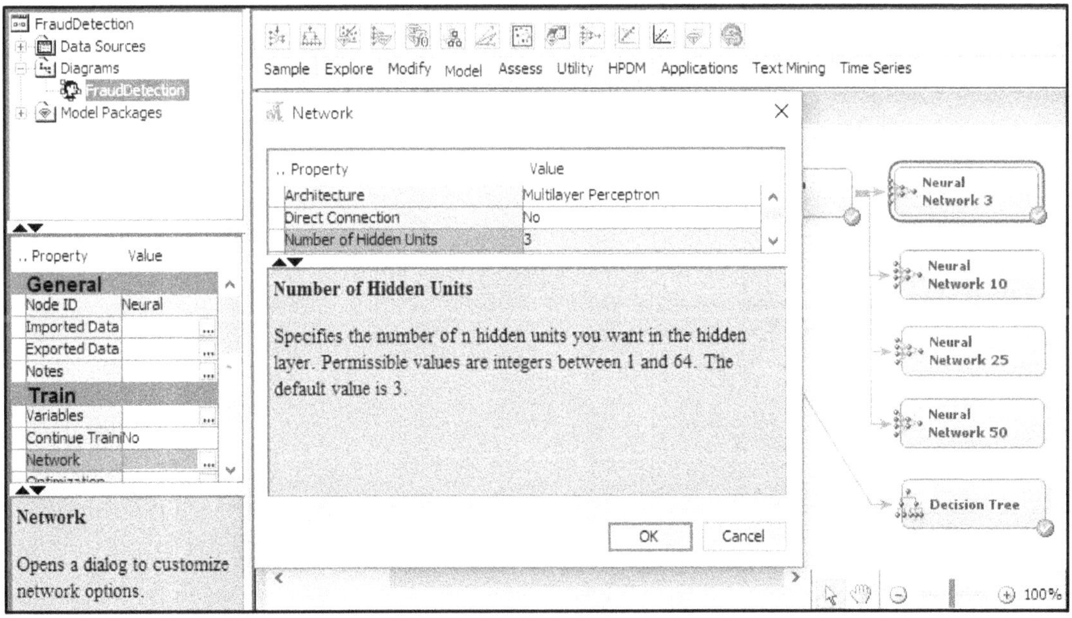

Step 35: Add an Ensemble node. The Ensemble node will compare the 3, 10, 25, and 50 hidden unit neural network models and decision tree model to determine the best result.

Figure 8.28 Ensemble Node

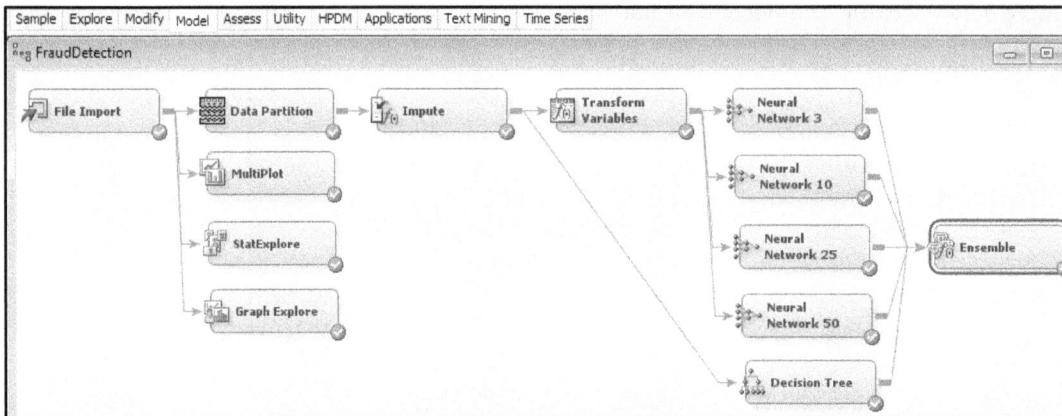

Step 37: Update the Ensemble node properties to use a Voting function when determining the class target. For this application, the Ensemble model will use the votes from each of the models to determine the fraud classification of yes or no.

Figure 8.28 Ensemble Node Properties

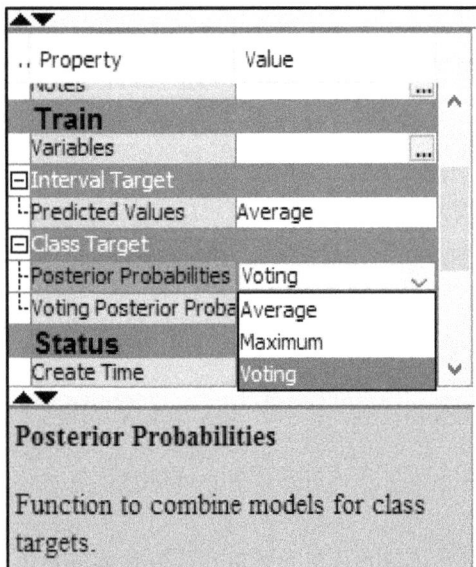

Step 37: Review the Model Results for the Ensemble node, each of the Neural Network nodes, and the Decision Tree Node to compare the results. Specifically, from the fit statistics, review the misclassification rate for the validation set for each model. With the Ensemble node, we can run several neural network models with our hidden unit adjustments and decision tree models to improve accuracy and reduce variance (Maldonado et al., 2014). For this data set, the Neural Network 10 is the parsimonious (or simplest-best)

model from the fit statistics misclassification rate. With considerations for future use, although the Neural Network 10 model and Ensemble model perform similarly when comparing misclassification rates, the Ensemble model might prove more accurate and robust when applying the model to a new data set and could be tested further. The evidence of reduced variability can be seen from the sum of squared errors (SSE), with the Ensemble model reporting the lowest SSE, which is the difference between the predicted and actual values. There are often tradeoffs between the models in terms of performance, interpretability, and error rates when deploying the final model.

Table 8.1: Model Results

Model	Validation Misclassification Rate	Validation Sum of Squared Errors (SSE)
Neural Network 3	1.38%	13.94
Neural Network 10	0.79%	8.31
Neural Network 25	2.17%	16.86
Neural Network 50	1.97%	13.93
Decision Tree	4.34%	39.25
Ensemble	0.79%	7.35

Model Summary

In summary, the neural network will take the form of 1 or more inputs and 1 target variable. The inputs can be interval, binary, or nominal. The target can also be interval, binary, or nominal. To evaluate the neural network, we can use error, lift, and misclassification rate for a binary, nominal or categorical target variable. We can also use the Ensemble model node to combine several neural network models into a final model to improve accuracy.

- Model: Neural Network node
- Neural Network: 1+ input and 1 target variable
- Input: Interval, Binary, or Nominal
- Target: Interval, Binary, or Nominal
- Evaluation: Error, Lift, Misclassification Rate

Experiential Learning Application: Hospital Readmissions

Executive Summary

Hospital readmissions are one of the most common measures of quality, and a common measure is if the patient is readmitted to the hospital within 30 days. Readmissions are an international healthcare issue, causing a strain on healthcare systems and leading to shortages of hospital beds. Patients with ongoing readmissions often experience psychological stress and financial hardships (Low et al., 2015).

Underlying risk factors for readmissions vary and have been found to be based on factors including age, race, provider of care, socio-economic status, surgery, co-morbidities, length of stay, previous admissions, and medications (Robinson and Hudali, 2017). A high readmission rate might indicate that the underlying conditions were not treated properly or the patient was discharged from the hospital early to save costs. In an effort to address readmissions, as part of the ACA, the Hospital Readmissions Reduction Program was also enacted, which as a CMS requirement, reduces payments between 1-3% for hospitals with excessive readmissions. As a result, hospitals must be able to identify those patients with a high risk of readmission both from a quality of care and financial standpoint (CMS, 2016c; Kulkarni et al, 2016). For the 2017 fiscal year, CMS has withheld over $500 million in Medicare reimbursements as part of the program, an increase from the previous year. A total of nearly 2,600 hospitals received a penalty at an average reduction of 0.73%, with 49 hospitals receiving the maximum 3% penalty (Punk, 2016; Murphy, 2016). Even with the many patient factors available for detection of readmission, healthcare providers often have poor predictive accuracy for patients at risk for readmission. Methods have been developed to help providers with identifying at risk patients; two of these tools are the HOSPITAL score and the LACE index (Robinson and Hudali, 2017).

The HOSPITAL score method uses seven components to identify high risk patients (Donze, et al., 2017).

- H - Hemoglobin level before discharge, a protein in red blood cells
- O - Oncology or cancer service discharge
- S - Sodium level before discharge, a mineral in blood
- P - Procedure performed during hospitalization
- I T - Index admission Type, such as emergency or elective
- A - Admissions in previous year
- L - Length of stay

The LACE index is a method recommended by the Institute of Health Improvement, which has moderate to high predictive value in identifying patients at risk for readmission. Although LACE method tools vary, four components are included to calculate the overall risk (low, moderate, high) of readmission (Besler, 2018).

- L - Length of stay (LOS) for initial admission, such as 1 day or 7+ days
- A - Acuity of admission, such as emergency or elective admission
- C - Co-morbidities, such as previous heart disease or diabetes
- E - Emergency department visits within the last six months, such as 0 or 4+

A 2017 study compared HOSPITAL to LACE as a method to predict 30-day hospital readmissions, and found that the HOSPITAL score outperformed LACE at a single hospital facility. Similarly, studies in Denmark and Switzerland found the HOSPITAL score to have better predictive performance. In contrast, larger nationwide Medicare studies in the U.S. have found no significant differences between HOSPITAL and LACE (Robinson and Hudali, 2017).

In one research study, authors used techniques to determine whether a patient would be readmitted to the hospital within 30 days of being discharged. The authors used a combination of model approaches (including neural networks, logistic regression, and decision tree) to find the best discriminating power and accuracy when using validation cases after training. The authors reviewed several factors for determining readmission, including the medical conditions, length of the hospital visit, care rendered during the stay, size of the medical facility, type of medical insurance, and discharge environment (Kulkarni et al, 2016; CMS, 2016c).

For this experiential learning application, you have been provided a data set of 112,749 records. Help your management team with the following objective.

Objective: develop a model for detecting hospital readmissions

Data Set File: 8_EL2_Readmissions.xlsx

Variables:

- RecordID, unique identifier
- DischargeDisposition, Discharge Setting of Hospital, Skilled Nursing Facility (SNF), Rehab, Other, Home
- Cohort, Cardiorespiratory, Cardiovascular, Medicine, Neurology, Surgery
- FacilitySize, Hospital size by number of beds
- InsuranceType, Managed Care, Medicare, Medicaid, Self-Pay, Workers Compensation, or Other
- Age, in years
- ICDCounts, count of diagnosis codes
- LengthOfStay, in days between admission and discharge
- Readmission, 1 indicator if patient was readmitted

Follow the SEMMA process for your experiential learning application and provide recommendations. A template has been provided below that can be reused across future projects.

Figure 8.30 SEMMA Process

Title	Hospital Readmissions
Introduction	Provide a summary of the business problem or opportunity and the key objective(s) or goal(s). Create a new SAS Enterprise Miner project. Create a new Diagram.
Sample	Data (sources for exploration and model insights) Identify the variables data types, the input and target variable during exploration. Add a FILE IMPORT Provide a results overview following file import: Input / Target Variables Generate a DATA PARTITION

Title	Hospital Readmissions
Exploration	Provide a results overview following data exploration Add a STAT EXPLORE Add a GRAPH EXPLORE Add a MULTIPLOT Summary statistics (average, standard deviation, min, max, and so on.) Descriptive Statistics Missing Data Outliers
Modify	Provide a results overview following modification Add an IMPUTE Add a TRANSFORM VARIABLES
Model	Discovery (prototype and test analytical models) Apply a neural network model and provide a results overview following modeling. Add a NEURAL NETWORK with 3, 10, 25, and 50 hidden units Model description Add an ENSEMBLE MODEL Model description Analytics steps Model results (Lift, Error, Misclassification Rate) Selection Model
Assess and Reflection	Provide overall recommendations to business Model advantages / disadvantages Performance evaluation Model recommendation Summary analytics recommendations Summary informatics recommendations Summary business recommendations Summary clinical recommendations Deployment (operationalization plan: timeline, resources, scope, phases, project plan) Value (return on investment, healthcare outcomes)

Learning Journal Reflection

Review, reflect, and retrieve the following key chapter topics only from memory and add them to your learning journal. For each topic, list a one sentence description/definition. Connect these ideas to something you might already know from your experience, other coursework, or a current event. This follows our three-phase learning approach of 1) Capture, 2) Communicate, and 3) Connect. After completing, verify your results against your learning journal and update as needed..

Figure 8.Key Ideas – Capture	Key Terms – Communicate	Key Areas - Connect
Government Anamatics		
Government Legislation		
HIPAA		
ARRA		
ACA		
AHCA		
BCRA		
Government Agencies		
Government Data Sharing		
Fraud and Abuse Interventions		
Neural Network		
Fraud Detection Application		
Readmissions Risk Application		

Chapter 9: Health Administration and Assessment

Chapter Summary

The purpose of this chapter is to develop administration and data assessment skills using SAS Enterprise Miner and the Assess capabilities within the SEMMA process. This chapter builds on the models from previous chapters. This chapter also includes experiential learning application exercises about health risk score and hip fracture risk. The focus of this chapter is shown in Figure 9.1. In healthcare, there are often provider assessments that are made based on quality improvements, value-based care, and health outcomes. Patient assessments of health also occur, for example, to identify potential risk factors. Similarly, for data mining, we can assess our models for improvement, comparison, and overall value. In this chapter, we will explore both assessment from a healthcare standpoint and assessment from a data mining standpoint. The two areas will be blended through an experiential learning application opportunity on health risk score. This chapter will combine and connect your knowledge throughout all the previous chapters to reinforce your learning and further build your learning pathways.

Figure 9.1: Chapter Focus - Assess

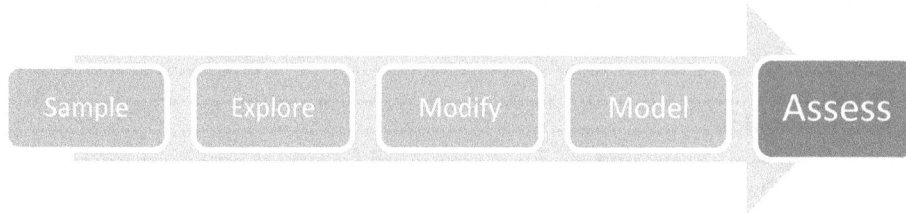

Chapter Learning Goals

- Define and describe the Assess process steps
- Understand the administration, privacy, and security principles of health anamatics
- Develop model assessment skills
- Apply SAS Enterprise Miner assess functions

Health Anamatics Administration

In comparing hospital administrative costs within the countries of U.S., Canada, U.K., Scotland, Wales, France, Germany, and the Netherlands, the U.S. had the highest cost, and Scotland and Canada had the lowest cost. In the U.S., administrative costs contributed to 25% of the total healthcare costs. If the U.S. were able to reduce administrative costs on par with Scotland or Canada, over $150 billion could have been saved each year. The share of administrative costs was also found to be higher in for-profit hospitals at 27% as compared with public hospitals at 23%. Consequentially, no link has been found between administrative costs and quality of care. Some of these administrative costs are attributed to the complexity of the U.S. health system, where billing multiple payers and negotiating rates with varying requirements and procedures requires additional administrative staffing (Himmelstein, 2014; Himmelstein et al., 2014).

Although the preceding numbers include only hospitals, billing and insurance-related administrative costs for the U.S. healthcare system were estimated at nearly $500 billion. It is estimated that a more simplified financial system in the U.S. would result in cost savings over $350 billion per year or 15% of healthcare spending. In a healthcare system, administration is required to improve efficiency and effectiveness of care. Nevertheless, the U.S. administrative costs as a percentage of healthcare have been increasing in the last 50 years, and more than doubled between 1980 and 2010. Private insurer's administrative costs have also more than doubled between 2001 and 2010. HIPAA, ACA, and ARRA were seen as achieving methods to further reduce costs. Due to increases in coverage, greater use of deductibles, and incentives, which all require additional administrative staff to process, costs might be expected to slightly increase. In Massachusetts, where universal coverage was implemented similar to the ACA, administrative staff increased (Jiwani et al., 2014).

Several legislative measures have been made over the last several decades to reduce the administrative cost burden. The 1996 HIPAA legislation included provisions for administrative simplification, including EDI (covered in Chapter 7) and code sets. Additional components to simplify administration included a standard Employer Identification Number (EIN) and National Provider Identifier (NPI). The EIN is distributed by the Internal Revenue Service (IRS) and is a unique 9-digit number. The NPI is distributed by CMS and is a unique 10-digit number for healthcare providers, including hospitals and individual physicians (HHS, 2017d). Other recommended standards, such as a national patient identifier, are still pending.

In a further effort to control the administrative costs, the 2009 ACA legislation created two mechanisms: Rate Review and 80/20 Rule. The rate review protects patients from unreasonable rate increases, and insurers must publicly explain any rate increases exceeding 10%. The 80/20 rule requires insurers to spend no more than 20% on administrative costs that include overhead and marketing, and 80% or more must be spent on direct medical care. The rule is intended to prevent insurers from raising rates to increase profits. The rule might also be called a medical loss ratio (MLR). If 80 cents of every dollar are used for direct medical care, the MLR is 80%. Insurers selling to large groups are required to have an 85% MLR. If these ratios are not met, rebates are provided to patients. Between 2011, when the 80/20 rule started, and 2015, approximately $2.8 billion in rebates have been distributed, with an average refund between $100-$150 per family (CMS, 2016a; Healthcare.gov, 2018).

Although spending on healthcare administration is higher in the U.S., across other categories of healthcare spending, it is also higher than in other counties. The other categories include provider care, pharmaceuticals, medical goods, and public health, with the U.S. spending 2.5 times more (on the average) on health expenditure per person. Research has shown that both the price of healthcare is higher and more services are provided than in other countries. As a seemingly conflicting result, given the higher costs and spending, life expectancy is lower in the U.S. than average (OECD, 2011).

As a result of aging populations, rising costs, quality exceptions, legislation, and ongoing changes, there are considerable differences found in the efficiency and effectiveness across hospitals and across countries. In identifying these differences, management (or administration) of the hospitals can play a role. One study investigated a set of 2,000 hospitals in Brazil, Canada, France, Germany, India, Italy, Sweden, the U.K. and the U.S. to determine whether management practices were associated with improvements in healthcare. The improvements where measured through clinical outcomes such as survival rates of heart attacks, as well as financial outcomes such as profit. The results found that within a single country, variation can be explained through hospitals with managers that have more clinical training, are larger in size, operate in competitive markets, are non-government owned, and are politically independent have the highest scores. Clinical training can influence management as clinically trained managers can be able to more easily communicate and understand both the business and medical sides of the hospital to improve decision-making. They can also experience greater levels of trust with medical staff. Some countries, such as the U.K., have gone a separate direction in having many CEOs from non-clinical business backgrounds. Also, growth in joint degrees, such as MD/MBA, can have an influence. Government-owned hospitals have lower scores as a result of profit motivations, political interference, and required patient mix and treatment offered. Competition increases management scores as a result of increasing efficiency to remain competitive, along with learning best practices from competitors. Size of the hospital can influence management, since more formalized practices are required for successful management. Another result of good management practices, is that hospitals can grow and gain more patients. Many of the management practices are fixed in cost such as human resource activities, which might give larger hospitals a cost advantage (Bloom et al., 2014).

Variation in outcomes between countries (such as the U.S., U.K., India, and Italy) can also be explained by accountability and governance. Countries have different requirements to report quality metrics such as survival rates and patient satisfaction. In Germany and Sweden, comparable data is widely available, although in other countries (such as Canada and France), only a limited data set exists, and in some countries (such as Italy), national quality data might be unavailable. For governance, countries (such as France and Italy) have appointment of hospitals made by politicians. In the U.S. and Germany, there is limited influence on CEO appointments as a result of privately owned hospitals. Interestingly, healthcare spending by country did not explain much of the variance, although Sweden spends 15% less than France as a percentage of GDP, their management scores are 18% higher (Bloom et al, 2014).

In the following sections. we'll discuss the major components of healthcare administration, including code sets, security, and privacy, many of which were developed or updated as part of HIPAA and other legislation.

Code Sets

One of the major components of healthcare administration are code sets. Code sets were briefly reviewed in earlier chapters, and, following our learning approach, will be reviewed in more detail to build connections to the concepts. A code set is any set of codes that can be used to assign values to various entities (such as terms, medical concepts, diagnostic codes, tests, treatments, supplies, or procedures). There can be clinical and non-clinical or non-medical code sets (CMS, 2017g). Using code sets is sometimes called medical coding or medical billing. The goal of coding is to take information (such as a diagnosis, procedure, and service) and translate that information into a standard code (such as an ICD-10, CPT, or HCPCS code). Typically, physician notes and medical records are used to translate the codes. The correct codes are necessary for accurate documentation as well as for billing purposes to ensure that the claim is paid (AAPC, 2017). Historical analysis shows that errors within coding are common. Known causes of error include focusing on quantity versus quality, following coding pathways, memorization of codes, limited documentation, diagnosis selection, secondary diagnosis, and DRG assignment. Capabilities to prevent coding errors include identifying edits, analyzing patterns in coding errors, and business intelligence systems. In one study of neurosurgical clinical coding, at least one coding error was found in 18.4% of the patient episodes reviewed, and in another study of codes reviewed for accuracy, the error rate was 19.7% (Orcutt, 2009; Royal College of Physicians, 2009; Haliasos, et al., 2010).

In Chapter 7, we discussed EDI, which for administration can be used as a strategic advantage along with other health anamatics capabilities. The standard code sets are transmitted within standard EDI transactions to improve efficiency and effectiveness of care through accuracy. With increased competition, organizations seek competitive advantage via price, product, and quality. Strategies against competition include product differentiation, market segmentation, and lowest cost. Technologies such as EDI and administrative methods such as use of standardized code sets can be used to address these strategies. In particular, EDI can be used to reduce cost and improve efficiency. Staged models have been introduced for EDI, where a company goes from beginning EDI (discovery), to regularly using EDI (operation), and to using EDI as a strategic advantage (innovation). Timely imaginative technology and investment in the innovation stage allows for competitive advantage to be realized in an otherwise equal environment. Examples of innovation in other mature EDI environments outside of healthcare include made-to-order automobiles and blue jeans, in which a customer enters as input information (such as options or measurements), and through EDI, the applications create and deliver the customized product. An example within healthcare includes the concept of personalized medicine (Woodside, 2013e).

Within HIPAA, several code sets are defined including International Classification of Diseases 10th edition (ICD-10), Healthcare Common Procedure Coding System (HCPCS), Current Procedure Terminology (CPT), Code on Dental Procedures and Nomenclature (CDT), National Drug Codes (NDC), and Laboratory and Clinical Codes (LOINC) (CMS, 2012b).

International Classification of Diseases (ICD)

ICD is used to classify diseases and health conditions and is used for reporting trends and statistics globally. ICD was started in 1893 by the International Statistical Institute, and in 1948, the World Health Organization released the 6th edition, ICD-6, which included morbidity. ICD has been translated into 43 languages and is used by over 100 countries (WHO, 2017). ICD codes are now in the 10th edition and is commonly called ICD-10 as the set of codes for medical diagnosis. Any services that were provided on or after October 1, 2015 will be coded with ICD-10 and applies to all HIPAA-covered entities including those billed to Medicare and Medicaid. Although ICD-9 can still be used if not required by a specific payer such as Medicare and Medicaid, it is in the best interest of the provider to use ICD-10 since ICD-9 is no longer being maintained (CMS, 2017h).

ICD-9, or the ICD 9th edition, is now over 30 years old, and many systems, user training, and process have been designed to accommodate ICD-9 throughout the last several decades. Although ICD-10 was expected to reduce errors as compared with ICD-9, the initial transition period was more challenging. The deadlines for use of ICD-10 had been postponed several times, and organizations had already spent considerable time preparing for the transition (CMS, 2012a). Although the healthcare industry has endured similar transitional challenges in the past (including Y2K, APC reimbursement, UB04 migration, and DRG reimbursement), the migration from ICD-9 to ICD-10 was estimated to have an even greater impact based on the human, process, and technical aspects required to successfully transition (Armstrong et al., 2010).

Even though the industry has successfully navigated the initial ICD-10 transition, difficulties with mapping ICD-9 to ICD-10 have created issues with contracts, revenue, reimbursement, edits, policies, coding, and so on. The changes also cause system updates, testing, training, denials, payment delays, reduced reimbursement, productivity, penalties, financial impacts, and coding errors. There are also significant benefits that will be available as a result of ICD-10, including improved quality measurement, disease reporting, data mining analysis, organizational performance monitoring, and reimbursement. The ICD-9 codes did not allow for new disease detection, biomedical informatics, genetics, and international integration. ICD-10 fully replaces ICD-9 to improve diagnostic coding, with a level of detail that has increased from approximately 18,000 codes to 140,000 codes (Armstrong et al, 2010). Although ICD-10 was first endorsed in 1990, it took many years to full implement. Despite the earlier challenges transitioning to ICD-10, a new version, ICD-11 is planned for release in 2018 and was first proposed in 2011 (WHO, 2017).

HCPCS Level I (CPT Procedure Codes) and Level II (HCPCS Service Codes)

The Healthcare Common Procedure Coding System (HCPCS) contains two sets known as HCPCS Level I and Level II. Level I includes the CPT codes and is maintained by the American Medical Association (AMA). CPT is a coding system used primarily for medical services and procedures completed by physicians and health professionals. Level II of HCPCS is a coding system maintained by CMS and used for those services outside of CPT, namely ambulance services, durable medical equipment (DME), prosthetics, orthotics, and supplies (DMEPOS). Since the additional items are also covered by most insurers, both types are needed, and level II was introduced in the 1980s. To more easily distinguish between level I and II, level II codes follow the format of a single alphanumeric code followed by 4 numbers versus CPT codes that contain only numbers. HCPCS codes can also occasionally contain a two-digit modifier code to provide additional information or to indicate a special circumstance. For example, there is a UE and NU modifier that is used with certain HCPCS codes. The UE modifier indicates whether an item is used equipment, whereas a modifier of NU modifier would indicate new equipment (CMS, 2011; CMS, 2013).

Code on Dental Procedures and Nomenclature (CDT Code)

The Code on Dental Procedures and Nomenclature (CDT Code) is primarily used for dental treatments. CDT is intended to provide a uniform, consistent, and specific method to accurately document dental treatments for both billing purposes and documentation such as within an EHR. CDT is included within HIPAA, and claims complying with HIPAA use CDT codes (ADA, 2017).

National Drug Code (NDC)

The National Drug Code (NDC) is primarily used for drugs and has been maintained by the FDA since the passing of The Drug Listing Act of 1972. All drugs that are manufactured prepared, propagated, compounded, or processed for commercial distribution are included and given a unique 10 digit, 3 segment number known as the NDC. The 3 segments identify the labeler, product, and trade package size. The labeler can include the company that manufactures, repackages, relabels, or distributes the drug. The product code contains the drug strength, dosage and formulation, with different strengths or formulations, each having a different product code. The package code segment contains the package size and type. As a result, the 10-digit NDC can be in one of the following combinations: 4-4-2, 5-3-2, or 5-4-1 (FDA, 2017c).

Logical Observation Identifiers Names and Codes (LOINC)

Logical Observation Identifiers Names and Codes (LOINC) is a code set used primarily for laboratory tests and results started in 1994. LOINC codes can be exchanged electronically and is currently in version 2.61, with nearly 2,000 codes. Like other code sets, internal codes must be coded in the LOINC standard to be understood across systems. LOINC can be used as part of the HIPAA EDI standards through the claim attachment standards. The healthcare attachment EDI transaction was proposed separately from the initial set of HIPAA EDI transactions (Regenstrief Institute, 2017; Mayo Clinic, 2017a).

Security

The healthcare environment is a prime target for data and identity theft due to the available patient, financial, and clinical content. Mobile devices can provide unwanted access to a variety of patient data including contacts, texts, calls, email, calendars, internal systems, credit card information, and health data. Increases in breaches can be tied to regulation requirements, automation growth, social media development, and human errors. The economic burden created by these data breaches in healthcare is estimated at $7 billion annually, with $1 million per organization annually in cases of a breach. Security is a major priority for healthcare organizations, where patients entrust their detailed information. When security monitoring systems are in place, because the historical threat information is not always up-to-date, it can generate false positives for threats, and the security detection results vary between vendors. With the increased usage of healthcare information technology (such as e-prescribing, electronic health records (EHRs), personal health records (PHRs), social media networks, health information exchanges (HIEs), and mobile devices), the potential risk and data information loss has increased (Woodside and Florea, 2015). For hackers, electronic health information is ten times more valuable than financial information, such as a credit card number, as determined by monitoring underground data exchanges where hackers sell information to others. The electronic health information is used to create fraudulent bills for patients or seek and obtain health services and equipment. The Federal Bureau of Investigation (FBI) has warned healthcare providers after a large U.S. hospital, Community Health Systems, had hackers steal information about 4.5 million patients. Hackers are targeting the healthcare industry due to the potential value, which are vulnerable by information systems that are typically aging and are without all the available security updates (Humer and Finkle, 2014).

Despite the priority and potential impacts, healthcare security breaches have become a common catastrophe. In a recent study, 94% of healthcare organizations suffered a data breach in the last two years, and nearly half experienced more than five data breaches in the last two years. In 2010, a flash drive with protected health information (PHI) of 280,000 members was stolen from a health plan. In 2006, a laptop and disc with PHI of 26.5 million veterans was stolen from an employee's home. Since the enactment of the Health Insurance Portability and Accountability Act (HIPAA), over 11,000 HIPAA violations have been reviewed and 7 million patients have been impacted (Woodside and Florea, 2015). In 2015, Anthem Health reported a data breach affecting up to 80 million patients, which increased the urgency and security scrutiny by organizations. In 1996, HIPAA passed a comprehensive set of security rules to improve various controls and methods. Even with these guidelines, there are still existing vulnerabilities along with new threats, and security spending must be approached deliberately to ensure that risks are addressed (Commins, 2015). In May of 2017, hospital and physician office systems in the U.K. experienced a wide-spread failure due to a cyberattack, which also impacted other countries around the world including Spain. The attackers requested payment for unlocking healthcare files. The attack led to confusion within the health system, causing operations to be canceled, emergency services to be reduced, and backup methods (such as hand-written notes) to be used. These types of attacks are becoming increasingly common due to hackers' ability to infiltrate computer networks (Witte and Adam, 2017).

Healthcare organizations must implement safeguards to address new security concerns as a result of increased electronification, as well as to comply with enacted laws (CMS, 2007). The primary components of a security defense include physical, technical, and organizational safeguards. Physical safeguards include protecting facility access against unauthorized entry, as well as security workstations, transportation, and storage of media and information. Physical safeguards also apply well behind the walls of an organization. Stanford's Lucille Packard Children's Hospital announced a breach of 57,000 patients' information due to a stolen laptop from a physician's vehicle. Gibson Hospital in Indiana also reported a breach of 29,000 patients' information due to a stolen laptop from an employee's home. Technical safeguards include unique user identification, automatic logoff, encryption, having a responsible person to authorize and verify passwords, strong passwords, locking accounts after invalid logins, and deactivating employee accounts after termination. From a mobile perspective, endpoint access should also be verified and permitted or prevented from accessing the network, including monitoring and notification of an unauthorized device. Wireless threats should be detected and prevented through security policies and location tracking. Records should be kept by administrators to verify personally owned and operated devices to ensure security compliance. Organizational safeguards include Business Associate Agreements (BAAs), customer requirements, and policies and procedures. BAAs should be updated to eliminate breaches and enforce liability that might not be covered under the law. A business associate is anyone who works on behalf of a healthcare entity and uses or discloses PHI (Woodside and Florea, 2015). The National Learning Consortium (NLC) has developed an Information Security Policy Template from the experience and knowledge of EHR implementations and through research and communities of practice. This template is available to download through the HealthIT.gov website. Components of the plan include email usage, passwords, encryption, telecommuting, training, background checks, and incident response (HealthIT.gov, 2011).

Privacy

Even though hospital security breaches and EMR data leaks are traced to hackers, there have been a number of high-profile cases that involved hospital employees who reportedly accessed celebrity medical records without authorization. In 2007, 27 workers at Palisades Medical Center in New Jersey were suspended for one month after reviewing George Clooney's medical record. In 2011, UCLA Health System paid $865,000 to address possible privacy violations of employees accessing information about two celebrity patients (Ornstein, 2015; Jayanthi, 2015). Privacy is an individual's right to control the acquisition, use, or disclosure of their identifiable health data. Privacy is protected by HIPAA and applies to all covered entities such as health plans, health care clearinghouses, and any health care provider that transmits health information (OCR, 2017).

Protected health information (PHI) includes any personally identifiable information about an individual such as a patient. In order to protect privacy, one key component of the HIPAA privacy rule is termed the minimum necessary standard. The standard holds that private and protected information should be used only when necessary for carrying out one's job duties. A patient's full medical record should not be sent if only the patient's insurance ID is required. Healthcare organizations are responsible for identifying those who are required to have access to information, and have set up policies, procedures, and protocols for accessing protected health information. In non-routine cases, best judgment is used to determine the disclosure of minimum information, as required (HHS, 2013). Another way to protect the privacy of PHI is through de-identification, which allows for the data to be modified such that no records can be individually identified. The de-identification practice is common for healthcare research, where detailed data is required although personal identifiers are not required. Some challenges to de-identification are that the de-identified data (when combined with other data sources, or with access to certain variables in the data set) would allow someone to reverse identify individual records. If an address is listed even though the patient name is not listed, this would allow someone to still easily identify the individual. Key variables that should be removed to ensure de-identification include street address, city, county, ZIP code, census tract code, specific dates such as date of birth, admission, discharge, death, ages, numbers such as telephone, fax, social security, medical record, insurance, account, vehicle, email address, internet protocol address, biometrics, photos, and any other unique characteristics (UW Medicine, 2013). A random number that is assigned to each de-identified record and is maintained separately with a crosswalk (or matching record equivalent) by an entity might be used to re-identify information where needed.

Despite the belief in the importance of privacy, implementation by administration is often challenging. In the U.S., 60% of respondents say they wouldn't share their email contacts. However, smaller studies have found conflicting results. One study of 3,000 Massachusetts Institute of Technology (MIT) students found that 98% of the MIT students were willing to provide their friends' personal email contact information when offered an incentive of free pizza. A few 6% gave fake emails to both earn the free pizza and protect their friends' privacy. The result also follows with security usage by many individuals. Although most agree that security is important, many still knowingly use weak passwords that are easy to remember or other methods that compromise security (Paul, 2016). In 2009, the American Recovery and Reinvestment Act (ARRA) was enacted, which also included components from the Health Information Technology for Economic and Clinical Health (HITECH) Act. The HIPAA enforcement for privacy and security safeguards improved along with increased penalties and accountability to increase security and privacy. To ensure privacy, administrative best practices include designating a privacy officer that is responsible for implementation, designates a contact for any privacy complaints or concerns, and establishes ongoing training for all employees including contractors, volunteers, and associates. Many times, privacy is considered a legal team area, and security is viewed as a technical team area. Typically, these are handled by different areas and, in some cases, ignored altogether as an issue that is handled by another area of the organization. There are overlaps between security and privacy, and security must be implemented to ensure privacy. The HIPAA privacy rule has a requirement of minimum necessary access to PHI, whereas the security rule requires administrative, physical, and technical safeguards. Security can be viewed as the process or action, and privacy is the successful result or unsuccessful consequence. Privacy is an individual's right to control access to PHI, whereas security is an organization's responsibility to protect and control access to PHI (Beaver and Herold, 2004).

Not all incidents are caused by hackers or malicious employees. In some cases, they might simply be accidents. In one reported case on New Year's Eve in the state of Washington, a driver crashed into a utility pole. Even though the driver survived, the crash caused an outage in the Epic EHR system at Jefferson Healthcare from New Year's Eve through New Year's Day, as well as outages in the 911 services in the county, creating a potential security issue (Dietsche, 2017). Therefore, administrators must take precautions to guard against all potential disruptions to service and develop plans to quickly recover from any potential issues and keep the organization running in business as usual mode.

Experiential Learning Activity: HIPAA Administration

HIPAA Administration

Description: In the healthcare environment of today, individuals are increasingly connecting using mobile devices such as tablets and smartphones. Clinicians and patients demand current information at their fingertips during all phases of healthcare delivery in order to save time, reduce errors, and improve outcomes. As a result of the growth of mobile technology in healthcare, this presents a challenge to administrators of the environment to ensure high levels of security, privacy, and control. With the growth of mobile technology, one study showed that over 60% of breaches occurred due to mobile devices that were lost or stolen. Risks of breaches are expected to continue to grow along with mobile technology usage. As a sub-component of health informatics, Intelligence and Security Informatics (ISI) is a cross-disciplinary field defined as the development of advanced information technologies, systems, algorithms, and databases for international, national and homeland security related applications, through an integrated technological, organizational, and policy-based approach. Mobile devices can provide unwanted access to a variety of data including contacts, texts, calls, email, calendars, internal systems, credit card information, and clinical and personal data (Woodside and Florea, 2015; Woodside and Johnson, 2015).

Researchers have said that no federal agency has the ability to appropriately regulate the quickly changing environment of mobile health and telehealth systems. A report made in Ponemon's Third Annual Benchmark Study on Patient Privacy and Data Security revealed that 81% of healthcare organizations permit employees and medical staff to use their own devices to connect to their organization's networks or enterprise systems, and 54% of these people said that they were not confident that the devices that they were using were secure. At the same time, 66% of the nurses declared that they used their smartphones for clinical communication, and 95% of them said that the hospital IT departments did not support their devices due to potential security risks. In addition, mobile devices are able to access cloud-based platforms for storing information and are used for daily operational items including notes, documents, messages, and other information that often contains PHI in a healthcare environment, where clinical staff might take notes on a patient or post clinical documents for clinical review and monitoring. Despite initial security risks and concerns over the use of personal devices, the growth in bring your own device (BYOD) and bring your own technology (BYOT) has been increasing and is driven primarily by cost reductions, productivity improvements, and employee satisfaction through allowing the use of personal smartphone devices. There are also many benefits to healthcare mobile usage. At ASAN Medical Center in Seoul, the staff is using smartphones and laptops to improve productivity, increasing diagnostics and problem-solving up to three times faster by keying and accessing information directly on their mobile devices. Mobile and telehealth patient data sent through devices and networks must be secured and private to ensure trust by patients and providers, and methods are available to reduce the risks. Four primary security and privacy risks for mobile and telehealth are 1) telehealth and mobile services fall largely outside of the HIPAA Act of 1996 and HITECH Act of 2009 due to services outside of a covered healthcare setting, 2) mobile devices and apps may share patient information with third parties such as advertisers, 3) systems might contain security flaws and be compromised by hackers, 4) Privacy

HIPAA Administration

laws such as the Computer Fraud and Abuse Act and Electronic Communications Privacy Act of 1986 protect privacy although they might not apply to patient information (Pittman, 2014; Hall and McGraw, 2014; Woodside and Florea, 2015).

As a Compliance Officer for Blue Cross Blue Shield, you are responsible for developing an updated plan for mHealth (mobile health).

Compare and contrast the HIPAA Privacy Rule versus the HIPAA Security Rule.

Why are health records more valuable than financial records?

List the 3 types of security rule safeguards to include within each area for mHealth:

1.

2.

3.

List the 3 types of privacy rule safeguards for mHealth:

1.

2.

3.

List the 4 types of primary security and privacy risks for mHealth:

1.

2.

3.

HIPAA Administration

4.

Find and describe an outside example of a mobile health application that is available today.

Select a security and privacy issue that can impact your selected mHealth application, and provide a recommendation or safeguard to help prevent the issue.

Who owns the mHealth data? If it's a data security or privacy breach, who is responsible?

Now that you are familiar with code sets, security, and privacy requirements for healthcare administration, we'll continue with our SEMMA process using available administrative data to review the Assess process step.

SEMMA: Assess

Assess Process Step Overview

During the Assess process step, the data mining models applied to the training and validation data sets are compared. Whereas previously individual model such as regression, decision tree, and neural networks were run and interpreted separately, during the Assess step the models are compared and the best model is selected. In addition, new data sets using the selected model can be used to score new data using the previously run model selection. In this chapter we will build on your previous knowledge of the SEMMA process steps.

Model Tab Enterprise Miner Node Descriptions

Model Comparison Node

The Model Comparison node is used to compare different models and select a final model. The final model might be selected based on various criteria such as error rates or statistical tests. The Model Comparison node is associated with the **Assess** tab in SAS Enterprise Miner.

Figure 9.2: Model Comparison Node

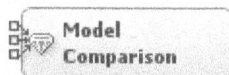

Score Node

The Score node is used for evaluating a new set of data based on a previously selected model. Thus far, we have reviewed historical data where the target is known. For new data sets where the target is not yet known, the data set can be scored or the target determined based on our historical data sets. The Score node is associated with the **Assess** tab in SAS Enterprise Miner.

Figure 9.3: Score Node

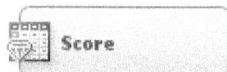

Now that we have covered the Assess process step and nodes, let's continue with an experiential learning application to connect your knowledge with a health application on health risk.

Experiential Learning Application: Health Risk Score

The top 1% of patients are often identified by the healthcare industry as a small number using up a large percentage of care and costs. In some studies, 1% of the patients accounted for 30% of the costs. Typically, the first step is to gather data sources, which include claims, demographics, eligibility, health assessments, biometric data, care management participation and intervention. The next step involves using the knowledge in the data warehouse to develop predictive models, gaps in case, utilization measures, and new engagement and interventions. The last step is to predict future healthcare costs, engage members, and intervene and impact future care and usage (McNeill, 2014).

Health Risk Assessment

Current healthcare risk assessment uses methods to assess relative risk of individuals within the population, with the relative risk predicting costs. The assessment can be carried out using various forms of data and typically includes claims, pharmacy, and self-reported survey information. The information has been used by the federal government to adjust payments to health plans, by employers in determining employee contributions to health coverage, by researchers in measuring outcomes of treatment methods, by policy makers for tracking access to care and quality of care, and by health plans for integrated care management, case management, disease management, quality improvement, payments to providers, and underwriting activities. Growth in consumer-driven health plans requires the need for improved risk assessment accuracy. Because patients and employees of sponsored plans have more options in selecting their benefit plans, this results in increased variability among the healthcare plan populations. Risk adjustment allows the health plan to determine healthcare outcomes appropriately or make equal payments to promote quality improvements rather than population selection, and ensure comparative price and consumer choice (Cumming et al., 2002; Rector et al., 2004; Woodside, 2010b). Blue Cross Blue Shield of Michigan (BCBSM) has an online health assessment for patients to determine risk and develop a customized plan to improve any identified areas of risk. The online assessment takes approximately ten minutes and contains digital health assistants, health trackers, community message boards, and videos and articles to help improve patient health (BCBSM, 2017).

Health Risk Adjustment

Within the ACA, to address varying health risk, there are three programs (reinsurance, risk corridor, and risk adjustment) that are intended to keep costs stable. The reinsurance program aggregates payments from all that offer health insurance coverage and reinsure those with high-cost claims. Due to the ACA provision that no longer allow exclusions based on pre-existing conditions, there will be a need to support those providing coverage to the high-cost enrollees in the early years. The risk corridor is similar in that insurers are likely to experience losses in the early years that they support the new plans and would be reimbursed. The last component includes risk adjustment and is a long-term program. Although the ACA prohibits coverage decision based on risk, insurers might be able to avoid high-risk enrollees through plan design, networks, and marketing techniques. The risk adjustment distributes revenues for insurers with low-risk enrollees to those insurers with high-risk enrollees. Various programs to accomplish risk adjustment techniques have been previously used in Medicare, Medicaid, and European programs. The current model used is the Medicare model although is modified for the entire population versus those who are only over 65. Enrollees receive a risk score based on factors including age, gender, ICD-10 codes, and condition categories. Enrollee factors are then used to estimate a risk score and plan expenditures. There is some criticism of this model, including that the model provides a disadvantage for small and new plans that have low premiums. Also, the costs do not reflect accurately for new enrollees when compared to the general set of enrollees. In addition, the model is recommended to include medication usage to increase accuracy. As a result, many ongoing changes have been made to the model to increase accuracy and address recommendations (Jost, 2016).

Here is an example of risk adjustment factors: An estimated 25% of patients in the U.S. have more than one chronic condition, and by the age of 65, the percentage of patients in the U.S. that have more than one chronic condition increases to 75%. Underlying causes include the aging population, increased life expectancy, and factors such as tobacco use and physical activity. Chronic conditions are also linked to healthcare and prescription costs. In the U.S., over 70% of the spending is for patients with multiple chronic conditions, and, among the Medicare patients, this number increases to over 90%. The most common chronic conditions among Medicare patients include high blood pressure, high cholesterol, heart disease, arthritis, and diabetes. For those under 65, the chronic conditions were similar with the exception of depression and asthma with higher than average expectancy. Chronic conditions also vary by gender. Women are more likely to have arthritis whereas men are more likely to have heart disease (CMS, 2012b; CDC, 2016b). Another example of a risk adjustment factor is geographic location, Northeastern states have a higher per capita health spending. D.C., Delaware, Massachusetts, and Vermont have per capita spending in excess of $10,000. Midwestern states (such as Indiana, Illinois, and Iowa) and Northwestern states (such as Montana, Oregon, and Washington) have a per capita spending of approximately $8,000. Southern states (such as Alabama, Louisiana, and Texas) have per capita spending of approximately $7,300. Western states (such as Arizona, Colorado, and Nevada) have the lowest per capita spending of approximately $6,500. One outlier based on location is Alaska with the second highest per capita spending after D.C. (Kaiser Family Foundation, 2017).

Health Risk Reduction

Health anamatics administration in the context of integrated care management includes the combination of analytics and informatics and is a well-defined and coordinated set of targeted services delivered to individuals by cooperating care providers across organizations, supported by applications and technologies used to gather, capture, access, consolidate, and analyze information to improve decision-making. Integrated care management can improve quality of care, quality of life, customer satisfaction, efficiencies, outcomes, and costs. The care impacts patients with issues that are complex, long-term, and run across many services and providers. Integrated care might appear in forms such as shared care, continuing care, case management, disease management, transmural care, comprehensive care, intermediate care. The primary goal of integrated care management programs is to use a delivery network across the continuum to improve clinical outcomes. The case (or disease) manager determines patient requirements along with risk and develops a care plan for the patient. Care systems in Western countries were historically focused on single disease state approaches, whereas in Europe integrated care enjoys more attention through conferences, journals, policies, and regulations. Integrated care is necessary when patients require multiple products by varying individuals and organizations to meet the patient needs. Previous research indicates that coordinated care can improve access to services, quality of care, and cost of healthcare. Nevertheless, Medicare and others do not fully support these integrative practices. Increasingly, knowledgeable patients are becoming more demanding in terms of treatment and provider options. In addition to population and financial changes, healthcare organizations are shifting to preventative care and patient education (Woodside, 2013b). Disease management programs (such as those offered by Amerigroup, Cigna, Evolent Health, Trizetto, UnitedHealth and others) use medical and pharmacy claims along with other information about a healthcare member (such as demographics such as age, gender, location, and so on) and apply their algorithms to determine conditions of a member to predict the likelihood that they will get the disease (NCQA, 2018). The disease management companies typically have nurses calling the members that have been rated as certain risks and either talk to them on a scheduled basis for managing their disease or as claims are brought in that indicate that the member has visited a physician, change in diagnosis, and so on. Nurses assist the members by providing information about the disease and educating the member on approaches for self-management. Health coaches will discuss with members how to change their lifestyle so that the predicted outcome (future disease) is not realized.

There are also alternatives to reducing risk besides patient interventions. In the U.S., a consortium of 38 companies including American Express, Macy's, and Johnson & Johnson was formed and aimed at using their combined negotiating power to lower healthcare costs. The nonprofit group has labeled themselves as the Health Transformation Alliance, and includes partnerships with CVS Health Corp for prescriptions, UnitedHealth Group Inc. for provider networks, and International Business Machines (IBM) to leverage the Watson software. The alliance estimated $600 million in savings over three years as compared with their current plans (Walker, 2017).

Health Risk Score

To understand the calculation of a health risk score, let's first review the more well-known financial credit scores. Most of us are familiar with credit score that range from 300-850, with higher scores meaning less of a credit risk for lending, and those with a score over 700 are given improved financing terms and rates. Following our learning approach, we connect new concepts to existing concepts that we already know to reinforce learning and build stronger mental models. On a positive note for individual financial credit scores, these have been improving steadily since the last economic downturn in 2009. In the second quarter of 2017, the average FICO (Fair, Isaac, and Company) score has now surpassed 700 for the first time. More importantly, the lowest score ranges have been decreasing, and the largest percentage category is now those with an 800-850 score at nearly 21% of the population (Dornheim, 2017). Currently, the FICO credit score is calculated from several factors, each having a weight: payment history at 35%, amounts owed at 30%, length of credit history at 15%, types of credit used at 10%, and new credit at 10%. The payment history includes accounts paid, collections, past due or delinquent accounts and past due time. Amounts owed includes the balances on accounts and credit lines used. Length of credit history includes time since accounts were opened with the longer time history the better. Types of credit include items such as mortgages and credit cards, with a variety being preferred. New credit includes recently opened accounts, credit inquires, and number of new accounts. Other credit scores such as the Vantage score use a similar method with different weighting of 32% payment history, 23% utilization of available credit, 15% credit balances, 13% length of credit history, 10% new credit, and 7% available credit (Skowronski, 2015; Wickell, 2016).

Instead of predicting if someone is a credit risk, we'll create and predict a health risk score on a scale. For the factors, these will include medical history, utilization of wellness and preventative services, length of health management, personal characteristics, and geographic location. The factors will also be weighted. However, instead of prescribing a fixed percentage to each, we could also use a more flexible model method such as a neural network that would automatically weight the factors to identify the optimal model for determining health risk. As with a FICO score, using historical health data and patient risk, we can create a scale with these factors. With current healthcare processes, most patients monitor their health at most only once per year unless they are ill, and, even then, many patients forego an annual checkup. Conversely, many consumers monitor their credit scores monthly and many with real-time alerts. Many consumers are so frightened at the possibility of daily fraudulent transactions, credit theft or individuals stealing their money! Yet surprisingly patients have much less concern at the daily possibility of stealing their own life. Shouldn't one's health have the same relevance and be of greater importance than finances? The creation of a health score will allow an easy to understand and trackable methodology for patients to compare and assess their health scores. Similar to credit scores, patients can also be offered incentives or interventions based on their placement in the health score continuum.

Table 9.1: Financial versus Health Risk Factor Weighting Comparison

Factor %	Financial	Health
35%	Your Payment History	Your Medical and Pharmacy Costs History
30%	Amounts You Owe	Chronic Conditions
30%	Length of Your Credit History	Personal Demographics and Family History
10%	Types of Credit Used	Preventative Testing and use of Wellness Services
10%	New Credit	Geographic Location

Once the health risk score or credit scores are calculated, the individual is placed within a range category indicating the level of their risk. If an individual is in the 800-850 credit score range, the individual is considered an exceptional borrower. Similarly, if an individual is in the 800-850 health score range that they are considered in exceptional health. A summary of the score category ranges is shown in Table 9.2.

Table 9.2: Financial versus Health Risk Score Category Comparison

Risk Score Category	Financial	Health
800-850	Indication of an exceptional borrower	Indication of exceptional health
750-799	Indication of a very good borrower	Indication of very good health
700-749	Indication of a good borrower	Indication of good health
650-699	Indication of an acceptable borrower	Indication of acceptable health
600-649	Indication of a below average borrower	Indication of below average health
550-599	Indication of a below average subprime borrower	Indication of below average and increased risk health
500-549	Indication of a borrower with poor credit	Indication of poor health
300-499	Indication of an unacceptable borrower	Indication of morbidity, or unhealthiness within a population and increased risk of mortality. In the U.S., seven of the top 10 causes of death are preventable chronic diseases (Johnson, 2014).

The experiential learning application data set contains two files. First, we will run through our Assess steps to generate a final model. The files contain average claim and pharmacy (or RX) spending costs (CMS, 2016b), preventative screening indicator, number of chronic conditions, age groups (0-18, 19-44, 45-64, 65-84, and 85 and over), gender (males and females), insurance type source of funding (private health insurance, Medicare, Medicaid, out-of-pocket, and all other payers and programs), region (Midwest, Northeast, Northwest, Southeast, Southwest), and health risk score category (300-499, 500-549, 550-599, 600-649, 650-699, 700-749, 750-799, and 800-850). We will use the second data set to calculate the health risk score category for a new set of data or a new patient population.

Objective: develop a model for overall patient risk for actionable healthcare

Data Set File Training: 9_EL1_Health_Risk.xlsx

Data Set File Score: 9_EL1_Health_Risk_Score.xlsx

Variables:

- PatientID
- AnnualClaimsCost
- AnnualRXCost
- PreventativeScreening

- ChronicConditions
- Age
- Gender
- InsuranceType
- Region
- HealthRiskScoreCategory

Step 1: Sign in to SAS OnDemand for Academics.

Step 2. Open SAS Enterprise Miner (Click the SAS Enterprise Miner link).

Step 3. Create a New Enterprise Miner Project (Click **New Project**).

Step 4: Use the default SAS Server, and click **Next**.

Step 5: Add the project name, and click **Next**.

Figure 9.4: Create New Project

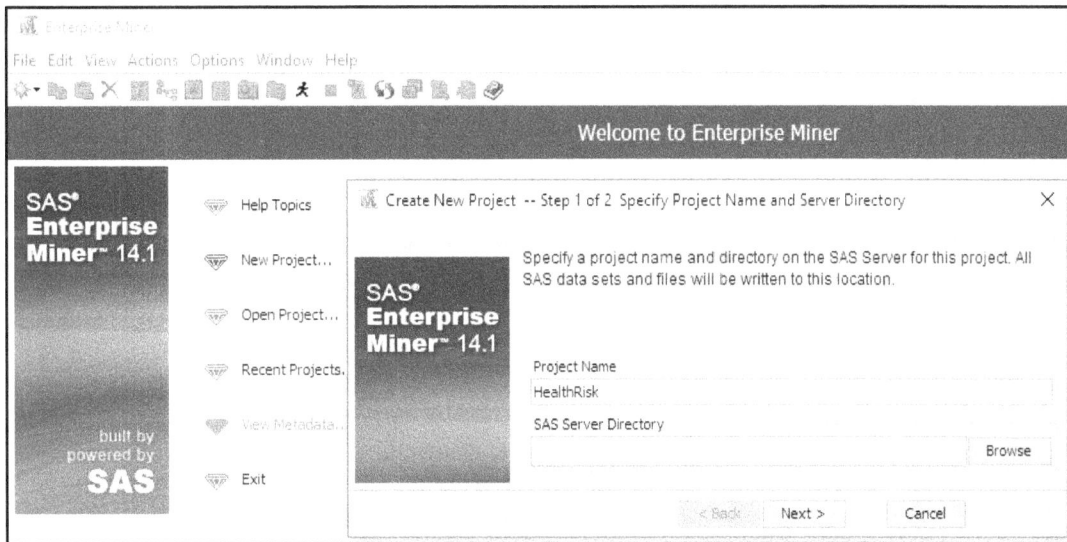

Step 6: SAS will automatically select your user folder directory (If you are using the desktop version, choose your folder directory), and click **Next**.

Step 7: Create a new diagram (Right-click **Diagram**).

Figure 9.5: Create New Diagram and Add Name

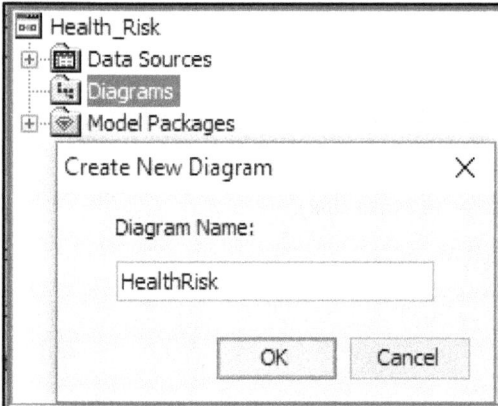

Step 8: Add a File Import node (Click the **Sample** tab, and drag the node onto the diagram workspace).

Step 9: Click the File Import node, and review the property panel on the bottom left of the screen.

Step 10: Click **Import File** and navigate to the *9_EL1_Health_Risk.xlsx* Excel file.

Figure 9.6: File Import

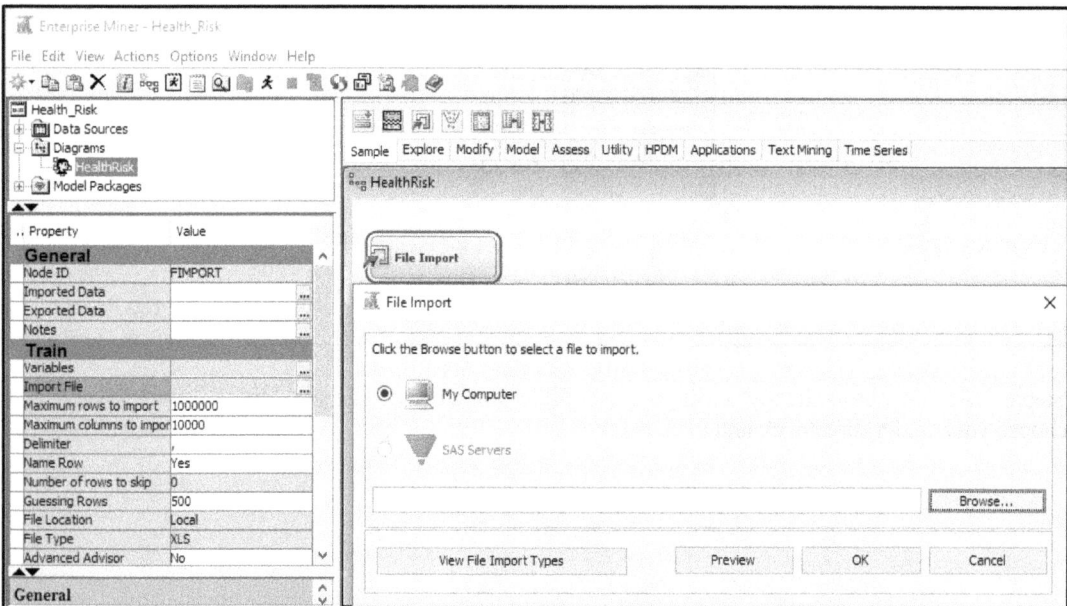

Step 11: Click **Preview** to ensure that the data set was selected successfully, and click **OK**.

Step 12: Right-click the File Import node and click **Edit Variables**.

Step 13: Set HealthRiskScoreCategory to the Target variable role, set PatientID to the ID role, and set all other variables to the Input role. Explore and set the remaining variables according to their nominal, interval, or binary levels. To review an individual variable to verify its role and level assignment, click the variable name and click **Explore**. After you have finished setting all variables, click **OK**.

Figure 9.7: Edit Variables

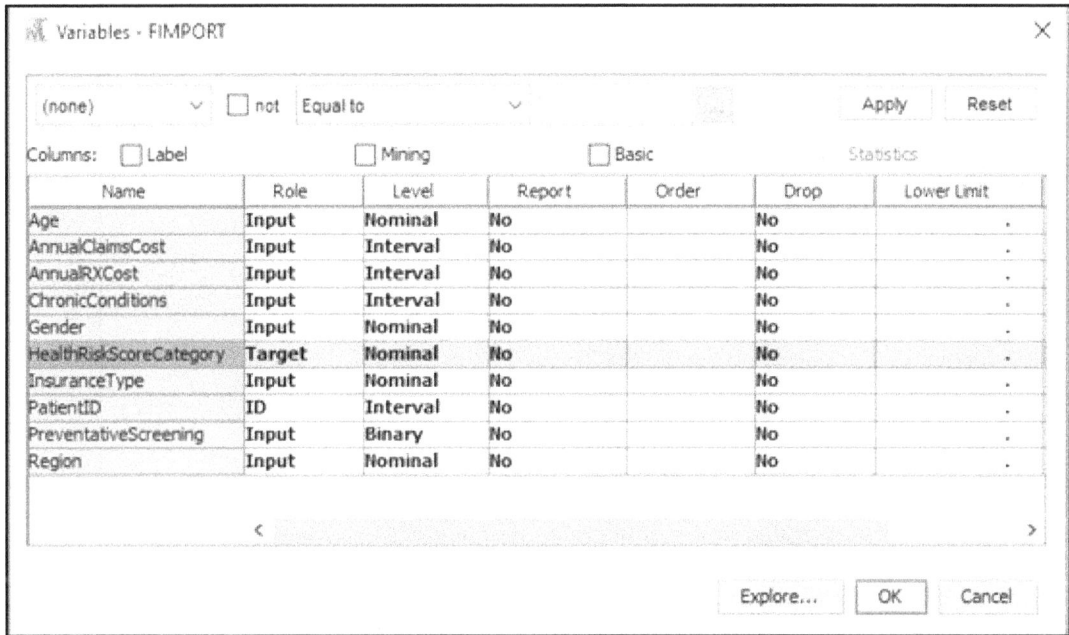

Name	Role	Level	Report	Order	Drop	Lower Limit
Age	Input	Nominal	No		No	.
AnnualClaimsCost	Input	Interval	No		No	.
AnnualRXCost	Input	Interval	No		No	.
ChronicConditions	Input	Interval	No		No	.
Gender	Input	Nominal	No		No	.
HealthRiskScoreCategory	Target	Nominal	No		No	.
InsuranceType	Input	Nominal	No		No	.
PatientID	ID	Interval	No		No	.
PreventativeScreening	Input	Binary	No		No	.
Region	Input	Nominal	No		No	.

Step 14: Add a Data Partition node (Click the **Sample** tab, drag the node onto the diagram workspace). Set the Data Partition Property Data Set Allocations to **60.0** for Training, **40.0** for Validation, and **0.0** for Test. Review the Partition Results.

Figure 9.8: Data Partition Node

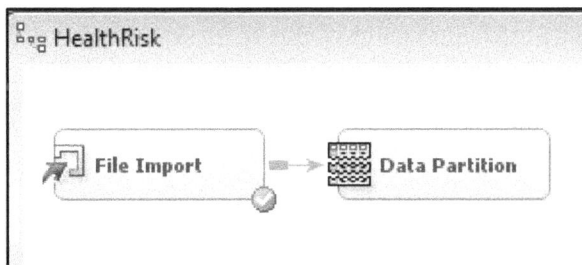

Step 15: Add a Stat Explore node, Graph Explore node, and MultiPlot node (Click the **Explore** tab, and drag the nodes onto the diagram workspace). Set the Graph Explore Property Sample Size to **Max**.

Figure 9.9: StatExplore, Graph Explore, and MultiPlot Nodes

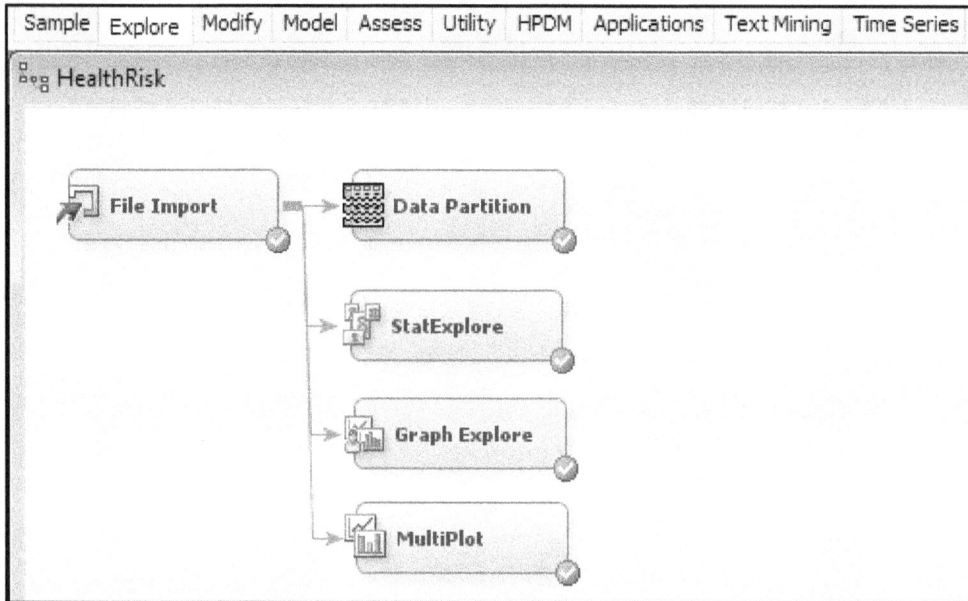

Step 16: Review the results. From the StatExplore results, we find that there is a total of 16,000 records, there is no missing data across almost all variables, with the exception of the Preventative Screening variable that has 52 missing records. The skewness and kurtosis for the AnnualClaimsCost and AnnualRXCost is outside acceptable ranges of -2 to 2, indicating potential outlier values. Initially, our reaction might be to correct or modify the outlier values. However, for claim and pharmacy costs, it likely there are outlier values in terms of healthcare costs. In these cases, the values likely are acceptable and should be reviewed within the exploration and ideally within the medical record source.

Figure 9.10: StatExplore Results

```
Output
39    Data Role=TRAIN
40
41                                          Number
42    Data                                    of                         Mode                    Mode2
43    Role       Variable Name        Role   Levels  Missing   Mode    Percentage   Mode2      Percentage
44
45    TRAIN      Age                  INPUT     5       0      0-18       20.00     19-44        20.00
46    TRAIN      Gender               INPUT     2       0      F          50.00     M            50.00
47    TRAIN      InsuranceType        INPUT     5       0      Medicaid   20.00     Medicare     20.00
48    TRAIN      PreventativeScreening INPUT    3      52      1          53.34     0            46.34
49    TRAIN      Region               INPUT     5       0      Midwest    20.00     Northeast    20.00
50    TRAIN      HealthRiskScoreCategory TARGET 8       0      650-699    15.44     600-649      14.38
51
52
53
54    Distribution of Class Target and Segment Variables
55    (maximum 500 observations printed)
56
57    Data Role=TRAIN
58
59    Data                                          Frequency
60    Role          Variable Name        Role   Level      Count    Percent
61
62    TRAIN      HealthRiskScoreCategory  TARGET  650-699    2471    15.4438
63    TRAIN      HealthRiskScoreCategory  TARGET  600-649    2301    14.3813
64    TRAIN      HealthRiskScoreCategory  TARGET  700-749    2292    14.3250
65    TRAIN      HealthRiskScoreCategory  TARGET  800-850    2274    14.2125
66    TRAIN      HealthRiskScoreCategory  TARGET  550-599    1915    11.9688
67    TRAIN      HealthRiskScoreCategory  TARGET  750-799    1911    11.9438
68    TRAIN      HealthRiskScoreCategory  TARGET  300-499    1512     9.4500
69    TRAIN      HealthRiskScoreCategory  TARGET  500-549    1324     8.2750
70
71
72
73    Interval Variable Summary Statistics
74    (maximum 500 observations printed)
75
76    Data Role=TRAIN
77
78                                   Standard      Non
79     Variable      Role    Mean    Deviation   Missing   Missing   Minimum   Median   Maximum   Skewness   Kurtosis
80
81    AnnualClaimsCost  INPUT  14889.31  23513.48   16000      0        0       6156    222335    3.047909   10.63456
82    AnnualRXCost      INPUT  5286.501   8493.471   16000      0        0       2001     80990    3.105797   11.72232
83    ChronicConditions INPUT  2.503875   1.749872   16000      0        0          3         6    0.088491   -1.11052
```

In Table 9.3, the percentage of the population is shown based on the standard financial credit risk and score health results. The financial credit score changes over time, and the average from two time periods were used as a benchmark (Wickell, 2016; Dornhelm, 2017). The category is skewed toward the higher range for financial credit score.

Table 9.3: Financial versus Health Risk Score Category % Comparison

Risk Score Category	Financial %	Health %
800-850	16%	14.2%
750-799	24%	11.9%
700-749	18.5%	14.3%
650-699	14.5%	15.4%

Risk Score Category	Financial %	Health %
600-649	10.5%	14.4%
550-599	8.5%	11.9%
500-549	6%	8.2%
300-499	2.5%	9.4%

Step 17: Review the results. From the StatExplore results, we find that the top three variables with a target of HealthRiskScoreCategory include ChronicConditions, AnnualClaimsCost, and AnnualRXCosts. The variable worth might also help develop a parsimonious model (or the simplest-best model) by including only a few variables in the final data set and model, which would improve performance of the model with similar accuracy.

Figure 9.11: StatExplore Results Variable Worth

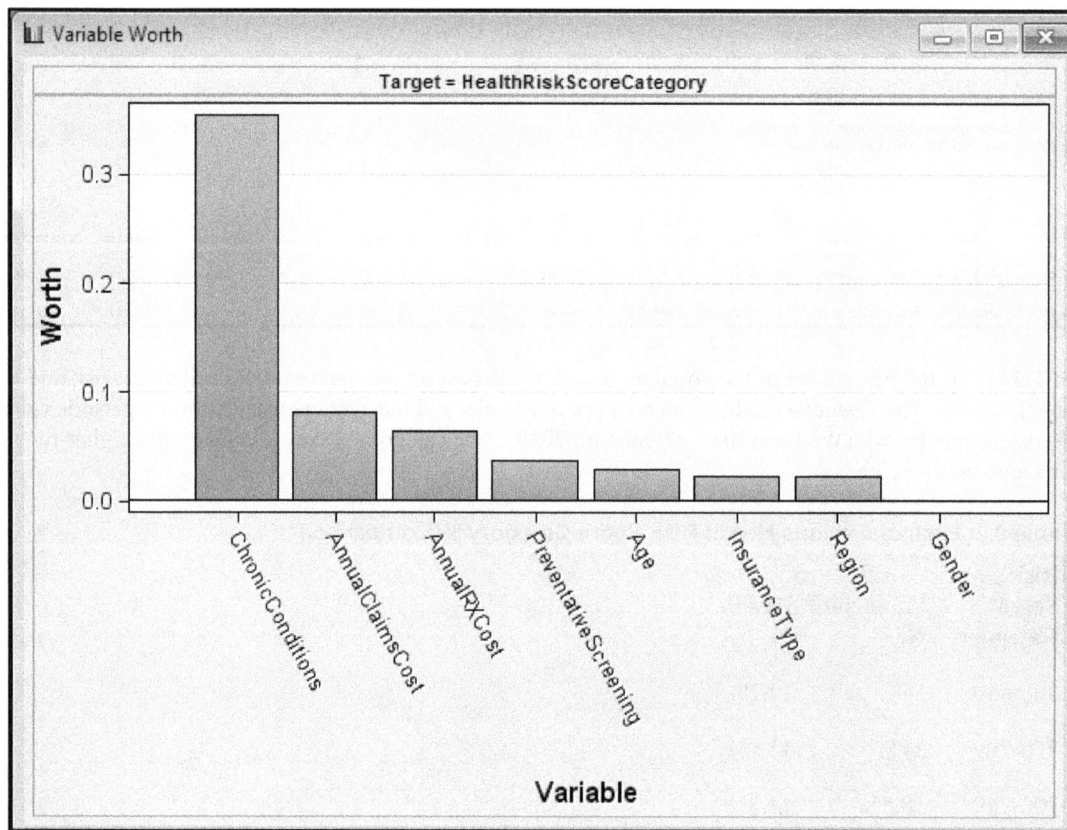

Step 18: Add an Impute node (Click the **Modify** tab, and drag the node onto the diagram workspace). Verify that the Impute Property is set to **Count** for Class variables and **Mean** for Interval variables.

Figure 9.12: Impute Node

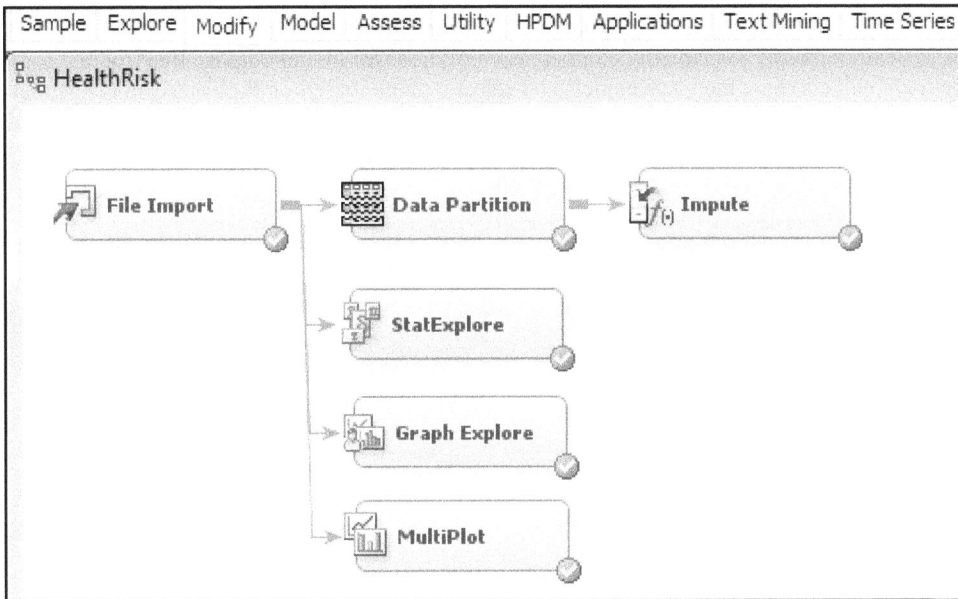

Step 19: Add a Regression node, a Decision Tree node, and Neural Network node (Click the Model tab, and drag the nodes onto the diagram workspace).

Figure 9.13: Model Nodes

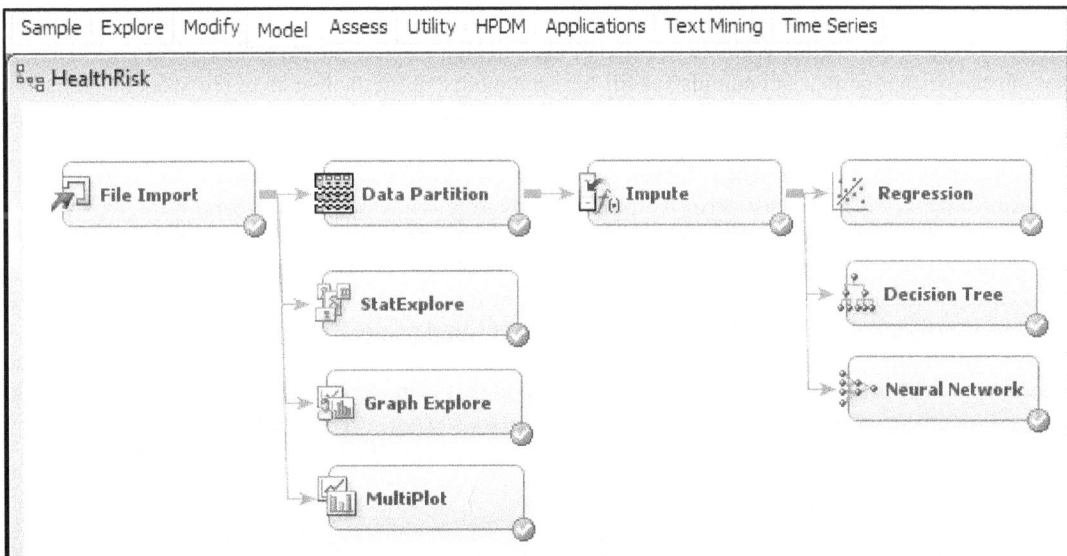

Step 20: Right-click the Regression node and Decision Tree node and click **Run**.

Step 21: Expand the Output Window Results for each model. Review the Fit Statistics for each, including the misclassification rates, and cumulative lift for each model.

Step 22: Add a Model Comparison node (Click the **Assess** tab, and drag the node onto the diagram workspace). Instead of having to manually compare each of the three model outputs, the Model Comparison node will compare the three models and select a model based on the criteria in the Model Comparison node properties. By default, we will use the validation misclassification rate to select the best model or use the model with the lowest event classification error percentage from our validation data set. Right-click the Model Comparison node and click **Run**.

Figure 9.14: Model Comparison Node

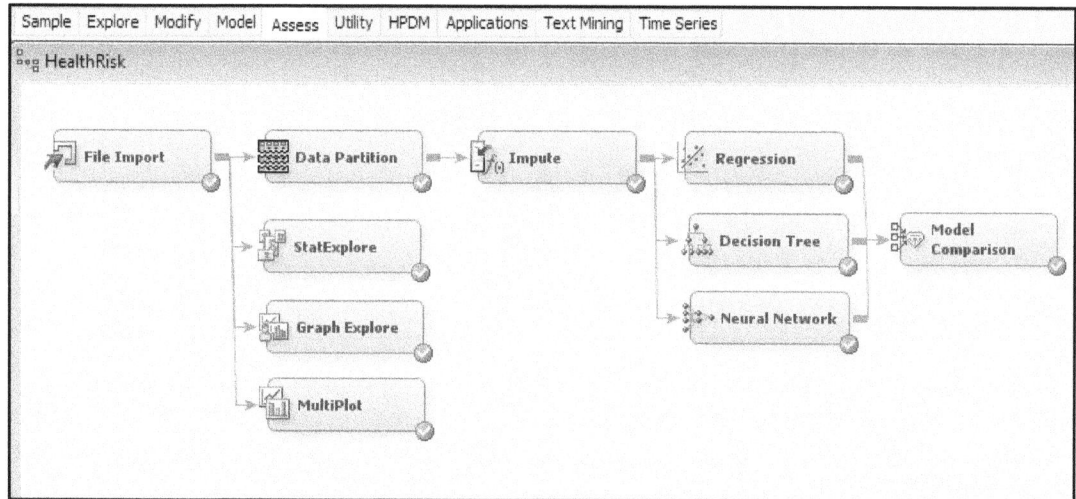

Step 23: Expand the Output Window Results for each model. Review the Fit Statistics for each, including the misclassification rates and cumulative lift for each model. With the results of the Model Comparison node, all three model results are included at once. In the Cumulative Lift chart, the Regression and Neural Network model outperform the Decision Tree model through the first 5% of the records although performance is equalized thereafter. For the validation misclassification rate from the fit statistics, the Decision Tree model is the best performer, followed by Regression and lastly Neural Network. The Decision Tree model might perform best due to missing data within the data set and split capabilities of variables. The misclassification rate near 30% for all models demonstrates that there remain some difficulties in accurately determining health risk from historical data. In addition, new government legislation makes it illegal to restrict insurance coverage based on pre-existing conditions. These restrictions also are applicable with financial lending institutions. In determining loan repayments, despite many decades of data and experience lending, current global loan repayments in some countries such as Italy are estimated at a 17% default rate (Friedman, 2016). Various events might cause models to lose accuracy and require additional variables for a more accurate assessment.

Figure 9.15: Model Comparison Nodes Cumulative Lift

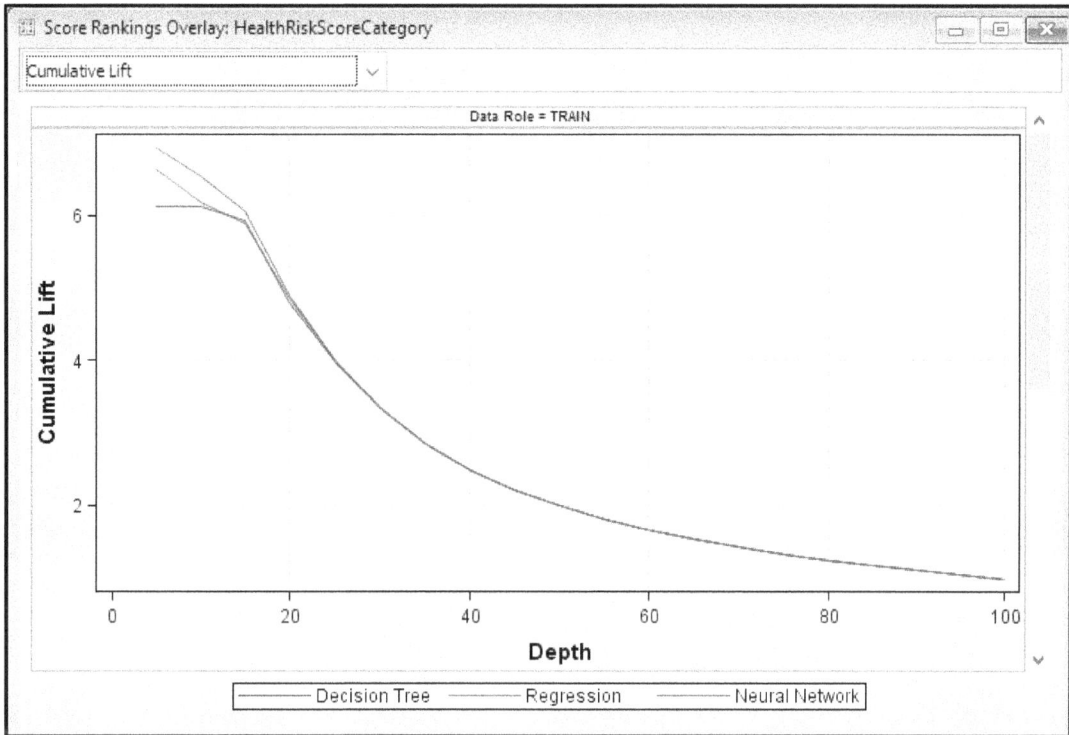

Figure 9.16: Model Comparison Nodes Fit Statistics

Selected Model	Model Node	Model Description	Selection Criterion: Valid: Misclassification Rate
Y	Tree	Decision Tree	0.288011
	Reg	Regression	0.351077
	Neural	Neural Network	0.40665

Step 24: Now that we have developed and selected a model for determining health risk category, we want to generate the health risk categories for a new set of data. We call this data set score data. To accomplish this, we will import a new file to score and a score node that will use the results of our previous model training to make a health risk category determination. To start, add a new File Import node and a Score node. The File Import node is accessible from the **Sample** tab, and the Score node is accessible from the **Assess** tab). Rename the File Import node to File Import Score. Connect the File Import Score node to the Score Node. Connect the Model Comparison Output to the Score node.

Figure 9.17: Score Node

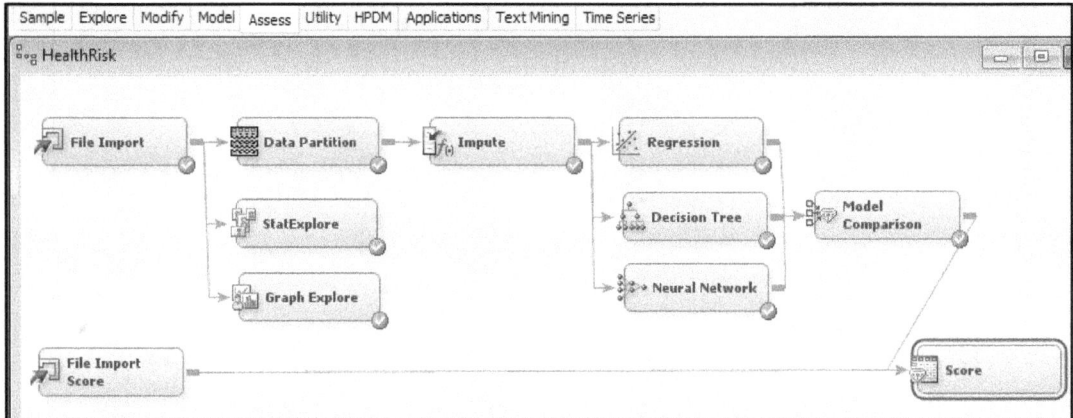

Step 25: Click the File Import Score node, click the Property panel, and set the Score Role property to **Score**. Typically, we use the Train role. However, since we have already trained the model, we now want to score new data.

Figure 9.18: File Import Score Property

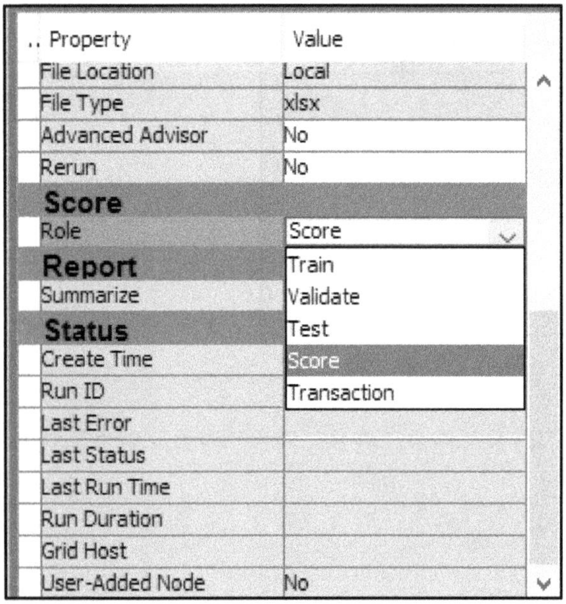

Step 26: Click the File Import Score node, and review the property panel on the bottom left of the screen. Click **Import File** and navigate to the *9_EL1_Health_Risk_Score.xlsx* Excel file.

Step 27: Click **Preview** to ensure that the data set was selected successfully, and click **OK**. Right-click the File Import node and click **Edit Variables**. Set the variables according to their roles and levels. Note the target variable is not included because we want to determine the target output from the Model Comparison node output and Score node.

Figure 9.19: File Import Edit Variables

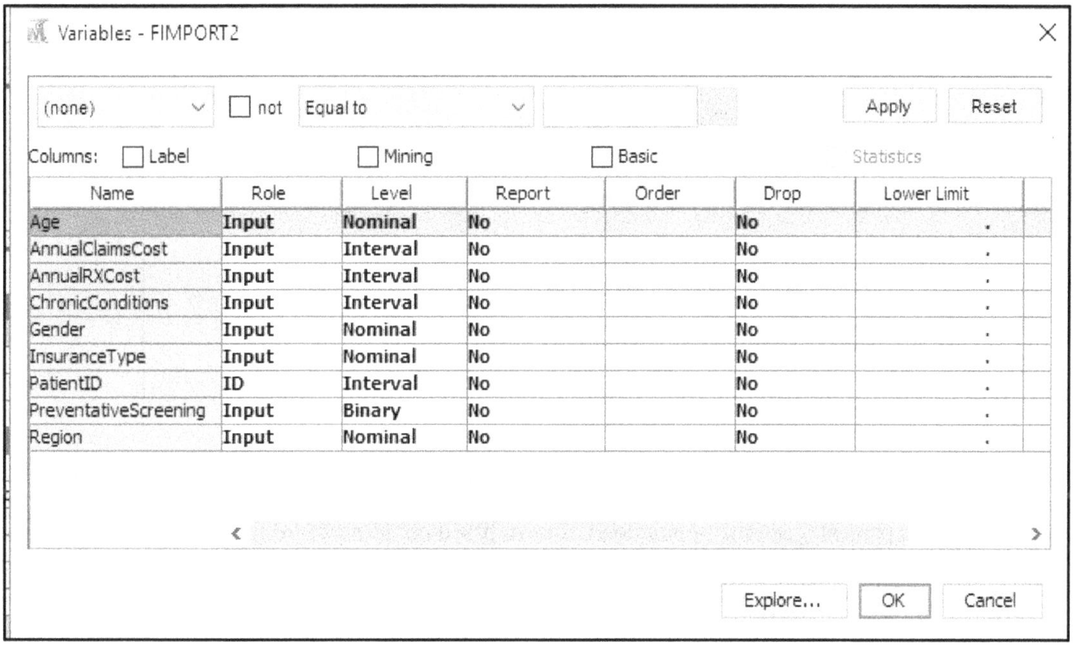
Variables - FIMPORT2
(none)
not
Equal to
Apply
Reset
Columns:
Label
Mining
Basic
Statistics
Name | Role | Level | Report | Order | Drop | Lower Limit
Age | Input | Nominal | No | | No | .
AnnualClaimsCost | Input | Interval | No | | No | .
AnnualRXCost | Input | Interval | No | | No | .
ChronicConditions | Input | Interval | No | | No | .
Gender | Input | Nominal | No | | No | .
InsuranceType | Input | Nominal | No | | No | .
PatientID | ID | Interval | No | | No | .
PreventativeScreening | Input | Binary | No | | No | .
Region | Input | Nominal | No | | No | .
Explore...
OK
Cancel

Step 28: Click the Score node and run the Score node. Review the results. From the results, we find the Frequency Count and Percent breakdown by HealthRiskScoreCategory. As one category, 21 records were classified under HealthRiskScoreCategory 300–499 (or 10.5% of the records). We see a distribution across HealthRiskScoreCategories. The scoring might be useful, for example, if assessing a new employer group that is requesting coverage. Using historical data, you could analyze and determine the health risk categories for the employees. As a next step, use of historical data would assist with the determination of premiums and the development of incentives for improving health risk score.

Figure 9.20: Score Data Results Classification

```
🗗 Output
  91    Class Variable Summary Statistics
  92
  93    Data Role=SCORE Output Type=CLASSIFICATION
  94
  95                              Numeric    Formatted   Frequency
  96          Variable             Value       Value       Count     Percent
  97
  98    I_HealthRiskScoreCategory     .       300-499        21        10.5
  99    I_HealthRiskScoreCategory     .       500-549        28        14.0
 100    I_HealthRiskScoreCategory     .       550-599        16         8.0
 101    I_HealthRiskScoreCategory     .       600-649        24        12.0
 102    I_HealthRiskScoreCategory     .       650-699        38        19.0
 103    I_HealthRiskScoreCategory     .       700-749        26        13.0
 104    I_HealthRiskScoreCategory     .       750-799        39        19.5
 105    I_HealthRiskScoreCategory     .       800-850         8         4.0
```

In Table 9.4, the percentage of the population is shown based on the standard financial credit risk and score health results. For our score data set, there is a larger percentage at the lower end, indicating higher health risk. By contrast, the category is skewed toward the higher range for financial credit score. The skewness might be caused through analysis of a high-risk health population. There are also relatively few members of the population with exceptional health. This could be true since there are factors outside of one's control that might influence health, including disease outbreaks or environmental factors, that still create health risk.

Table 9.4: Financial versus Health Risk Score Category % Comparison

Risk Score Category	Financial %	Health %
800–850	16%	4%
750–799	24%	19.5%
700–749	18.5%	13%
650–699	14.5%	19%
600–649	10.5%	12%

Risk Score Category	Financial %	Health %
550–599	8.5%	8%
500–549	6%	14%
300–499	2.5%	12%

Step 29: Click the Score node, and then click **Exported Data** under properties. Next, click **Score** and then click **Explore**.

Figure 9.21: Explore Score Data

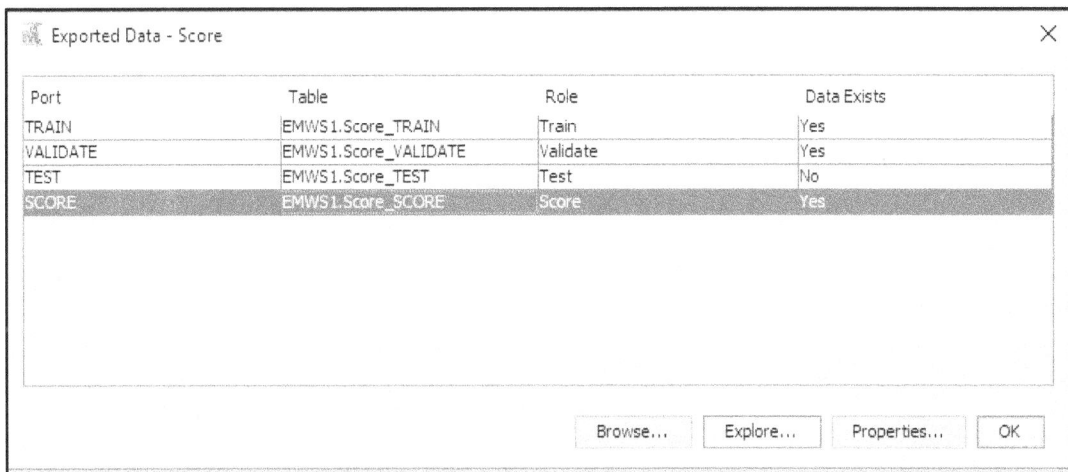

Step 30: Review the results. The display includes the individual patient records and the right-most column will show the Prediction for HealthRiskScoreCategory.

Figure 9.22: Explore Score Data Detail

Obs # ▲	PatientID	AnnualCl...	AnnualRX...	Preventati...	ChronicC...	Age	Gender	Insurance...	Region	Prediction for HealthRiskScoreCategory
1	3000727	76375	16068	1	6	45-64	M	Private heal...	Northwest	500-549
2	3015573	42823	6898	1	1	0-18	F	Private heal...	Northwest	700-749
3	3022565	2412	1440	0	0	85+	F	Other Payer...	Southeast	750-799
4	3031223	149237	27780	1	5	45-64	M	Private heal...	Northwest	500-549
5	3038675	102901	16963	1	4	65-84	F	Medicare ...	Midwest	550-599
6	3056705	29575	9541	0	2	85+	M	Out-of-Pock...	Southwest	650-699
7	3075184	59243	9533	0	5	45-64	M	Out-of-Pock...	Southwest	300-499
8	3097540	10565	879	1	1	19-44	F	Medicare ...	Midwest	750-799
9	3131546	77012	12881	0	3	19-44	F	Medicaid ...	Northeast	600-649
10	3165436	128818	17423	1	5	85+	F	Medicare ...	Midwest	500-549
11	3178384	25946	1994	1	2	0-18	F	Medicaid ...	Northeast	650-699
12	3181677	147649	19614	0	4	65-84	M	Out-of-Pock...	Southwest	550-599
13	3197039	1265	837	1	2	65-84	M	Medicaid ...	Northeast	650-699
14	3228327	82033	12472	0	3	85+	M	Private heal...	Northwest	600-649
15	3235501	11967	1295	1	2	0-18	F	Other Payer...	Southeast	650-699
16	3242993	7694	3034	0	2	0-18	M	Medicaid ...	Northeast	650-699
17	3244446	50251	16575	1	3	45-64	M	Medicare ...	Midwest	600-649
18	3251156	44041	9700	0	2	45-64	M	Medicare ...	Midwest	650-699
19	3290649	1288	1100	1	1	0-18	M	Medicare ...	Midwest	750-799
20	3295392	136906	32377	1	6	65-84	F	Other Payer...	Southeast	500-549
21	3297695	71537	6445	0	5	65-84	F	Out-of-Pock...	Southwest	300-499
22	3300711	181947	23705	0	6	85+	F	Private heal...	Northwest	300-499
23	3325683	4940	53	1	1	85+	F	Other Payer...	Southeast	750-799
24	3333949	8676	3390	1	1	85+	F	Out-of-Pock...	Southwest	750-799

Assess Summary

In summary, the Assess process step combines all the preceding steps in the SEMMA process. In this step, multiple models might be used and results interpreted individually, or model comparison and scoring might be used for a final model. All known model parameters might be used for evaluation, interpretation, selection, and recommendations. A Score node might also be added to evaluate a new data set from the selected model.

- Input: Interval, Binary, or Nominal
- Target: Interval, Binary, or Nominal
- Model: Regression node, Decision Tree node, and Neural Network node
- Evaluation: Assess Model Comparison node, or individual nodes evaluation: R^2, Adjusted R^2, Error, Lift, p-value, Odds Ratio, English Rules, Misclassification Rate
- Scoring: Evaluate a new data set using selected model

Experiential Learning Application: Hip Fracture Risk

Executive Summary

As human life expectancy increases, the occurrence of osteoporosis is increasing. One result is an increase in hip fractures, which have a high rate of increased morbidity up to 33%. By 2025, an estimated 2.6 million hip fractures will occur worldwide, with 4.5 million by 2050, creating health and cost impacts. Researchers have attempted to identify the risk factors to prevent hip fracture. Previous studies have reviewed factors such as bone mineral density, age, gender, chronic conditions, previous falls, physical activity, smoking, drinking, vision, medications, calcium and vitamins, BMI, and other factors such as geography, ethnicity, and culture. The researchers evaluated methods and models, such as a logistic regression and neural network, to evaluate results. Falls leading to hip fracture might be a combination of individual and environmental factors. However, in the past, researchers have found that individual factors might play a greater role (Tseng et al., 2013).

For the variables, milk intake indicated consumption of at least six times per week. Walking difficulty indicated the need for assistance such as a walker. Significant fall indicated a major fall more than once in the last year. Low education indicated lower than junior middle school. Smoking indicated more than a half pack per day. Vision impairment indicated blurred vision during walking. Exercise indicated at least four times per week, and Height and weight were used for BMI calculation (Tseng et al., 2013).

In the National Taiwan University Hospital study, the hip fractures included first-time, low-energy hip fractures, with is a fracture caused by injury caused by a fall at standing height or less than standing height. The study population includes 217 pairs of 149 women and 68 men, with and without hip fracture, all with age over 60 years. Participants were provided with a standard questionnaire. The researchers found many variables that achieved significance including low BMI, milk intake, walking difficulty, and significant fall at home. In their models, the neural network performance exceeded the logistic regression (Tseng et al., 2013).

For the experiential learning application, you have been provided a data set of 434 records and 15 variables. Help your management team with the following objective.

Objective: develop a model for hip fracture risk

Data Set File: 9_EL2_Hip_Fracture_Risk.xlsx

Variables:

- RecordID, unique identifier
- Gender, (F=female, M=male)
- BMILessThan21.4, (1=True, 0=False)
- VegetarianDiet, (1=True, 0=False)
- MilkIntake, (1=True, 0=False)
- WalkingDifficulty, (1=True, 0=False)
- SignificantFallInLastYear, (1=True, 0=False)
- MultiStoryDwelling, (1=True, 0=False)
- LowEducationLevel, (1=True, 0=False)
- CurrentSmoking, (1=True, 0=False)
- PreviousFractures, (1=True, 0=False)
- VisionImpairment, (1=True, 0=False)
- LessThan2MajorDiseases, (1=True, 0=False)
- RegularExercise, (1=True, 0=False)
- HipFracture, (1=True, 0=False)

Follow the SEMMA process for your experiential learning application and provide recommendations. A template has been provided below that can be reused across future projects.

Title	Hip Fracture Risk
Introduction	Provide a summary of the business problem or opportunity and the key objective(s) or goal(s). Create a new SAS Enterprise Miner project. Create a new Diagram.
<u>S</u>ample	Data (sources for exploration and model insights) Identify the variables data types, the input and target variable during exploration. Add a FILE IMPORT Provide a results overview following file import: Input / Target Variables Generate a DATA PARTITION
<u>E</u>xploration	Provide a results overview following data exploration Add a STAT EXPLORE Add a GRAPH EXPLORE Add a MULTIPLOT Summary statistics (average, standard deviation, min, max, and so on) Descriptive Statistics Missing Data Outliers
<u>M</u>odify	Provide a results overview following modification Add an IMPUTE
<u>M</u>odel	Discovery (prototype and test analytical models) Apply a regression, decision tree, and neural network model and provide a results overview following modeling. Add a REGRESSION Add a DECISION TREE Add a NEURAL NETWORK Model description Analytics steps Model results (Lift, Error, Misclassification Rate) Selection Model

Title	Hip Fracture Risk
<u>A</u>ssess and Reflection	ADD a MODEL COMPARSION Provide overall recommendations to business Model advantages / disadvantages Performance evaluation Model recommendation Summary analytics recommendations Summary informatics recommendations Summary business recommendations Summary clinical recommendations Deployment (operationalization plan: timeline, resources, scope, phases, project plan) Value (return on investment, healthcare outcomes)

Learning Journal Reflection

Review, reflect, and retrieve the following key chapter topics only from memory and add them to your learning journal. For each topic, list a one sentence description/definition. Connect these ideas to something you might already know from your experience, other coursework, or a current event. This follows our three-phase learning approach of 1) Capture, 2) Communicate, and 3) Connect. After completing, verify your results against your learning journal and update as needed..

Key Ideas – Capture	Key Terms – Communicate	Key Areas - Connect
Assess Process Step		
Health Anamatics Administration		
Administrative Code sets		
ICD		
CPT		
HCPCS		
NDC		
CDT		
LOINC		
Security		
Privacy		
Health Risk Application		

Key Ideas – Capture	Key Terms – Communicate	Key Areas - Connect
Health Risk Application		
Hip Fracture Risk Application		

Chapter 10: Modeling Unstructured Health Data

Chapter Summary

The purpose of this chapter is to develop unstructured data modeling skills using the text mining capabilities of SAS Enterprise Miner. Text mining capabilities are an extension to the SEMMA process, and it build on previous chapters with sample, explore, modify, model, and assess capabilities. This chapter includes exploring unstructured data through text mining. This chapter also includes experiential learning application exercises on U.S. presidential speeches and healthcare legislation tweets. The text mining focus of this chapter is detailed in Figure 10.1.

Figure 10.1: Chapter Focus - Text

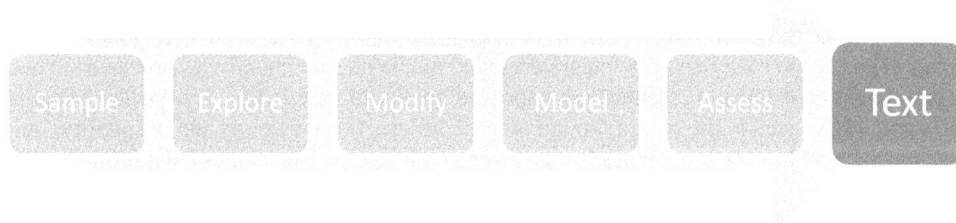

Chapter Learning Goals

- Describe the text mining process steps
- Understand unstructured data sources
- Develop text anamatics skills
- Apply SAS Enterprise Miner text functions
- Master the SAS text mining capabilities

Unstructured Health Anamatics

One of the greatest management challenges is found in the complexity and fluidity of the healthcare system, requiring advanced clinical and technical systems for decision making, particularly in the world of growing unstructured data sources. Health anamatics capabilities are used to increase power of workers to make decisions on factual data. Typically, a significant portion of knowledge resides with employees from personal experience, and is not directly codified. As a result, information systems are required to capture, store, and disseminate this data. Today, health care organizations are capturing increasing amounts of data in various places. The right information in the hands of patients, providers, case managers, and others can lead to improved patient outcomes and quality of care (Woodside, 2013). Increasingly, data is being stored in unstructured formats. According to some estimates, 80% of data is unstructured although most organizations are using only the structured 20% for analysis. Therefore, this chapter is devoted to unstructured data mining. In order to perform data mining on unstructured data, we first transform the data into a structured format. Once the data is structured, we are able to run data mining methods, as previously shown. In earlier chapters, we focused on structured data or data that is provided in nice and neat formats such as rows and columns. Structured data is typically found within an Excel spreadsheet, standard database tables, or delimited files. In this chapter, we'll move to unstructured data. Unstructured data is commonly found in the healthcare industry, and large amounts of unstructured data is included in sources such as instrument results, radiology images, clinical notes, and social media sites (including blogs, microblogs, social networking sites, virtual worlds, audio, video, and applications). Blogs enable patients to keep in contact with family and friends while they undergo long-term care. Microblogs such as Twitter can be effectively used during emergency situations, status updates, and receiving feedback from patients. Social networking sites like Facebook are online communities that can be used for support of specific conditions and diseases. Virtual worlds allow for personalized avatars for use in patient education while offering an entertaining method of delivery. Audio-based deliverables and videos are used to distribute educational information such as policies and procedures for visually impaired and hearing-impaired individuals. Apps have been developed for locating services, medication management, risk assessment, and symptom relief (Woodside, 2012)

Social Media

Social media is one of the most common data sources of unstructured data. Early adopters of social media in the healthcare industry have successfully integrated social media into their analytics processes in order to measure the effectiveness of social media through improved quality of care, efficiency, loyalty of patients, and increased revenue. As seen in the past, although other industries (such as retail and hospitality) quickly adopted social media, healthcare adoption has been slower. Healthcare opportunities exist for patients, providers, payers, and governments. Children's Hospital Boston has created a Facebook page (with over 700,000 likes) that uses disease-related support groups for current and prospective patients, helping patients with community support and serving as a referral network for the hospital. Another online community for physicians and healthcare providers is called SERMO, and has over 800,000 physician members from 150 countries who share information for learning and treatment to improve patient outcomes (Sermo, 2017). Blue Cross Blue Shield has set up Facebook and Twitter sites to increase the relationship with insured members through improved online web support and social media customer service. In one instance, a member tweeted a negative comment that their new insurance card was confusing. In response, a community service manager reached out to the member. The next day the member re-tweeted the encounter and issued a positive tweet that all the concerns were resolved. Government agencies such as CDC are also using social media for health education and public health information about epidemics (Anderson et al., 2012). The Centers for Disease Control (CDC) has been a leader in social media, and the CDC has collaborated with social networking sites for physician communication and information exchange, as well as sites that connect family and friends during critical illness. Increasingly, individuals are turning to their social networks for information about health issues, and CDC is demonstrating social media strategies to help individuals lead healthier, safer lives. Using the online networks, physicians can meet to discuss questions and issues. Researchers are also able to learn regarding treatment plans and adverse effects. Prior to widespread web information, in the late 1970s, offline social networks had displayed improvements in mortality rates between 2–4.5 times better. The social network trend continues today with online social media information access rivaling that of individual physicians (Sarasohn-Kahn, 2008).

Social media in healthcare can lead to improved health outcomes, reduced costs, and improved productivity and employee health. A stable and supportive social network can improve health outcomes. A majority of patients indicate that online information has helped with treatment decisions, physician decisions, healthcare maintenance, physician dialog, pain management, stress management, diet management, and exercise management. Regardless, adoption of healthcare social media has been slow and the full benefits have not been realized. The extent of social media use varies, and there is still much room for improvement and growth in healthcare social media as compared with traditional forms of social media. Patients of healthcare are seeking increased access to quality, cost, and value information, and they are increasingly using online sources for self-management (such as reviewing diseases and conditions, looking for better treatment options and providers, comparing costs, providing other individuals with assistance and support, and collaborating with clinicians). Individuals under the age of 49 were the most dissatisfied with their healthcare experience, as they expect faster, simpler and more accessible healthcare service. These individuals are also more tech savvy and are increasing their demands on healthcare organizations to provide streamlined services. In addition to having younger audiences, social media and online communities are also drawing older audiences, with those aged 50–64 among the fastest growing segment, and segments over 65 are also growing. Social media is currently used by patients to share information about healthcare experiences and to provide comments on physicians and hospitals. Social media is being used to support health behaviors, improve ability to cope with diseases, participate in treatment decisions, navigate health systems, motivate groups, follow emergency measures, provide physician interaction, and support policy development. Many healthcare patients actively search for individualized information, and nearly a quarter of patients have read another patient experience on a website, blog or news group, and reviewed provider or hospital rankings. Others receive updates or listen to podcasts. A smaller number, less than 10%, have developed new content including posting information such as photos, videos, comments, or provider and hospital reviews. Over half of the patients indicate that online information influenced decisions around treatment of an illness, condition, physician questions, or health maintenance. Nearly half of patients reported that they are changing their thoughts on diet, exercise, stress, and chronic conditions or pain management (Woodside, 2012).

For social media activities, it is important to understand which activities are most frequent and then to track those activities and decisions to determine which have a greater impact on improving outcomes. Using this information, a healthcare organization (such as a hospital, insurer, or managed care organization) might be able to set specific methods to achieve results. If patients with specific conditions more frequently use online sources, this might reduce costs of care, improve behaviors, or facilitate an improved quality of life for patients. Another example might include a microblog (Twitter) reminder about upcoming flu shots and tracking the online community response and resulting outcomes. Secondarily, the knowledge developed by online communities often exceeds that of individuals, such that collective knowledge might be used to improve healthcare at a rate faster than individual interactions (Woodside, 2012).

Experiential Learning Activity: Social Media Policy

Social Media Policy

Description: Over 1,200 hospitals are included on 4,200 social media sites, and participation is growing. In light of security and privacy concerns, most healthcare organizations have limited access to social media sites within on-site workstations, and they require employees to limit their social media usage and abide by specific guidelines. In one case, a surgeon who was proud of their suturing technique, posted an online photo as well as a summary of the medical information. Soon after, hospital administration was quickly notified of the privacy violation. Many organizations have developed specific social media policies for their employees and visitors to their websites. The following is a typical policy example: (Selvam, 2012; Cleveland Clinic, 2017; AHA, 2017; Mayo Clinic, 2017b).

Social Media Policy:
If you post on any public facing social media site or site that allows user-generated content including but not limited to Facebook, LinkedIn, Pinterest Twitter, and YouTube, you must conform to the following guidelines:

1. Must be at least 18 years of age
2. Prohibited from posting protected health information including images and videos
3. Prohibited to provide medical advice, medical referrals or medical commentary
4. Agree to all laws including intellectual property, trademarks, or unauthorized advertising
5. Agree to allow your content to be reproduced, distributed, and edited
6. Agree to be professional, respectful, use good judgment, and be accurate and honest in communications
7. As an employee receive approval for any social media project
8. As an employee, social media use should not interfere with any work commitments
9. As an employee friending of patients on social media websites is strongly discouraged
10. As an employee if not authorized in an official spokesperson capacity, include a disclaimer when sharing personal opinions and beliefs, such as the views do not reflect the views of my employer

As a patient and as an employee how do you feel about the social media policy.

Recommend 3 components to include in your organizational social media policy.

Social Media Maturity

Although the implementation of strategies using social media has been extensively studied in marketing literature and applications in industry, strategies that identify the specific practices, criteria, and methodology for healthcare are limited. Technology and consumer behavior are driving the development of new business models using social media for delivering healthcare, behavior support, and health information. This new environment creates opportunities for improving overall care. The effective adoption of social media into the business model requires that businesses reach certain levels of knowledge, sophistication, and integration of their social media strategy. A healthcare organization's position relative to the progression of capability levels reflects its social media maturity. A social media maturity model can be an important tool that healthcare managers can use to determine where their organization stands in terms of its social media capabilities, what strategies are appropriate to its maturity level, and how to move to the next level. At the broadest level, the building blocks of social media maturity can be stated as follows: Social Media Maturity: Content + Community + Integration => Coherence. Content includes the sources of data, types of data, and presentation of data. Community includes the audience, tools, resources, partnerships, trust, and interaction. Integration includes overall social media inclusion in the organization's operations, plans, and strategies, and use of social media to support patient-centered care. As the organization approaches social media maturity, its community relationships, its operations, and its vision become more seamlessly meshed. It moves symbiotically with its health consumers and achieves Coherence. A general social media model and social media strategies include a range of progressions or levels to social media maturity (Thomas and Woodside, 2015).

- Level 1: Social media presence
- Level 2: Knowledge of and engagement with the audience
- Level 3: Connection to a strategy or objectives
- Level 4: Corresponding monitoring and metrics through analytics
- Level 5: Integration with processes through anamatics

Social media and social media tools represent a set of internet-based tools for sharing and discussing information or activities that integrate technology and social interaction. The importance of social media has emerged along with the social revolution in consumer decision-making and information seeking that has been brought about by changing technologies and is played out through social behaviors. Based on current usage, it is predicted that social media is becoming one of the two most important forms of engagement with employees and customers, second only to face-to-face interactions, which have important implications for healthcare. Related to the paradigm shift in consumer behaviors is the concept of the social data revolution in which new data sources are emerging, and marketing and business strategy is driven by studying consumer connections and sharing. These data sources can be mined in new ways to help businesses provide forums for consumers to discuss ideas and help other consumers make decisions. Specific to the healthcare industry, topics of research and practice include the health-seeking behavior of individuals, development of social media strategies for healthcare organizations, and descriptions and criteria for defining healthcare delivery's evolving versions. A key component of empowering health consumers is to be able to assess their health literacy level and to assist them in learning so that they can make truly informed decisions. It is important to offer tools and content that correspond with the health consumer's health literacy level in guiding them along the knowledge path toward decision-making (Thomas and Woodside, 2015).

Text and Email Messaging

Healthcare information can also be disseminated in the form of unstructured text and email messages. Similar to Twitter's limit of 140 characters, text messages are 160-character messages on mobile phones and are available on 98% of all mobile devices. With over 91% of U.S. adults owning a mobile device, text and email provide a low-cost, convenient mechanism to reach a large percentage of the population with health messages. Over 1.5 trillion text messages are sent annually, and users receive more text messages than phone calls each month. Text messaging has also been shown to improve healthcare outcomes such as medication compliance (CDC, 2011; CDC 2014; CDC, 2016d). Below are a few examples of health-related text messages:

- Test your smoke alarms and carbon monoxide detector when u turn your clocks back on Nov 1.
- Cover cough and sneezes to protect others.
- Thanksgiving is Nat'l Family History Day. Talk to UR family about health conditions that run in UR family. Learn more http://m.cdc.gov/family.

Unstructured text and email messaging can also be used for healthcare administrative functions such as sales, marketing, and customer service to improve engagement. Aurora Health Care, a hospital system with 15 hospitals and 150 clinics in Wisconsin, is using Salesforce for their marketing campaigns for current and future patients. Salesforce is the world's leading cloud-based Customer Relationship Management (CRM) system for sales, marketing, and customer service functions, and includes various healthcare-related plug-ins through their application market including Evariant, which is a selected healthcare CRM and analytics solution. This model increasingly treats the patient as a consumer of care. With a health system patient population of over 1 million, the marketing campaign has been successful with less than a 1% unsubscribe or stop messaging rate and an email open rate of between 30–50%, with the hospital system already implementing over 40 marketing campaigns. In addition, the system sends text and email message reminders for routine health checks such as annual physicals and mammograms. The hospital system takes a customer-centered approach, even comparing themselves to the Amazon.com approach to healthcare (Sutner, 2017; Evariant, 2017; Salesforce, 2017).

Experiential Learning Activity: Dr. Google

"Dr. Google"

Description: The increase in online information and social media sites have led to the term "internet doctor" or "Dr. Google." Increasing access to information by patients is changing the dynamics between patients and providers. Online communities and information sources are debating the role of the internet in providing medical information for treatment and care and the qualifications of those providing advice online. Despite concerns, there is also research to support that the internet might be a valuable source of information by improving health literacy and self-management capabilities, such as through chronic condition communities that lead to improved health outcomes. Combined with EHRs, PHRs, and other health informatics technology, the storage and use of online information might have a transformative effect on the patient-physician relationship (Glynn, 2014). Watson, by IBM, is also being used in projects at major hospital systems such as Cleveland Clinic. Amazon is also working with HealthTap in their version of "Dr. Google" for Alexa (Maney, 2017).

List an advantage and a disadvantage to "Dr. Google":

Given the growth in online information, what are your thoughts on the evolution of the doctor-patient relationship for the next 5–10 years? How will patient expectations change as customers of healthcare? Will any regulatory or business process changes be required?

Text Mining

Text Mining Process Step Overview

In general, a text mining process consists of three steps: 1) collect the documents, 2) parse and transform the text, and 3) analyze patterns. The first step is establishing the corpus (or the set of text documents) that is used for discovery. The second step is to parse and transform the text, where each word is separated into terms, and the relationship between the terms and documents is measured in terms of frequency. During this step, standard and expert-driven stop words are removed to improve differentiating value. The output includes each unique term (excluding stop terms), each unique document, and the frequency count of each term within each document. The last step of pattern analysis relies on methods including clustering and classification models such as regression, decision tree, and neural network. Clustering is a data mining process that allows the set of mission statements to be segmented into natural groupings based on similar characteristics.

Text Tab Enterprise Miner Node Descriptions

Model Application Examples

Text mining has been used in detecting coronary artery disease (CAD) (or heart disease) that is the leading cause of death worldwide. The disease is due to plaque that accumulates in the coronary arteries and blocks the flow of blood. Unstructured data within EHRs can be used in the text mining, such as demographics, medical history, medication, allergies, immunizations, laboratory test, images, and other personal statistics. One of the challenges of the study was missing data, which was handled through imputation. In the study, a cohort of 296 diabetes patients were used to detect CAD (Jonnagaddala et al., 2015).

Another study used text mining to review and identify the communication needs for oncological (or cancer) patients. Interviews were text mined to support a patient-centered health communication system. The text mining capability in conjunction with other advancements in cancer research (such as early detection, screening, and treatment) has allowed increased life expectancy for cancer patients. Patient-empowered care through communication is a self-care strategy and is aimed at improving patients' understanding of and impact on their health. This all leads to improved health outcomes and improved quality of life (Falotico et al., 2015).

Many other studies use information contained in EHRs. Most EHRs are a combination of structured and unstructured data as free text. In one study that reviewed case diagnosis detection, it was learned that many studies use the structured data only, which might lead to quality concerns. Provider detection results were improved with the inclusion of unstructured data in detecting clinical conditions, leading to improved patient outcomes. In determining the reasons why clinicians use unstructured free text in their detection, some find structured coding to be limiting and nuances, others find the process for identifying and looking up the code to take longer than free text. Text is also used where a code does not specifically describe the diagnosis or a recording of unknown or uncertain diagnostic symptoms, for describing a range of diagnoses, for supporting evidence, and recording thought process and deduction (Ford et al., 2016).

In another study, authors used online physician reviews and mined the text-based user reviews. In the U.S., nearly half the patients have searched for their physician online, and 59% said that online reviews were important. In a study of European countries, 33–40% of users said that the online information was important. The online platform was the largest health community in China, containing more than 770,000 reviews of 100,000 physicians. The major areas identified were the medical treatment, the technical skills and bedside manner, and the process to locate the physician (Hao and Zhang, 2016).

Many physicians are looking to improve their online presence and practice. Sites such as Yelp, Google Reviews, and Healthgrades contain physician reviews. Physicians are hiring reputation management companies, such as Empathiq and IHealthSpot, to help improve their online image. In responding to reviews, healthcare laws create additional challenges, as private health information cannot be shared in the response. If a patient review indicates that the physician missed a diagnosis or ordered the wrong test, the physician can respond only in a general way. In these cases, it is recommended that the physician's response should be to ask the patient to contact them to discuss further. Larger organizations such as Banner Health are using natural language technology to review social media sites such as Facebook for comments and to create tickets for a call center, as needed, within their Salesforce CRM system (Wang, 2017).

Text Parsing Node

The Text Parsing node is used to split the unstructured text data into individual word parts (or terms). The Text Parsing node is associated with the **Text** tab in SAS Enterprise Miner.

Figure 10.2: Text Parsing Node

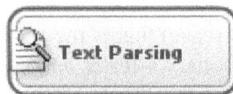

The Text Parsing node properties contain a number of additional options including Ignore, Synonyms/Stemming, and Filter. The Ignore property settings can be customized by clicking the ellipsis (…) in order to remove parts of the text that are not meaningful such as numbers and punctuation. The Synonyms/Stemming options combine similar terms in order to have a unique count. Instead of listing the terms physician, doctor, and provider, we could combine these as a unique term. In addition, instead of listing the terms treats, treatment, and treated, this could be combined as a unique term. The Filter removes additional text that is not meaningful called stop words. These words are a dictionary of commonly ignored terms and can be customized.

Figure 10.3: Text Parsing Properties

Property	Value
General	
Node ID	TextParsing
Imported Data	...
Exported Data	...
Notes	...
Train	
Variables	...
⊟ Parse	
┊ Parse Variable	
┊ Language	English ...
⊟ Detect	
┊ Different Parts of Speech	Yes
┊ Noun Groups	Yes
┊ Multi-word Terms	SASHELP.ENG_MULTI ...
┊ Find Entities	None
┊ Custom Entities	
⊟ Ignore	
┊ Ignore Parts of Speech	'Aux' 'Conj' 'Det' 'Interj' 'Part' ...
┊ Ignore Types of Entities	...
┊ Ignore Types of Attributes	'Num' 'Punct' ...
⊟ Synonyms	
┊ Stem Terms	Yes
┊ Synonyms	SASHELP.ENGSYNMS ...
⊟ Filter	
┊ Start List	...
┊ Stop List	SASHELP.ENGSTOP ...
┊ Select Languages	...
Report	
Number of Terms to Display	20000

Text Filter Node

The Text Filter node allows additional customization of the final word list or term list. The Text Filter node is associated with the **Text** tab in SAS Enterprise Miner.

Figure 10.4: Text Filter Node

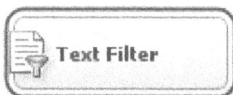

The Text Filter node properties allow Spelling verification, Weighting, Term and Document Filters. Under the Term Filters properties, in order for a term to appear in the final analysis, it must occur within at least four documents. This ensures more meaningful terms depending on the application.

Figure 10.5: Text Filter Properties

.. Property	Value
General	
Node ID	TextFilter
Imported Data	...
Exported Data	...
Notes	...
Train	
Variables	...
⊟Spelling	
├Check Spelling	No
└Dictionary	...
⊟Weightings	
├Frequency Weighting	Default
└Term Weight	Default
⊟Term Filters	
├Minimum Number of Documents	4
├Maximum Number of Terms	.
└Import Synonyms	...
⊟Document Filters	
├Search Expression	
└Subset Documents	...
⊟Results	
├Filter Viewer	...
├Spell-Checking Results	...
└Exported Synonyms	...
Report	
Terms to View	All
Number of Terms to Display	20000

Text Cluster Node

The Text Cluster node allows classification of the final word list or term list. The Text Cluster node is associated with the **Text** tab in SAS Enterprise Miner.

Figure 10.6: Text Cluster Node

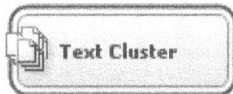

The Text Cluster node properties can be customized to provide a maximum number of cluster classes based on the cluster algorithm, or an exact number can be set. The number of clusters might be useful for certain applications; a set of five clusters might be desired for a prototype technique with five variations. The number of descriptive terms can also be customized. This is the number of words or terms that describe the clusters and helps group documents or records.

Figure 10.7: Text Cluster Properties

.. Property	Value
General	
Node ID	TextCluster
Imported Data	...
Exported Data	...
Notes	...
Train	
Variables	...
⊟ Transform	
┆-SVD Resolution	Low
┆-Max SVD Dimensions	100
⊟ Cluster	
┆-Exact or Maximum Number	Maximum
┆-Number of Clusters	40
┆-Cluster Algorithm	Expectation-Maximization
┆-Descriptive Terms	15

Text Mining Model Description

Text analytics is a relatively recent term and is an umbrella term that includes information retrieval, information extraction, data mining, and text mining. Text mining is an extension of data mining and is an automated method used to find and extract useful patterns, models, directions, trends, or rules from unstructured text (Woodside, 2016). Text mining includes many functions that are similar to other models, including data selection, data cleansing, data transformation, data mining, and results evaluation. Applications of text mining for information processing and analysis can use text summarization, link analysis, clustering and classification analysis (Hung, 2008; Abdous & He, 2011).

Data Preparation and Evaluation

Text mining is largely an exploratory process and requires a set of unstructured data. The unstructured data might be contained within a single file or within multiple files. For text mining, the text field will be assigned a Text role within SAS Enterprise Miner. The results of a text mining process such as clustering can be combined with the model nodes that we have covered in previous chapters including regression, decision trees, and neural networks. The combination allows the output of a text mining cluster analysis to be the input of a data mining model, and the methods of lift, error, and misclassification to be used for evaluation as shown in previous chapters.

Now that we have covered the text mining process step and nodes, let's continue with an experiential learning application to connect your knowledge with a health application on presidential speeches.

Experiential Learning Application: U.S. Presidential Speeches

The Centers for Medicare and Medicaid Services (CMS) programs, which has garnered the focus of Presidents in the U.S., has compiled a subset of presidential speeches that related to CMS and Medicare. The speeches include major policy as well as less formal discussions such as press conferences and public comments. As noted by CMS, some presidents spent more time and attention on Medicare. However, all have made continuous contributions to the program, which was developed over time. Additional material sources are available in the National Archives and Records Administration outside of CMS (CMS, 2017i).

Data Set File: 10_EL1_CMS_President_Speeches.xlsx

Variables:

- PRESIDENT
- DATE
- TOPIC
- SUBJECT
- TEXT

Step 1: Sign in to SAS OnDemand for Academics.

Step 2. Open SAS Enterprise Miner (Click the SAS Enterprise Miner link).

Step 3. Create a New Enterprise Miner Project (Click **New Project**).

Step 4: Use the default SAS Server, and click **Next**.

Step 5: Add a project name, and click **Next**.

Figure 10.8: Create New Project Step

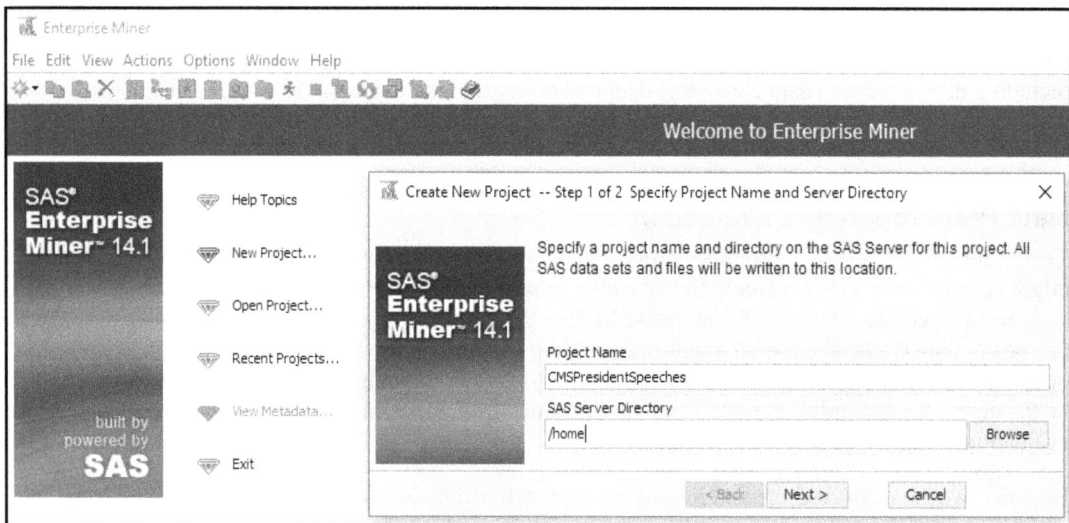

Step 6: SAS will automatically select your user folder directory that is using SAS OnDemand for Academics if using a local installation version, choose your folder directory), and click **Next**.

Step 7: Create a new diagram (Right-click **Diagram**).

Figure 10.9: Create New Diagram

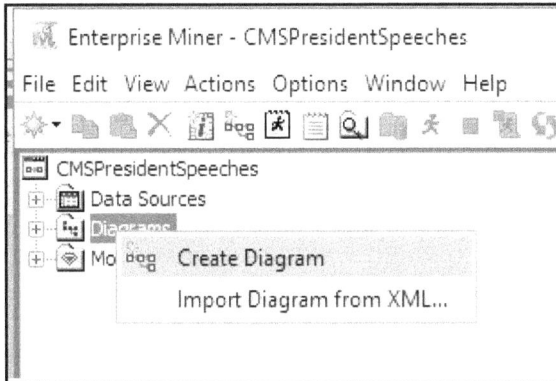

Figure 10.10: Add Diagram Name

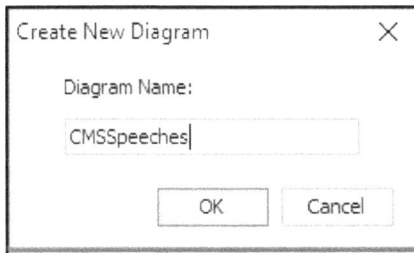

Step 8: Add a File Import node (Click the **Sample** tab, and drag the node onto the diagram workspace).

Figure 10.11: Add a File Import Node

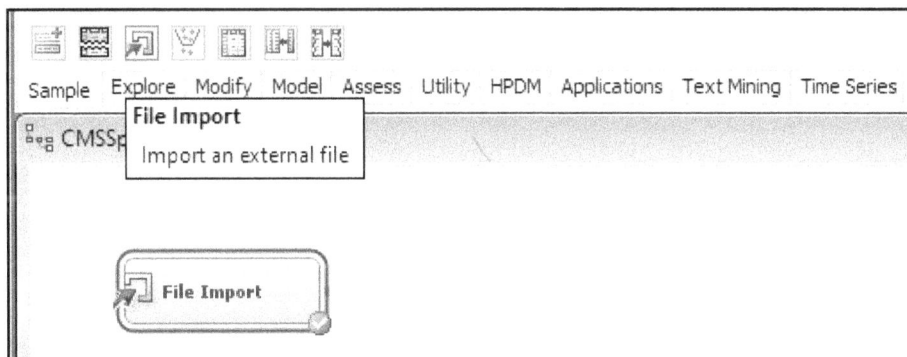

Step 9: Click the File Import node, and review the property panel on the bottom left of the screen.

Step 10: Click **Import File** and navigate to the *10_EL1_CMS_President_Speeches.xlsx* Excel file.

Step 11: Click **Preview** to ensure that the data set was selected successfully, and click **OK**.

Figure 10.12: File Import Preview

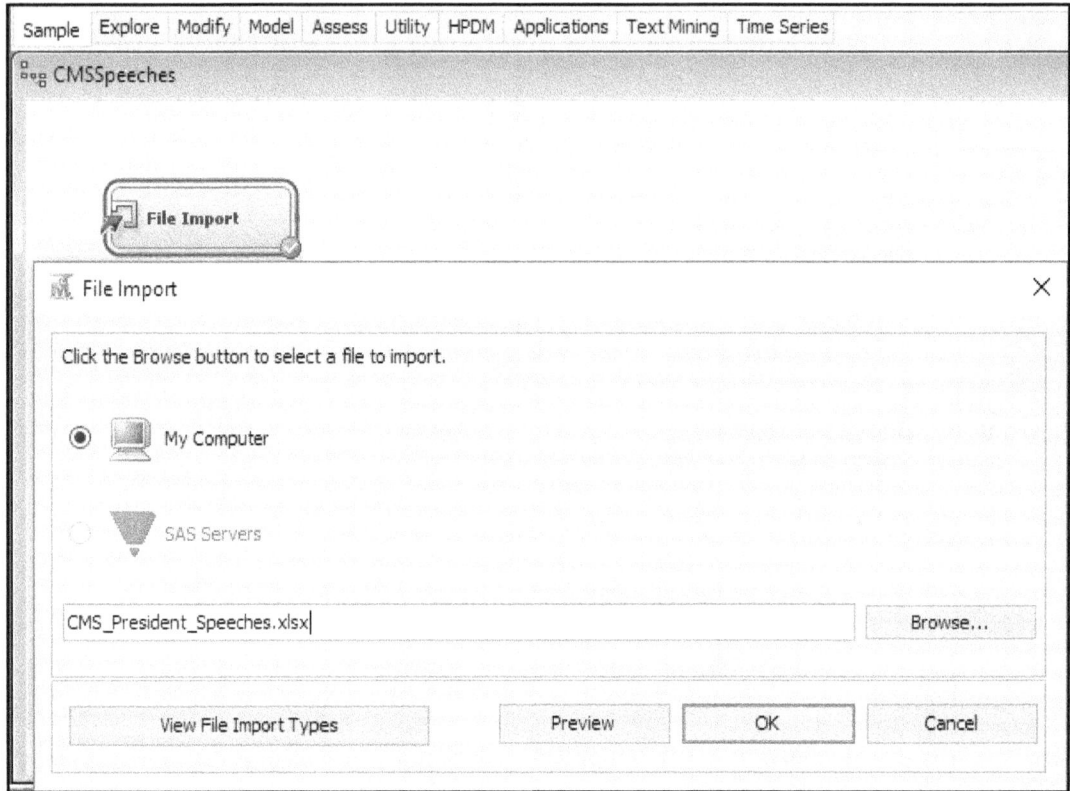

Figure 10.13: Preview File Import Node

PRESIDENT	DATE	TOPIC	SUBJECT	TEXT
Lyndon B. Johnson	09May1964	Medical Aid un...	REMARKS IN ...	There is a third program where ...
Lyndon B. Johnson	26Mar1965	Medical care I...	REMARKS TO ...	Ladies and Gentlemen:I have b...
Lyndon B. Johnson	08Apr1965	Social Security...	Statement by t...	Q: Mr. President, on another su...
Lyndon B. Johnson	09Jul1965	Social Security...	Statement by t...	The 22-year fight to protect the ...
Lyndon B. Johnson	30Jul1965	Social Security...	Remarks at th...	PRESIDENT TRUMAN. Thank y...
Lyndon B. Johnson	08Apr1966	The Medicare ...	Remarks at si...	So we come here now to sign th...
Lyndon B. Johnson	08Apr1966	Launching Me...	Letter to Secret...	Dear Mr. Secretary:I expect short...
Lyndon B. Johnson	30Jun1966	The inaugurati...	Statement by t...	Medicare begins tomorrow.Tom...
Lyndon B. Johnson	19Aug1966	Remarks at th...	Racial discrimi...	Last year your Congressmen a...
Lyndon B. Johnson	01Jul1967	The first anniv...	Statement by t...	The success of the Medicare pr...
Lyndon B. Johnson	29Jun1968	The second a...	Statement by t...	Tomorrow America celebrates t...
Lyndon B. Johnson	14Aug1968	Medicare	Remarks befor...	But my friends, the greatest bre...
Richard M. Nixon	23Mar1972	The President' ...	Special messa...	Growing old often means both d...
Richard M. Nixon	30Oct1972	Medicare	Radio address...	In addition, H.R. 1 will pay a spe...
Richard M. Nixon	06Feb1974	Comprehensiv...	Special messa...	The Medicare program now pro...
Gerald Ford	19Jan1976	Catastrophic ...	Address befor...	Hospital and medical services i...
Gerald Ford	21Jan1976	Catastrophic ...	Remarks at a ...	Q: Mr. President, last night you p...
Gerald Ford	21Jan1976	Catastrophic ...	Remarks on gr...	The second point I addressed, I...
Gerald Ford	09Feb1976	Catastrophic ...	Special messa...	I believe that the prompt enactm...
Gerald Ford	22Jul1976	Medicare Impr...	Special messa...	The proposed "Medicare Improv...
Jimmy Carter	05Mar1977	"Ask President...	Remarks durin...	MEDICARE: HEALTH CARE CO...

Step 12: Right-click **File Import** and click **Edit Variables**.

Figure 10.14: Edit Variables

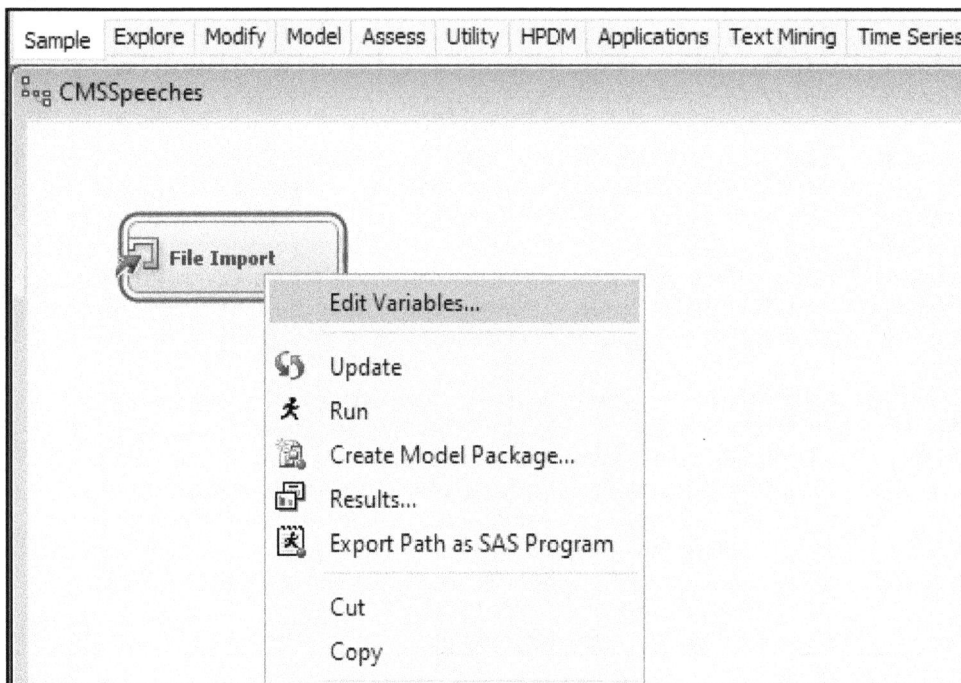

Step 13: Set DATE to the Time ID variable role. Set PRESIDENT to the Target role. Set SUBJECT and TOPIC to the Input role. Set TEXT to the Text role, and click **OK**.

Figure 10.15: Edit Variable Roles and Levels

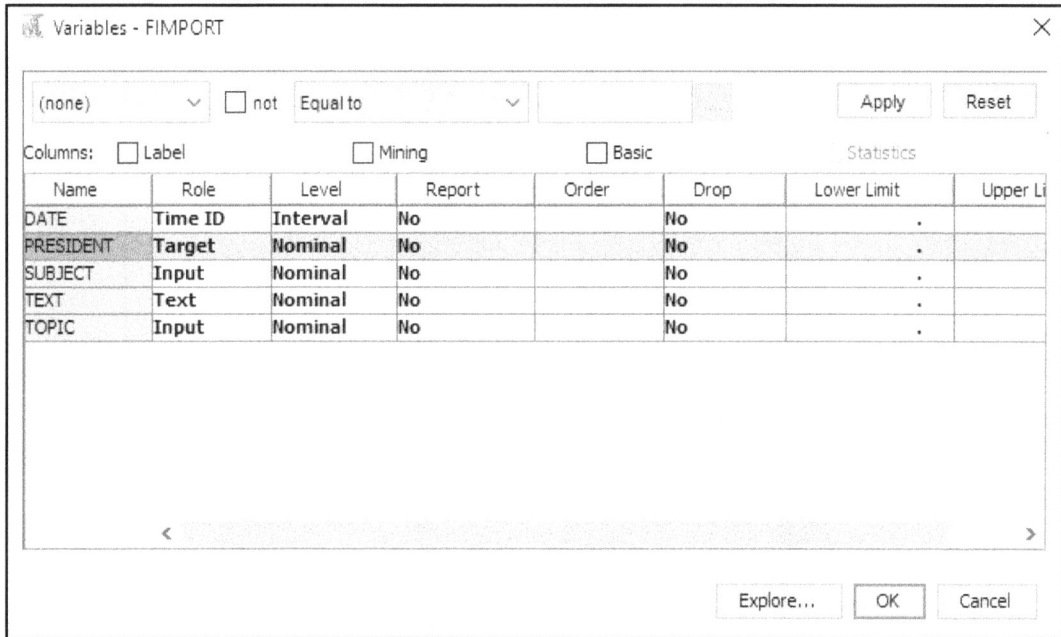

Step 14: Add the Text Parsing node from the **Text Mining** tab. The Text Parsing node splits the unstructured text data into individual parts to allow analysis.

Figure 10.16: Add Text Parsing Node

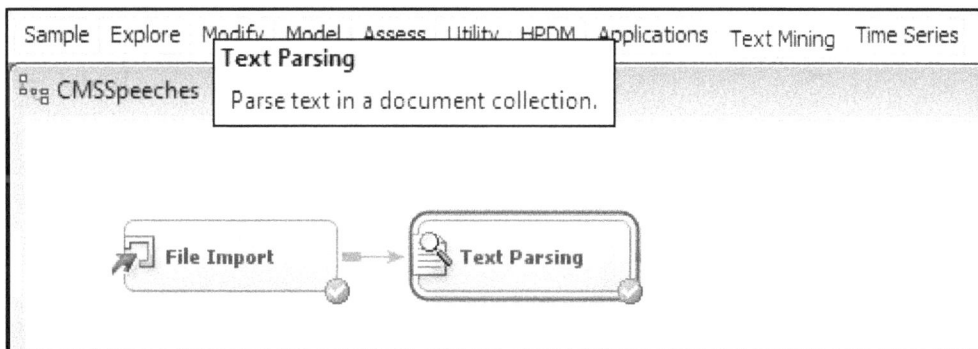

Step 15: Right-click Text Parsing node, and click **Run**.

Figure 10.17: Run Text Parsing Node

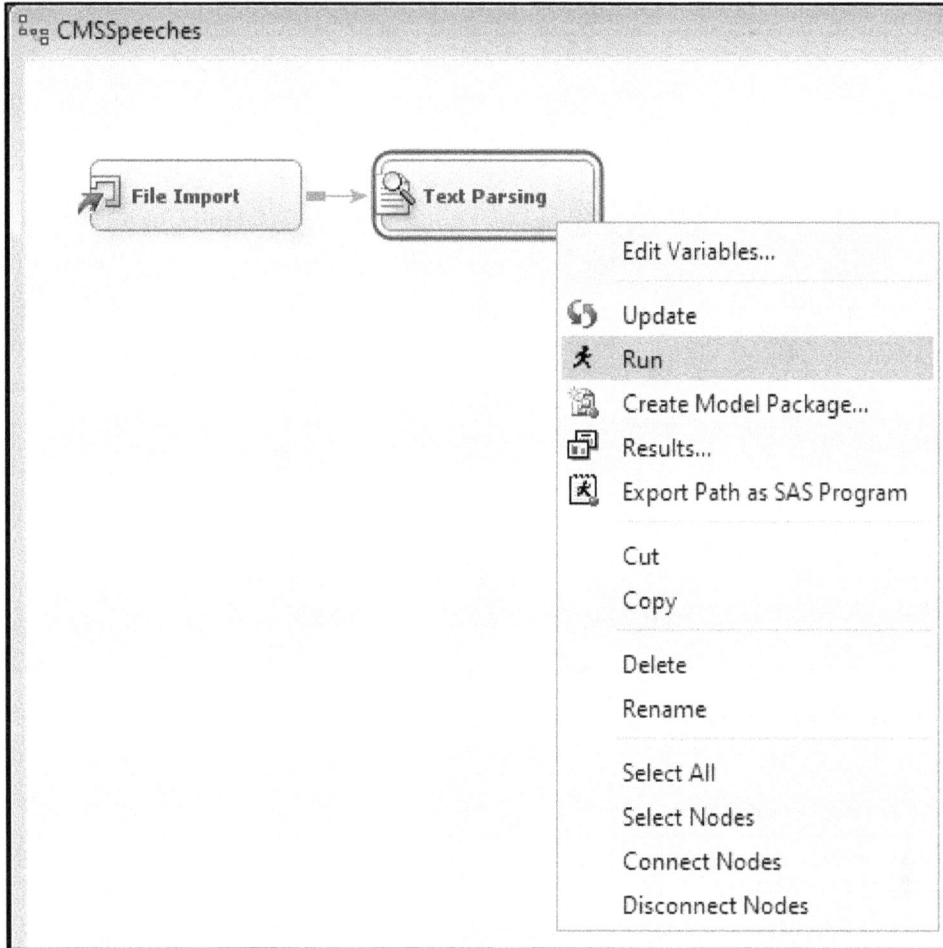

Step 16: Click **Results**. Review the Terms Window within the Results.

Figure 10.18: Sample Results

Term	Role	Attribute	Freq	# Docs	Keep	Parent/Child Status	Parent ID	Rank for Variable numdocs
+ be	... Verb	Alpha	2738	85N	+		13217	1
+ have	... Verb	Alpha	1224	85N	+		13121	1
care	... Noun	Alpha	813	83Y			5181	3
health	... Noun	Alpha	989	79Y			7194	4
medicare	... Noun	Alpha	1184	79Y			9188	4
now	...Adv	Alpha	352	76N			13313	6
s	... Noun	Alpha	842	76N			13069	6
+ year	... Noun	Alpha	456	75Y	+		6437	8
not	...Adv	Alpha	944	73N			13141	9
+ do	... Verb	Alpha	879	70N	+		13267	10
+ make	...Verb	Alpha	523	66N	+		13176	11
people	Noun	Alpha	544	65Y			3325	12

Step 17: Add Text Filter node from **Text Mining** tab. The Text Filter node reduces the data set by removing unwanted or unimportant terms, which are sometimes called stop terms. These are words that have little value in the final analysis. Examples are words such as 'a' or 'the' and can be removed without affecting the final results.

Figure 10.19: Add Text Filter Node

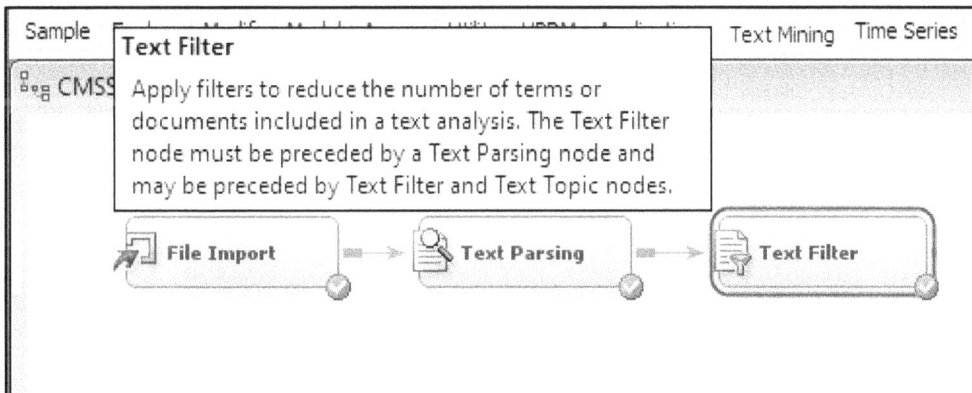

Step 18: Right-click Text Filter node, and click **Run**.

Step 19: Click **Results**.

Step 20: View results.

Figure 10.20: View Sample Results

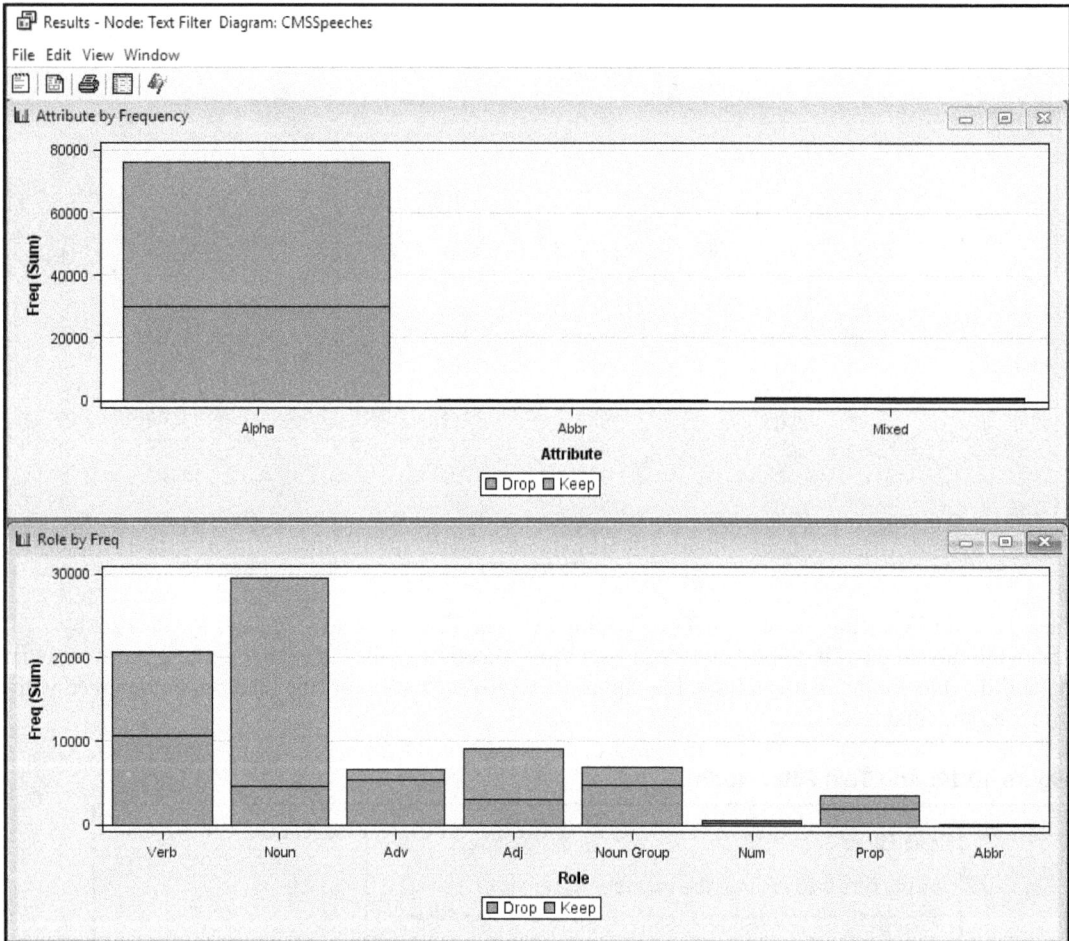

Step 21: Click Text Filter node, and in the **Properties** panel, click **Filter Viewer**.

Figure 10.21: Text Filter Viewer

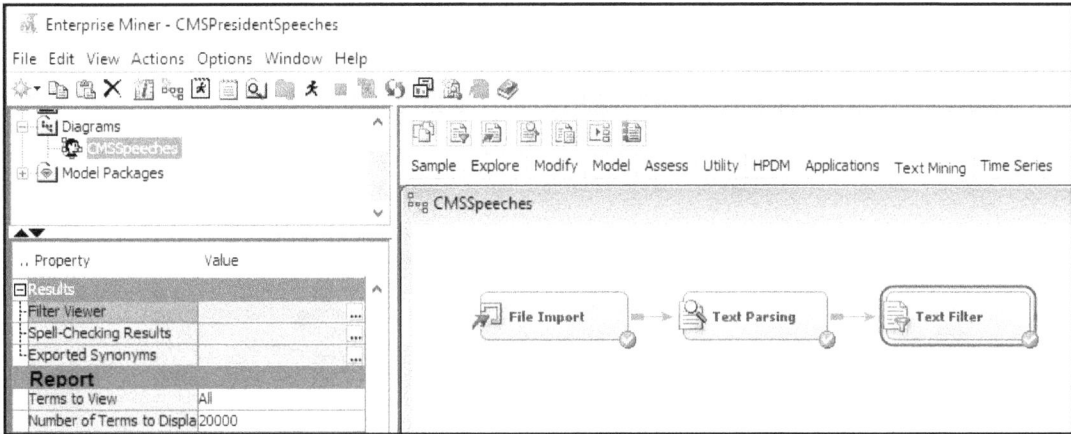

Step 22: Next, we want to further explore certain terms within the documents. For this example, we are interested in reviewing Medicare policy-related terms. In the Filter viewer, click the term, **medicare**, right-click, and click **View Concepts Lin**ks.

Figure 10.22: View Concept Links

Step 23: Review Results. Concept Linking is a user-friendly text method, which shows relationships between terms, with the thicker line indicating greater association between terms. Concept linking provides a method to display the terms that are related with the selected term from the Terms table. The terms that surround the selected term, medicare, indicate the greatest correlation.

Figure 10.23: Concept Linking

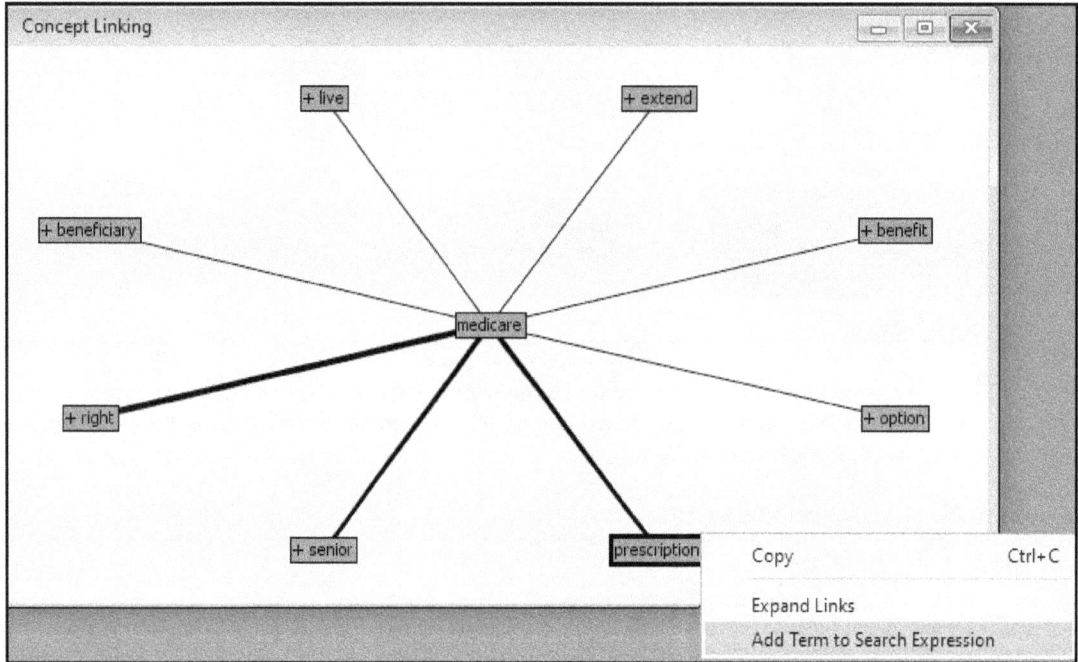

Step 24: From the concept links, we want to review the term, **prescription**, further. Right-click **prescription** and click **Add Term to Search Expression**, which will add **prescription** to the search box, and click **Apply**. Review the text snippets and dates for each of the lines that include **prescriptions**. Interestingly, while the cost of prescription drugs made headlines in 2016–2017 during the U.S. presidential elections, the cost of prescription drugs has been a topic of concern for many decades and presidential cycles.

Figure 10.24: Prescription Filter

Interactive Filter Viewer						
File Edit View Window						
Search : prescription			Apply	Clear		

TEXT	TEXTFILTER_SNIPPET	TEXT...	DATE	PRESIDE...	SUBJECT	TOPIC
I am excited to see so many	... 's got a **prescription** for health	0.034	1992-05-13	George H....	Remarks t...	Health car...
Mr. Speaker, Mr. President,	... the cost of **prescription** drugs.	0.034	1993-09-22	Bill Clinton	Address to...	Health car...
Dear Gentlemen:	... offers all Americans **prescription**	0.069	1993-10-27	Bill Clinton	Remarks o...	The Health...
Thank you, Ruth. I think she	... to overpayments for **prescription**	0.034	1998-01-06	Bill Clinton	Remarks a...	Medicare
Thank you. I would like to	... but also a **prescription** drug	0.034	1998-10-08	Bill Clinton	Remarks o...	HMO's and...
Thank you. I would like to	... lower rates for **prescription**	0.034	1998-12-07	Bill Clinton	Remarks o...	Medicare F...
Thank you very much.	... helping seniors with **prescription**	0.241	1999-03-30	Bill Clinton	Remarks o...	Social Sec...
Thank you very much, and	... our seniors afford **prescription**	0.345	1999-06-29	Bill Clinton	Remarks	Medicare ...
Dear Mr. Chairman: (Dear	... long- overdue **prescription** drug	0.034	1999-10-19	Bill Clinton	Letter to C...	Medicare r...
THE PRESIDENT: Thank you	... to better afford **prescription**	0.276	2001-07-12	George W....	Press brief...	Medicare I...
THE PRESIDENT: Thank you	... which must include **prescription**	0.138	2001-07-13	George W....	Address at...	Medicare ...
THE PRESIDENT: I want to	... plan that includes **prescription**	0.103	2002-01-28	George W....	Remarks t...	Medicare ...
THE PRESIDENT: Well,	... Medicare must provide	0.241	2002-05-17	George W....	Remarks t...	Medicare P...
THE PRESIDENT: Good	... Many seniors need **prescription**	0.172	2002-05-18	George W....	Radio addr...	Medicare P...
THE PRESIDENT: Thank you	... guarantee of a **prescription** drug	0.276	2002-07-11	George W....	Speech in ...	Medicare P...
Our second goal is high	... plan that provides **prescription**	0.034	2003-01-28	George W....	State of th...	Medicare ...
THE PRESIDENT: Thanks for	... help in buying **prescription** drugs.	0.31	2003-03-04	George W....	Addressin...	Medicare ...
In the coming weeks and	... long–awaited **prescription** drug	0.138	2003-06-06	George W....	Press release	Medicare ...
THE PRESIDENT: Good	... Medicare to offer **prescription**	0.276	2003-06-07	George W....	Radio addr...	Medicare ...
PRESIDENT BUSH: Thanks	... better benefits like **prescription**	0.345	2003-06-11	George W....	Speech to ...	Medicare ...
I commend the Senate for	... long– awaited **prescription** drug	0.069	2003-06-27	George W....	Praises Co...	Medicare ...

Step 25: Add a Text Cluster node from the **Text Mining** Tab. The node will group the text into similar segments or clusters.

Figure 10.25: Add Cluster Node

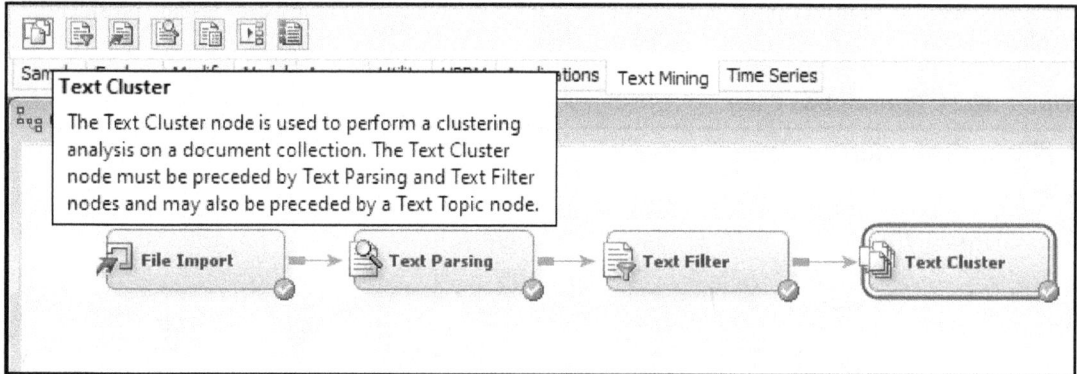

Step 26: View results. In the Distance Between Clusters window, note how Cluster IDs 2 and 4 and Cluster IDs 1 and 3 appear to be closer in distance. The clusters with near distance might be able to be combined to form better defined clusters. Cluster analysis is often iterative and exploratory. Let's run again with three instead of five clusters.

Figure 10.26: View Sample Results

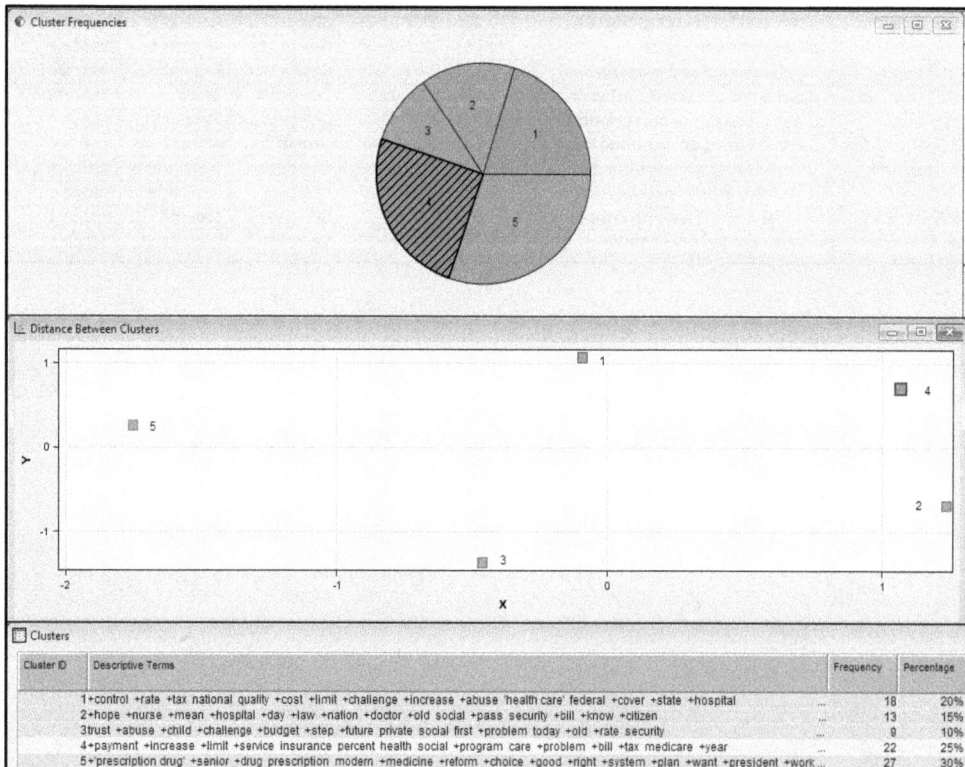

Cluster ID	Descriptive Terms		Frequency	Percentage
1	+control +rate +tax national quality +cost +limit +challenge +increase +abuse 'health care' federal +cover +state +hospital	...	18	20%
2	+hope +nurse +mean +hospital +day +law +nation +doctor +old social +pass security +bill +know +citizen	...	13	15%
3	trust +abuse +child +challenge +budget step +future private social first +problem today +old +rate security	...	9	10%
4	+payment +increase +limit +service insurance percent health social +program care +problem +bill +tax medicare +year	...	22	25%
5	+'prescription drug' +senior +drug prescription modern +medicine +reform +choice +good +right +system +plan +want +president +work	...	27	30%

Step 27: To reduce the number of clusters, click the Text Cluster node, and click the Property Panel. Under the Cluster property, set the **Exact** value for the Exact or Maximum Number property and **3** for the Number of Clusters.

Figure 10.27: Text Cluster Properties

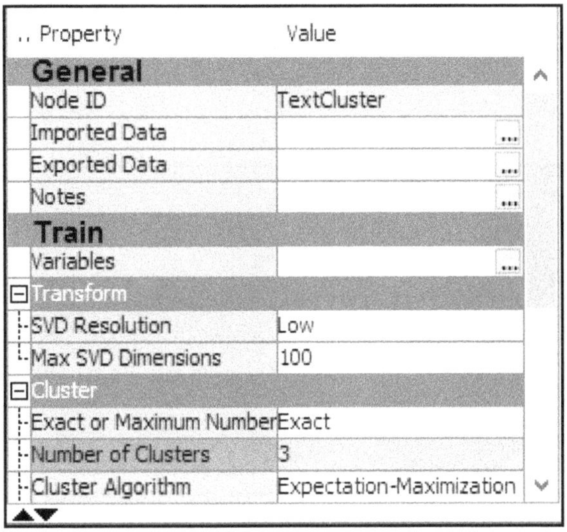

Property	Value
General	
Node ID	TextCluster
Imported Data	...
Exported Data	...
Notes	...
Train	
Variables	...
Transform	
SVD Resolution	Low
Max SVD Dimensions	100
Cluster	
Exact or Maximum Number	Exact
Number of Clusters	3
Cluster Algorithm	Expectation-Maximization

Step 28: View Results. Note how clusters have good distance from one another and how they represent different descriptive terms. For our next step, we attempt to label each of the clusters. As one technique, review the different descriptive terms within each cluster, and attempt to generate a word or phrase that best represents those terms:

- Cluster 1 Description: Social Security
- Cluster 2 Description: Prescriptions
- Cluster 3 Description: Medicare

Figure 10.28: Cluster Results

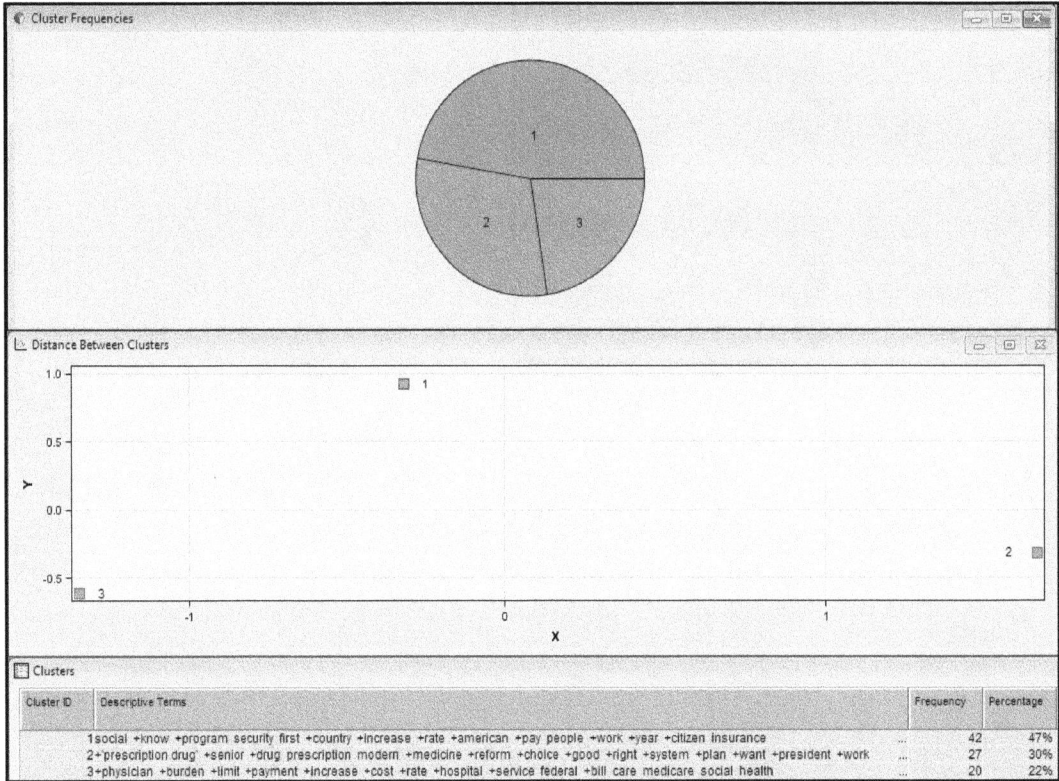

Step 29: To review the individual records and the assigned clusters, click the Text Cluster node. Click the Property panel and click **Exported Data**. Click the **Train** data and click Explore.

Figure 10.29: Cluster Export

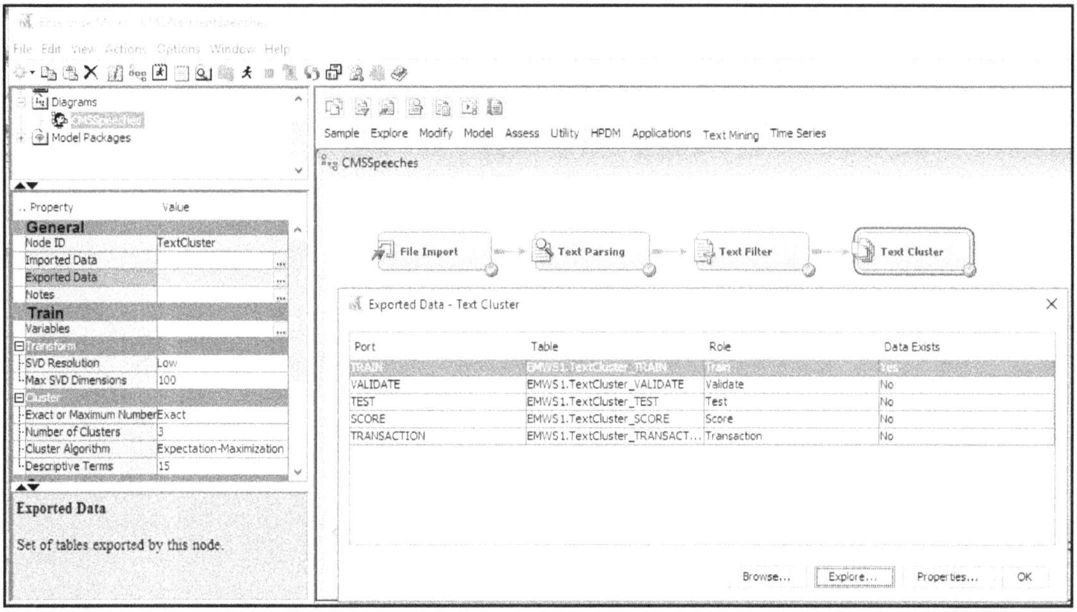

Step 30: The exported data shows clusters of text that is assigned by cluster and president. From earlier, our cluster descriptions were 1 - Social Security, 2 - Prescriptions, and 3 - Medicare. President Johnson, for example, has entries in clusters 1 and 3, which might indicate that President Johnson was focused on Social Security and Medicare, which were foundational issues in his presidency and was responsible for expanding Social Security and Medicare programs during his term.

Figure 10.30: Cluster Export Results

Obs #	TEXT	TextCl ▲	PRESIDENT	DATE	TOPIC	
1	There is a third program whe...	1	Lyndon B. Johnson	09May1964	Medical Aid under Social Security	...
2	Ladies and Gentlemen:I have...	1	Lyndon B. Johnson	26Mar1965	Medical care legislation	...
3	Q: Mr. President, on another ...	1	Lyndon B. Johnson	08Apr1965	Social Security Amendments of 1965	...
4	The 22-year fight to protect th...	1	Lyndon B. Johnson	09Jul1965	Social Security Amendments of 1965	...
5	PRESIDENT TRUMAN. Than...	1	Lyndon B. Johnson	30Jul1965	Social Security Amendments of 1965	...
6	So we come here now to sig...	1	Lyndon B. Johnson	08Apr1966	The Medicare Extension Bill	...
8	Medicare begins tomorrow.T...	1	Lyndon B. Johnson	30Jun1966	The inauguration of the Medicare prog...	
9	Last year your Congressmen...	1	Lyndon B. Johnson	19Aug1966	Remarks at the dedication of the Ellen...	
11	Tomorrow America celebrate...	1	Lyndon B. Johnson	29Jun1968	The second anniversary of the Medica...	
12	But my friends, the greatest b...	1	Lyndon B. Johnson	14Aug1968	Medicare	...
14	In addition, H.R. 1 will pay a s...	1	Richard M. Nixon	30Oct1972	Medicare	...
15	The Medicare program now p...	1	Richard M. Nixon	06Feb1974	ComprehensiveHealth InsurancePlan...	
22	We seem to have some happ...	1	Jimmy Carter	25Oct1977	Medicare-Medicaid Anti-Fraud and Ab...	
27	Q: Mr. President, I'm Marcella...	1	Jimmy Carter	12Sep1979	National Retired Teachers Associatio...	
29	Today I have signed H.R. 323...	1	Jimmy Carter	09Jun1980	Social Security Disability Amendment...	
32	None of the great achieveme...	1	Jimmy Carter	31Oct1980	Medicare and Social Security	...
33	MEDICAREQ: Mr. President, ...	1	Ronald Reagan	24May1982	Medicare	...
37	Second is Medicare. All our a...	1	Ronald Reagan	31Aug1984	Medicare	...
38	MEDICAREQ: What do you pr...	1	Ronald Reagan	07Nov1984	Medicare	...
39	I am excited to see so many ...	1	George H.W. Bush	13May1992	Health care reform	
40	Please be seated, and thank ...	1	George H.W. Bush	02Jun1992	Health care reform	

Step 31: As a second part to our activity, go back to review the main diagram. Instead of using the Text Cluster to separate the president speeches, we'll create a model that connects our knowledge from previous chapters with the text capabilities of SAS Enterprise Miner. Below the Text Cluster and connected to the Text Filter node, add the Regression node, Decision Tree node, and Neural Network Node.

Step 32: Set the Decision Tree properties to have a Minimum Categorical Size of **2**, Leaf Size of **1**, and Subtree Method to **Largest**.

Figure 10.31: Add Model Nodes

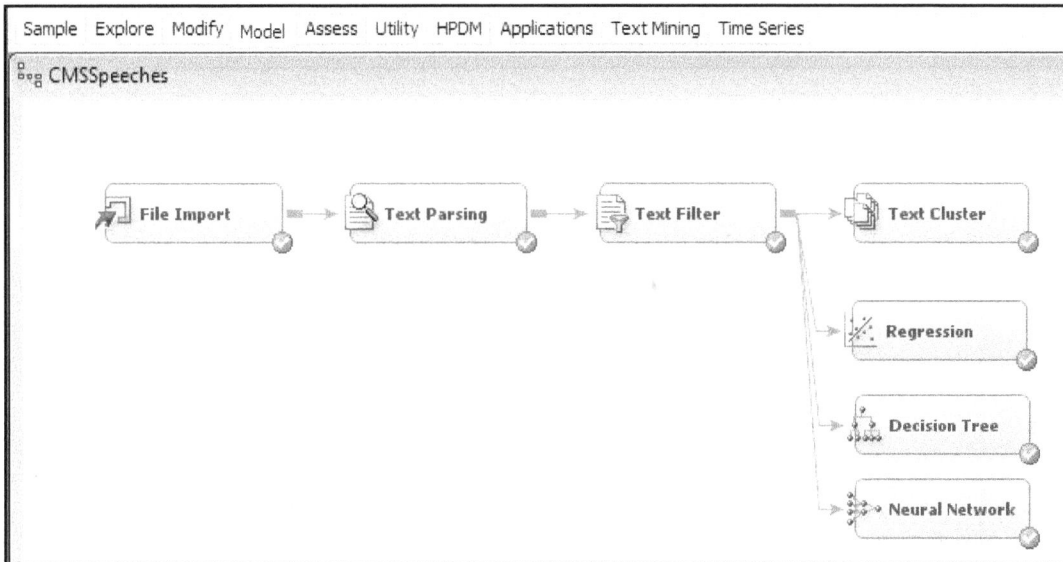

Step 33: Add the Model Comparison Node.

Figure 10.32: Add Model Comparison Node

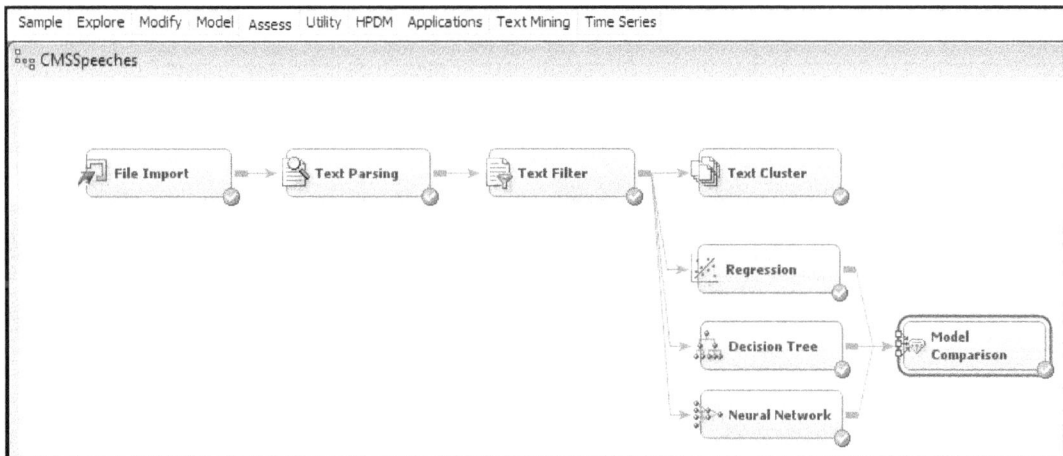

Step 34: Review Results. From the Cumulative Lift and Fit Statistics for the Training data set, the Regression and Neural Network models outperform the Decision Tree model. To further expand this example, we could also include a Data Partition node to verify a validation data set for our model. Also, we could include a score data set for unclassified president speeches in order to classify them into a category.

Figure 10.33: Model Comparison Node Results

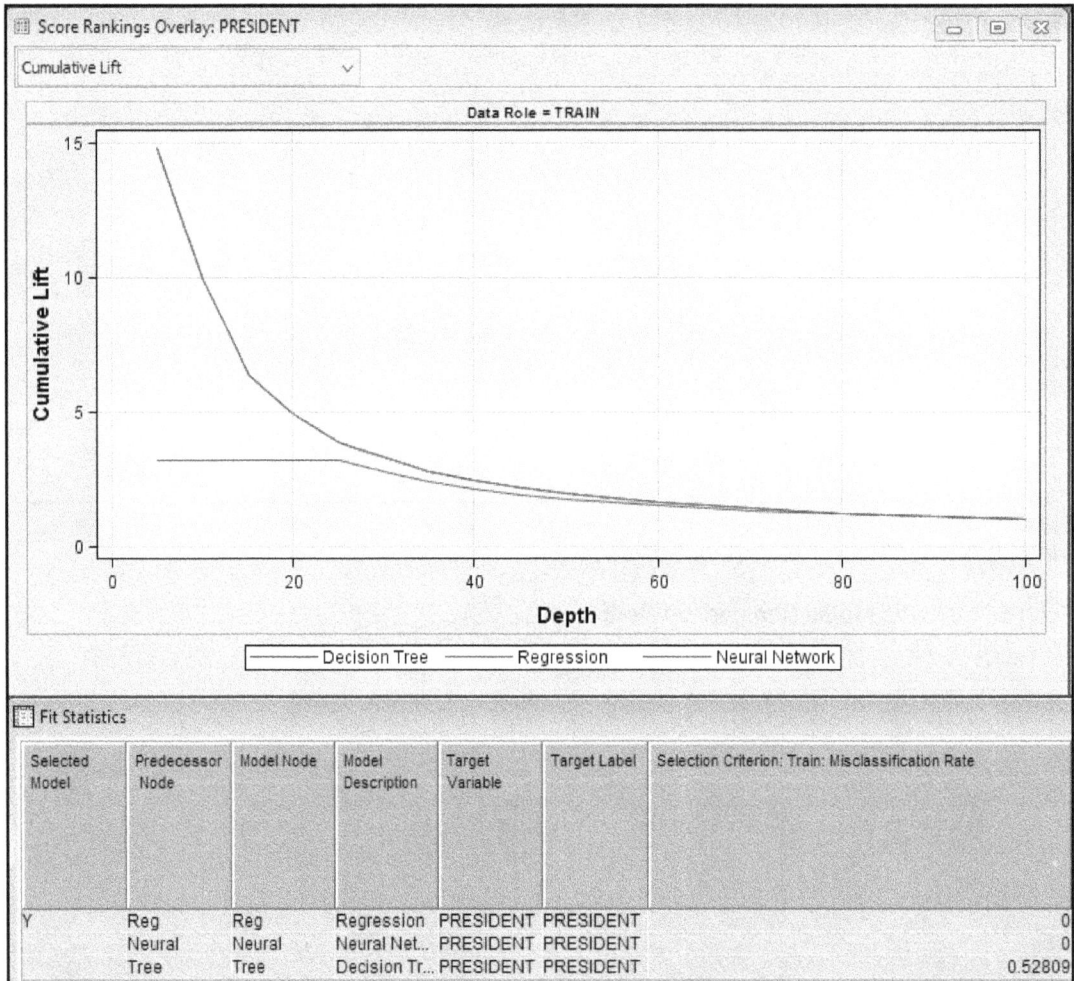

Step 35: Suppose that we want improve the results of the Decision Tree model. One new variable input that we could include are the results from the Text Cluster output. To use this additional cluster variable, add a second Decision Tree node and connect to the Text Cluster node. Connect the Decision Tree node to the Model Comparison node.

Figure 10.34: Text Cluster Input

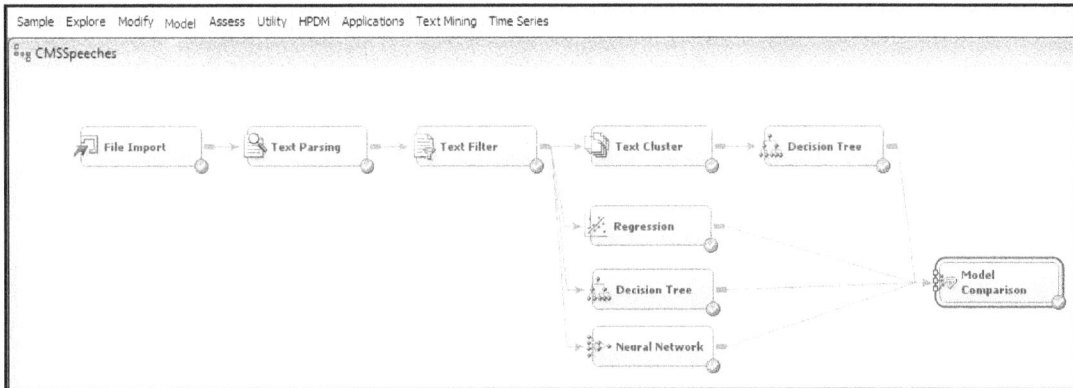

Step 36: Re-run the Model Comparison node. Note how Tree 2 now has an improved performance over the Tree using the Text Cluster input variable.

Figure 10.35: Text Cluster Model Comparison

Selected Model ▼	Predecessor Node	Model Node	Model Description	Target Variable	Target Label	Selection Criterion: Train: Misclassification Rate
Y	Reg	Reg	Regression	PRESIDENT	PRESIDENT	0
	Neural	Neural	Neural Net...	PRESIDENT	PRESIDENT	0
	Tree2	Tree2	Decision Tr...	PRESIDENT	PRESIDENT	0.258427
	Tree	Tree	Decision Tr...	PRESIDENT	PRESIDENT	0.52809

Model Summary

In summary, the text mining process consists of three steps: 1) collect the documents, 2) parse and transform the text, and 3) analyze patterns. The text mining output can be combined with the modeling and assess nodes in the SEMMA process. Input: Interval, Binary, or Nominal.

- Variable: Text
- Evaluation: Text Cluster node, Regression node, Decision Tree node, and Neural Network node

- Evaluation: Text Cluster descriptive terms, Assess Model Comparison node, or individual nodes evaluation
- Scoring: Evaluate a new data set using selected model

Experiential Learning Application: Healthcare Legislation Tweets

The opportunity to use real-time data from Twitter and other social media sites can provide early public health warning and alerts to hospitals and patients on infectious diseases. Nearly 90% of 18–29 year olds use social media sites such as Twitter, and usage has grown across all age groups. Although critics are quick to dismiss Twitter data as frivolous, scientists are finding that Twitter data is valuable, and, when combined with other real-time sources (such as environmental sensors or data from fitness apps), can provide alerts on chronic disease such as asthma and adverse drug reactions. Researchers have achieved a 50–60% accuracy rate using Twitter data alone and up to 75% when combining Twitter data with environmental sensors for determining asthma conditions. The average individual spends nearly three hours each day on their mobile device, generating real-time data through social media and similar sites, and in many cases are more willing to share health-related symptoms or behaviors on Twitter than they are with a medical professional. Social media users can help alert public health authorities on emerging concerns. As one emerging example, with ongoing issues of prescription opioid abuse, researchers found that nearly 70% of the opioid tweets referred to forms of misuse and unsafe behaviors (Kuehn, 2017).

Health-related text on social media sites such as Twitter has provided an opportunity to study a variety of public health areas such as chronic diseases, disease outbreaks, sleeping patterns, and quality of care. Advantages include real-time data, the open nature of the platforms, global data, and ease of data accessibility. By contrast, typical research studies with surveys or patient studies might take months or years to collect data on only a small subset of the population. In describing the open nature of social media platforms reviewed tweets on stomach problems and diarrhea, one study could trace these behaviors to food poisoning. Another study analyzed tweets on common medications and tweets describing side effects of those medications. The real-time location data can also be useful for allocating resources based on health needs (Twitter, 2015).

For this experiential learning application, you'll review a sample set of Tweets made on the AHCA, BCRA, and ACA legislation. The Tweets are available under the hashtags #ACA, #AHCA, and #BCRA.

Objective: Review and explore unstructured data patterns in legislative Tweets. Create a model to determine for which of the legislative items the Tweets were made (ACA, AHCA, or BCRA).

Data Set File: 10_EL2_Healthcare_Legislation_Tweets.xlsx

Variables:

- ID, unique identifier
- text, tweet
- favoriteCount, count from tweets favorite button
- sid, internal identifier
- statusSource, tweet source
- Topic, legislation

Follow the combined SEMMA and text mining process for your experiential learning application and provide recommendations. Note that the Explore and Modify steps have been combined because text mining is often an exploratory process where textual data is processed, modified, and explored together as a single process step. A template has been provided below that can be reused across future projects.

Title	Healthcare Legislation Tweets
Introduction	Provide a summary of the business problem or opportunity and the key objective(s) or goal(s). Create a new SAS Enterprise Miner project. Create a new Diagram.
Sample & Collect the Documents	Data (sources for exploration and model insights) Identify the variables data types, the input and target variable during exploration. Add a FILE IMPORT Provide a results overview following file import: Input / Target Variables Generate a DATA PARTITION
Exploration, Modify & Parse and transform the text	Text mining overview and description, steps Provide a results overview following data exploration and modification Add a TEXT PARSING Add a TEXT FILTER Add a TEXT TOPIC Add a TEXT CLUSTER
Model & Analyze Patters	Discovery (prototype and test analytical models) Apply a regression, decision tree, and neural network model and provide a results overview following modeling. Add a REGRESSION Add a DECISION TREE Add a NEURAL NETWORK Model description Analytics steps Model results (Lift, Error, Misclassification Rate) Selection Model

Title	Healthcare Legislation Tweets
<u>A</u>ssess and Reflection	ADD a MODEL COMPARSION Provide overall recommendations to business Model advantages / disadvantages Performance evaluation Model recommendation Summary analytics recommendations Summary informatics recommendations Summary business recommendations Summary clinical recommendations Deployment (operationalization plan: timeline, resources, scope, phases, project plan) Value (return on investment, healthcare outcomes)

Learning Journal Reflection

Review, reflect, and retrieve the following key chapter topics only from memory and add them to your learning journal. For each topic, list a one sentence description/definition. Connect these ideas to something you might already know from your experience, other coursework, or a current event. This follows our three-phase learning approach of 1) Capture, 2) Communicate, and 3) Connect. After completing, verify your results against your learning journal and update as needed..

Key Ideas – Capture	Key Terms – Communicate	Key Areas - Connect
Unstructured Data		
Social Media Policy		
Text Mining		
Text Mining Process		
Corpus		
Text Parsing Node		
Text Filter Node		
Text Filter Viewer		
View Concept Links		
Text Cluster Node		

Chapter 11: Identifying Future Health Trends and High-Performance Data Mining

Chapter Summary

The purpose of this chapter is to identify future trends in health anamatics and to develop high-performance data mining skills using SAS Enterprise Miner. The high-performance data mining (HPDM) focus of this chapter is detailed in Figure 11.1. This chapter builds on the models from previous chapters. This chapter also includes experiential learning application exercises on SIDS and lifelogs. This chapter is intended to prepare you for a future work environment and to continue developing the learning skills and knowledge for a successful long-term career. Individuals that obtain knowledge and viewpoints on emerging trends and technologies can help their organizations anticipate, prepare for, and navigate changes in the future. Organizations that understand and embrace the evolution of health anamatics can guard against obsolescence. Future trends can significantly impact all areas of health anamatics and stakeholders of healthcare including patients, providers, government, and administration.

There are five major trends identified and discussed within this chapter:

- Population and Consumer Changes
- AI and Robotics Automation

- Healthcare Globalization and Government
- Public Health
- Big Data Health Anamatics

Figure 11.1: Chapter Focus - HPDM

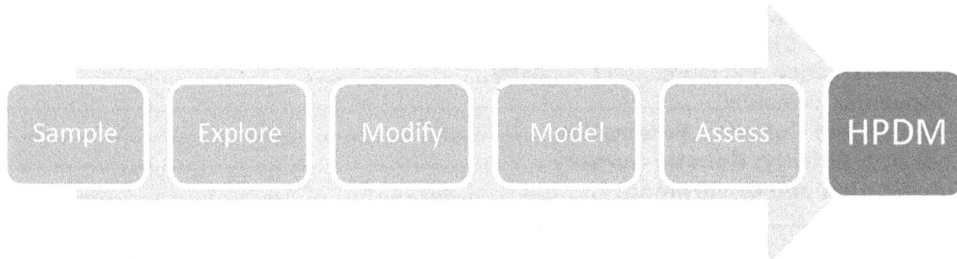

Chapter Learning Goals

- Describe future trends and their impact on health anamatics
- Describe high-performance data mining
- Understand Big Data sources
- Develop Big Data modeling skills
- Apply SAS Enterprise Miner HPDM data functions
- Master HPDM

Population and Consumer Changes

Growing World Population and Increased Longevity

The world's population doubled in the 40 years between 1959 and 1999 from 3 billion to 6 billion. The U.S. Census Bureau has projected that the world's population will continue to grow to 9 billion by 2044, which is about a 50% increase over 45 years, and grow to 9.3 billion by 2050. This growth implies that even though the population is growing, the growth rate has slowed from the previous century (US Census Bureau, 2016a; US Census Bureau, 2016b). Population growth and population centers are also anticipated to change through 2050. Some continents such as Europe are projected to shrink in overall world population. At the same time, the greatest population growth is projected in Asia and Africa, which might require redistribution of health resources (Roser and Ortiz-Ospina, 2017).

Figure 11.2: World Population

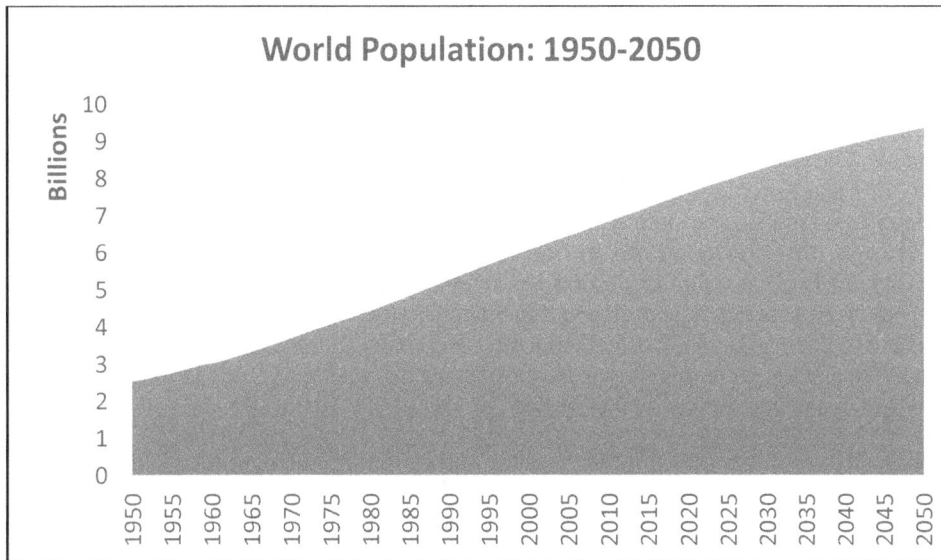

Each generation now, on average, lives a few years longer than the previous generation, and because of continuing improvements in health, life expectancy is expected to increase through 2030. South Korea is projected to be the leader in life expectancy by 2030, with a life expectancy for women of 90.8 years. For men, life expectancy rates are impacted by higher rates of smoking and alcohol consumption than for women even though the difference between men and women is projected to be close. For men, the top country for life expectancy in 2030 is expected to be Switzerland at 83.95 years of age. In reviewing high-income countries, the U.S. is expected to have the lowest life expectancy at 83.3 years of age for women, and 79.5 years of age for men. The mortality rate for high-income countries continues to be high between the ages of 40-60. Factors affecting both high-income and emerging country life expectancies include overall diet, obesity, homicides, accidents, and limited universal health coverage (Senthilingam, 2017a).

A life table is a method to display the probability of a population that lives to or dies at a set age. The Social Security Administration (SSA) develops life tables to review mortality rates in the population over time. Based on the SSA life tables, a newborn has a probability of death within the next year less than 1% and a remaining life expectancy of 76.28 for males and 81.05 for females. By contrast, at the social security full retirement age of 67, there is a 2% probability of death within the next year and a remaining life expectancy of 16.32 years for males and 18.73 years for females, reflecting a declining life expectancy (SSA, 2013). There are some concerns over usage of actuarial tables for healthcare decision-making. A major concern is that healthcare is treated as a simple product and that the cost of a product is being weighed against the cost of a life. Cases where an expensive treatment might prolong end of life might not be deemed "cost effective."

Increasing aging and life expectancy also has an effect on the cost of healthcare. Researchers have found that the mortality rate for patients in the U.S. with Alzheimer's disease has increased to 55% in the studied period between 1999 and 2014, with no known cure for the disease. Alzheimer's disease primarily occurs in those over 65 years of age and is the sixth leading cause of death in the U.S., with more than 5 million with the disease. The number is projected to increase to 16 million in the U.S. by 2050. As individuals live longer, their chances for Alzheimer's disease increases. Researchers suggest that the increase might be due to a number of factors including the increasing number of those over 65 years of age, improved earlier diagnosis, improved reporting by physicians, and a decrease in deaths caused by other diseases such as heart disease (Senthilingam, 2017b).

Further, end-of-life healthcare accounts for a disproportionate share of healthcare costs. In one study, inpatient costs were measured over the last two years of life for Medicare patients with heart failure, lung disease, cancer, dementia, vascular disease, kidney disease, and liver disease over a four-year period at 3,000 U.S. hospitals. The findings showed a wide range of costs between Mayo Clinic with an average cost per patient of $53,432 and New York University with an average cost per patient of $105,000. Since the patient fatal outcome was similar, the extra costs and care were attributed to additional hospital and intensive care unit days, physician visits, and specialist consultations. Although the costs were not adjusted for disease intensity, the larger remaining question was whether any cost savings caused an extension or improvement of any lives. Due to medical advances in treatment of conditions, it can be difficult to determine when end of life begins, and whether, ethically, everything should continue to be done to prolong life, including additional costs spent to improve end-of-life care and to deliver compassionate care even when a terminal patient outcome is already determined (Neubert, 2009).

The evidence of the high cost of end-of-life care has been shown from studies dating to 1960s (Scitovsky, 2005). Approximately 25% of Medicare spending occurs on end-of-life care for only 5% of the patients, with a total cost of $125 billion. This has an impact on the overall program costs for Medicare as well as out-of-pocket expenses for individuals. Medicare patients in the last five years of life averaged $51,000 for couples, exceeding financial assets 40% of the time. Hospice care focuses on end-of-life services, often in patients' homes. The focus is on comfort of care and less aggressive interventions, which can be delivered at a reduced cost over other options such as hospital stays. Researchers found that hospice care reduces Medicare spending in the last year of life up to $7,000 per patient (Wang, 2012).

Changing Consumer Demands

Another growing trend is patients who follow a consumer-driven approach to healthcare, in reviewing healthcare options and basing decisions on cost and quality. Patients expect greater customer service, delivery on requirements, and a value offering. Organizations (such as referralMD, Healthgrades, and Vitals) provide patients with the tools to measure cost and quality (Govette, 2017).

Consumer preferences on natural foods, viewpoints on personal health and wellness, and company commitment to sustainability are also changing healthcare. Panera Bread banned nearly 100 ingredients from their menu items, calling the new menu "clean," which means that it contained no artificial flavors, preservatives, sweeteners, or colors from nonnaturally occurring sources. The new menu required considerable changes through redesigning recipes, training employees, and educating customers, with the total initiative dating back nearly ten years (Jargon, 2017).

As another example of changing preferences, consumers in the U.S. now drink more bottled water than soda at 39.3 gallons per person versus 38.5 gallons per person, respectively. However, soda has a higher purchase cost, and therefore total sales revenues are nearly double that of bottled water despite consumption of soda at nearly a 30-year low. Impressively, between 1976 and 2015, bottled water consumption grew 27 times. Many of the largest soda companies, including both Coca-Cola and PepsiCo, own or have acquired bottled water brands after recognizing this trend. On a positive note, most of the bottled water, while originating from tap water, has been further filtered or has vitamins added. Conversely, soda leads to an estimated number of 184,000 deaths per year as a result of diabetes, heart disease, and obesity-related diseases (Cottrell, 2017).

Artificial Intelligence and Robotics Automation

Increasing amounts of data and new artificial intelligence (AI) brought to market in the last few years are projected to cause similar disruption to healthcare as with other industries within the next 5-10 years. In the U.S., nearly 10% of the total U.S. population of over 300 million, or 30 million people, have diabetes (Maney, 2017). If including the number of people in the U.S. with diabetes or prediabetes, this number jumps to 100 million (CDC, 2017g). The U.S. is not alone. In ten years, the number of people in China with diabetes will be greater than 100% of the U.S. population. In other words, over 300 million people in China will have diabetes. Diabetes is estimated to be the most expensive disease in the U.S. with up to $10,000 per person a year spent on medication, plus additional physician and hospital bills, as well as with lost productivity. There are nearly 130 tech-based start-ups involved in diabetes. One example is Virta Health, which was founded in 2014 and allows patients to enter their glucose reading, weight, blood pressure, activity, mood, energy levels, and hunger using a smartphone app and cloud computing. The app's AI analyzes the data to provide information to the patient and doctor on potential warning signals or symptoms along with recommendations on diet and medication. Early estimates are that the software increased productivity and decision-making by a factor of 10 and could save $100 billion in U.S. diabetes related costs. AI can also help automate processes. For example, IBM's Watson uses patient data, medical research papers, news, and diseases to help diagnosis medical conditions (Maney, 2017).

Robotic process automation (RPA) has the ability to code and computerize many simpler tasks that require physical effort or tasks with limited knowledge understanding and insight requirements. Employee tasks are increasing as a result of the growth in data, regulation, and bureaucracy. RPA will allow employees to spend their time on more meaningful, or value-added, activities to improve healthcare. An estimated 36% of all healthcare job activities can be automated. Healthcare and administrative industries still have one of the lower ranges in terms of industry automation potential, with 36% and 31%, respectively. In contrast, the food services industry has an automation potential of 75%. The most difficult activities to automate will be those involved with managing people, making decisions, or doing creative work. Those activities that involve direct patient care also have a lower automation potential. For example, nurses might have less than 30% of activities automated, and dental hygienists might have less than 13%. Other activities (such as information collection, reading radiological scans, or anesthesia) could be more easily automated (Lhuer, 2016; Chui et al., 2016).

In addition to process automation, robotics can be applied to other healthcare opportunities such as surgeries and patient mobility. The Carson Tahoe Regional Medical Center began using robotic surgical systems in 2017 and are expanding the use of robotics for hernia repair and prostate cancer treatment. The Chief Operating Officer at Carson Tahoe Health explained that the robotics are not intended to replace surgeons but rather to improve their abilities through high-definition 3-D images and smaller incisions made with robotic arms that allow a greater range of motion than the human wrist (Carson Tahoe Health, 2018). Another example of robotics in healthcare is the ReWalk, a wearable robotic exoskeleton that allows individuals with hip, knee, or spinal cord injuries to stand, walk, turn, and climb stairs. The ReWalk device is the first of its kind to receive FDA approval for use in the U.S. (ReWalk, 2017a). The ReWalk 6.0 has a list price of $77,000, but ReWalk Robotics is also developing a lower cost, lighter, and simpler exoskeleton suits for those who have suffered a stroke or have multiple sclerosis (Smith, 2017). In a recent court decision, Blue Cross and Blue Shield of Florida (Florida Blue) was required to provide coverage of the device for a patient with spinal cord injury. The ruling showed that the technology was no longer experimental, and was medically necessary. Although denial of coverage has occurred across the U.S., in over 80% of appeals, the coverage was granted (ReWalk, 2017b).

Experiential Learning Activity: Robotic Surgery

Robotic Surgery: For or Against

Description: Review the following summaries for and against robotic surgery use.

For Robotic Surgery:
In India, urology and gynecology are two early application areas of robotic surgery, with general surgery, cardiovascular, oncology and transplants on the rise. The first surgical robotic program in India began in 2006. Today, there are 19 surgical robotic systems in the country, which is a far cry from the number that is needed to serve the population of 1.2 billion. The da Vinci robot, manufactured by Intuitive Surgical, aims to assist with minimally invasive surgery. The robotic device was first approved by the FDA in 2000, and in 2017 had a total of 133 da Vinci surgical robotic systems that were sold (Thomas, 2017). Costs of the robotic systems are often cited as a concern, although as more options enter the market, the cost is expected to decrease. Costs can also be decreased through increased usage, typically on the order of 100-150 patient procedures per year per robotic system. Learning and developing expertise for medical professionals is also being added to existing centers of excellence to improve usage. Overall, the learning curve for robotic systems was less than for non-robotic methods and resulted in a reduced length of stay for patients (Desai et. al, 2015).

Against Robotic Surgery:
Although robotic surgery has been supported for many years by medical centers of excellence, it has lacked adoption by the general surgical community due in part to high costs and limited benefits. Advantages of robotic surgery include magnified 3-D vision, ergonomic movements with full range of motion, reduction in strain, precision, increased patient service, and minimization of hand tremors. Drawbacks include space required for the robot, increased operating room time, changing docking position of the operation table, lack of tactile feedback, and safety. Lack of tactile touch is seen as too high a price to pay for the technology. Complications of surgery or malfunctions are thought to be underreported. Robotic surgery has an initial investment of $2 million with annual maintenance of $400,000 along with disposable components. In comparing prostate cancer outcomes, open surgery resulted in a 5-year outcome rate of 80%, 78% for laparoscopic surgery, and 84-90% for robot-assisted laparoscopic surgery. For technology to be of benefit to countries, and in particular developing countries such as India, a five-step approach is recommended called the 5 As: Affordable, Acceptable, Accessible, Available, and Appropriate (Udwadia, 2015).

List a point for robotic surgery:

List a point against robotic surgery:

Robotic Surgery: For or Against

Which point of view do you agree with? Discuss your justification:

Discuss Global Robotic Surgery Readiness in Terms of the 5As:

Acceptable:

Affordable:

Accessible:

Available:

Appropriate:

Based on the preceding factors, which countries do you foresee as the leaders in robotic surgery? What are the barriers and how can these be addressed? Will robots render healthcare professionals obsolete?

Describe one robotic procedure and technology. Are robotic surgeries safe? What is the government's role in robotics?

Healthcare Globalization and Government

Pricewaterhouse Coopers (PwC) identified a few of the top health issues as part of their industry report. The overall themes trended around value of care, including adapting, innovating, and building value-based care. Key areas around adapting include uncertainty of the healthcare legislation on Affordable Care Act (ACA) and incentive-based payments. Key areas around innovation include emerging technologies such as artificial intelligence and virtual reality, along with population health management. Key areas around building value-based care include reducing medication prices and improving education in a new value-based healthcare world (PwC, 2017). Given the ongoing legislation of healthcare proposals in the U.S. (including the repeal and replacement of the ACA), some advocates are calling for a more dramatic change to a single-payer healthcare system. A similar system is already in place in Canada, France, the U.K., and Australia. This model provides universal coverage through government-based funding or hybrid public/private setup. In the U.S., the states of California and Illinois have discussed legislation on a single-payer system, and even CEOs of health insurance companies such as Aetna have discussed publicly a single-payer system. The Physicians for a National Health Program also supports a single-payer insurance model, also known as Medicare for all, and would give coverage to everyone in the U.S. for medically necessary services including physician, hospital, preventative, and long-term care, as well as mental health, reproductive health, vision, prescriptions, and medical supplies. Co-pays, deductibles, restrictive coverage networks, and premiums would be eliminated. The program estimates that 95% of households would save money. Also, the program would be paid for by primarily replacing the inefficiencies that are currently in place, which, along with changes in taxes, would result in an estimated $400 billion in savings. Even under the ACA mandate for universal coverage, still some 30 million remain uninsured. Those without coverage would continue with the American Health Care Act (AHCA) plan. In addition, both the ACA and AHCA add administrative costs and healthcare spending through new legislative changes, which is an estimated $1.1 trillion over the next decade (Bryan, 2017; PNHP, 2017).

International healthcare growth in overseas markets and changes in governmental coverage and control are driving insurers to provide healthcare coverage and services internationally. In China, 90% of the population lack coverage, compared with 16% in the U.S. The market in China is expected to grow to an estimated $56 billion by 2020. Although healthcare costs have increased, they are still far less than in developed markets, which is attractive to insurers. Latin America has also experienced a rapid influx of multi-national corporations that provide healthcare coverage. During the period of 1996-1999, revenues of multi-national healthcare corporations increased faster in Latin American than in U.S. revenues. As the U.S market becomes saturated, companies are seeking new market locations for sustained growth (Lim 2006; Ran, Waitzkin, Merhy 2004; Woodside, 2008). In international studies, setup costs were identified as a major barrier to electronic system implementations. Healthcare services that occur outside the country of residence require a lengthy manual process in order to gain reimbursement. Trust factors (such as information security, infrastructure, education, government, and culture) influence adoption rates. Security issues include data theft, corruption, and personal or confidential data. Trusted authorities are also a key component to security and trust. Global certificate and intermediary services that improve security are being developed internationally. Culture might also influence the communication language that is selected or a willingness to engage in nonpersonal contracts between systems (Angeles, Corritoreb, et al. 2001; Hassan, Pans, Collins 2003; Hennick 2007; Woodside, 2008).

Health Tax

Government initiatives that intend to improve healthcare and to encourage healthier behaviors have also drawn concerns. Similar to car insurance safe-driving discounts, some government initiatives have proposed the offer of healthcare discounts or penalties according to food consumption, genomic sequencing results, or preventative care completion. Critics consider healthcare monitoring to be an invasion of privacy and micromanagement of one's life by government. However, precursors to the creation of larger government regulations have been tested through wellness programs and taxation initiatives. One example of an initiative is the reduction of soda consumption via a tax. Several states and cities (such as New York, Seattle, and Chicago) have proposed taxes on soda and related drinks although nearly 45 such proposals have been defeated. However, Philadelphia approved a tax of 1.5 cents per ounce tax on diet soda, regular soda, iced tea, energy drinks, juice drinks with under 50% juice, and other sports drinks. This was Philadelphia's third attempt at the soda taxes to encourage healthier drinking and to raise funds. Critics such as the American Beverage Association argue that the tax is regressive and impacts the poorer population disproportionately. Since implementation in the beginning of 2017, the city has reported greater-than-expected revenue of nearly $5.7 million in the first month. However, retailers have reported nearly a 50% decrease in sales, leading to layoffs, which have been blamed on the tax. The city had previously projected a 27% decrease in sales as a result of the tax. Court cases on repealing the tax are ongoing (Premark, 2016; Zwirn, 2017).

Similar to the soda tax, initiatives on genetic testing have also been proposed for improving healthcare management. As a government establishment, the Food and Drug Administration (FDA) approved genetic testing through companies such as 23andMe to allow patients to learn their risks for ten diseases: Parkinson's, Alzheimer's, Celiac, Alpha-1 antitrypsin deficiency, Primary dystonia, Factor XI deficiency, Gaucher disease type, Glucose-6-phosphate dehydrogenase deficiency, Hereditary hemochromatosis, and Hereditary thrombophilia (Pirani, 2017). A recently introduced bill allows employers to require employees to complete genetic testing and allows employers access to the information, with the risk of penalties for employees not completing the testing. Currently, the genetic privacy and nondiscrimination (GINA) law passed in 2008 and the 1990 Americans with Disabilities Act prohibit this testing and access. Employers argue that the testing is needed for workplace wellness programs that are not covered by the current laws (Begley, 2017).

Life Sciences

Recently, governments have focused on the life sciences aspect of healthcare, namely on rising drug costs and research of new medicines. Life sciences incorporate pharmaceutical, biotechnology, medical device, vaccines, and research organizations that bring therapies to market. The combination of safety, patents, and approvals led to an estimated cost of $1.2 billion for each new therapy (Handelsman, 2014). Analytics has been used in the discovery process of new treatments. The mapping of the human genome in 2001 was a major discovery process that led to new treatment opportunities. Analytics can also be applied to the development of trials to measure safety and effectiveness, which is a form of descriptive analytics. Analytics is also applied to manufacturing to address shortages, identify demand, and improve the supply chain. Six Sigma has been used in reducing the number of defects during manufacturing. In 2007, the FDA released the report titled *Pharmaceutical Quality for the 21st Century* to identify manufacturing quality improvements using analytics and other methods in order to prevent recalls and safety concerns. Sales and marketing can also use analytics to determine physician targeting, marketing mix, and rebate optimization. Healthcare reform can have an impact on the life sciences through cost controls and differentiated treatment requirements that target new areas of need rather than replications of existing therapies (Handelsman, 2014).

Social Inequality

Another growing area of importance of government and public health is income inequality and social inequality. Top incomes continue to grow while the percentage of those in poverty continues to grow. Poverty is calculated by comparing annual income based on family household size, number of children, and age. Poverty rates varied by state from 8.5% to 21.9%. Based on the latest U.S. Census Bureau data, there were nearly 44 million people who live below the poverty line, which is approximately $25,000 for a family of four (Bishaw and Macartney, 2010, U.S. Census Bureau, 2017). In 2011, the CDC released the Health Disparities and Inequality report, with a follow-up report in 2013, which highlighted disparities and inequality in healthcare based on gender, race, ethnicity, income, education, disability, and additional factors. A health disparity is a health difference that is connected with an economic, social, or environmental disadvantage. The CDC findings showed that inequalities in income and health demonstrate that gaps are evident. Individuals with low levels of income generally experienced increased rates of mortality, morbidity, and decreased rates of access to healthcare and quality of healthcare. U.S. Department of Health and Human Services (HHS) and other agencies have launched strategic initiatives to reduce health disparities including an increased access to healthcare and healthcare coverage, recruitment of underserved areas for healthcare workers, and work to target specific diseases that show higher levels of disparities such as childhood obesity, maternal health, and cardiovascular disease (CDC, 2015).

Public Health

Environmental Factors

Nearly one quarter of deaths worldwide are due to addressable environmental factors, including physical, chemical, and biological elements. Components that can be directly modified include air pollution, water quality, sanitation, ultraviolet radiation, working conditions, and ecosystem change. There is the potential to prevent up to 12.6 million deaths, with over 100 diseases attributed to the environment. These diseases can include respiratory infections, asthma, cardiovascular, injuries, infections, malaria, and cancers. The environment disease burden is higher in lower income and developing nations (Pruss-Ustun, 2016). A 2017 cholera outbreak in Yemen had an estimated number of 200,000 cases and 1,300 deaths, qualifying it as the worst cholera outbreak in the world, up to that time. Still, an estimated number of 3-5 million cases of cholera in the world occur each year. Cholera is a diarrheal illness caused by bacteria in drinking water, causing those infected to lose fluids, leading to dehydration. The United Nations International Children's Emergency Fund (UNICEF) and World Health Organization (WHO) worked with local response teams to educate households on clean drinking water. The cholera outbreak was, in part, a result of several years of conflict in the country. The conflict led to limited funding for healthcare, which led to a lack of clean water and sanitation systems (CDC, 2016c; Lake and Chan, 2017).

Product Recalls

Product recalls are another area of public health and occur regularly. Ten individuals in California contracted botulism that was traced to nacho cheese sauce as reported by the Sacramento County Department of Health and Human Services (Andone, 2017). Botulism is a rare illness with 145 cases reported per year in the U.S. Symptoms begin 18-26 hours after consuming the food and include blurred vision, slurred speech, and paralysis. The disease can be fatal in 3-5% of cases (Andone, 2017). After three consumer complaints of metal within the packages, 200,000 pounds of hot dogs were recalled, although no reports of injuries or reactions were found (AP, 2017). In June 2017, nearly 2.5 million pounds of breaded chicken products were recalled, as the products might have contained milk, a known allergen, which was not declared on the product labeling. The supplier of the bread crumbs indicated that the bread crumbs might have potentially contained undeclared milk. The U.S. Department of Agriculture (USDA) has three recall classifications. Class I is a health hazard situation where a reasonable probability exists that the use of the product would cause serious, adverse health or death. Class II is a health hazard situation, which has a remote probability of adverse health. Class III is when no adverse health conditions would be caused (Medina, 2017). The USDA's Food Safety and Inspection Service (FSIS) announced the recall of bread crumbs as a Class I, with no confirmed cases of any adverse reactions to the products. In 2018, an E. coli outbreak in 11 states caused the CDC to issue a consumer warning to avoid all romaine lettuce for a second time in the same year. Of striking interest to researchers and health officials was that approximately 70% of the illnesses were reported by women. This follows a similar pattern to 2016 where 73% of those ill from an alfalfa sprouts outbreak were also women. In analyzing the data, researchers have found that diet can be the greatest contributing factor, with women's diets including more vegetables. Others suspect another factor is that the willingness to report symptoms to their doctors can occur at a higher rate for women than for men. Despite these recalls associated with lettuce, health officials stress that this health outbreak is not an excuse to stop eating vegetables (Rossman, 2018).

Health Tracking

One method of detecting public health occurrences is through geographic information systems and mapping. These geographic (or spatial) types of tools can help public health officials monitor and deploy resources more easily to affected areas. Walgreens developed the Flu Index, which is a weekly report that provides state-level and market-level detail incidence of flu. The index is calculated based on prescriptions filled for influenza. In a 2017 report, several of the top states were concentrated in a southeastern section of the U.S., including North Carolina, South Carolina, Kentucky, Tennessee, and Virginia (Walgreens, 2017). In a 2018 report, the top ten designated market areas with flu activity were all within the single state of Texas, demonstrating how the patterns might change over time (Walgreens, 2018). A spatial epidemiological approach to healthcare studies provides significant insight into evaluating health intervention and decision-making and offers organizations the opportunity to analyze unusual geographical patterns of disease. Factors such as neighborhood economics, social fragmentation, and rurality can also affect individual risk factors. Routine, aggregated healthcare data that is stored in health systems can be used to identify disease clusters (or utilization patterns). A typical advantage is that routine service data is readily available and, in some cases, data can be available in real time. Cluster analysis can be used to study whether diseases such as lung cancer or causes of death such as suicide are spatially clustered (Woodside and Sikder, 2010; Woodside and Johnson, 2015; Johnson et al., 2016).

Big Data Health Anamatics

Big Data Health Anamatics is truly an interdisciplinary topic that spans the entire organization and areas of management. Given the high failure rates of big data projects, most managers do not focus on the organizational span or focus solely on skills-based learning. Managers must apply Big Data Health Anamatics to build both short-term skills along with lifetime learning capabilities for their organizations and their members. The learning capability aligns and connects with our knowledge of mental models for short-term learning (mental-model maintenance) and long-term learning (mental model building). Mental models influence behavior and create a basis for reasoning, which improves human decision making. By allowing pre-defined models, which speed information processing speed is improved when supplemented by Big Data Analytics (Vandenbosch and Higgins, 1995). Learning theories have evolved from memorization to understanding of information and have often influenced organizational practices from simple structured learning domains to complex unstructured learning domains. Knowledge can be defined as the repository of intellectual assets that have been combined from experience, learning, and practice. Knowledge is often described in terms of data and information. However, knowledge can only be distinguished through an individual's personalization and interpretability to others. Increasing amounts of information and Big Data are useful only when processed by an individual through learning processes (Nonaka, 1994).

Big Data is characterized by Vs: volume, velocity, variety, veracity, variability, and value of the data. Health analytics seeks to uncover hidden patterns and information in Big Data that is retrieved from health systems, with methods of analytics including descriptive, diagnostic, predictive, prescriptive and cognitive (Hughes, 2011; Heudecker, 2014). The current volume of healthcare data is estimated at 150 exabytes and is soon growing to zettabytes and yottabytes. Cognitive computing systems are aimed at using this data and systems to enable clinical decision support, allow knowledge sharing between teams, use the latest evidence-based guidelines, increase preventative care, and improve health (Govette, 2017). Benefits of Big Data include reducing cost to analyze and use evidence-based practices. Given that Big Data in healthcare generates one-third of the world's data and is estimated at $300 billion over the next decade, many new companies and recent IPOs are in healthcare and analytics-related areas. Today, healthcare organizations are becoming increasingly computerized, thereby capturing increasing amounts of data in various places. In other industries, the use of information systems can be tied to improved quality and competitive advantage. Extracting, formatting, analyzing, and presenting this data can improve quality, safety, and efficiency of delivery within a healthcare system. Big Data analytics has been used in radiology to address the complex, poorly integrated, and functionally limited legacy systems as practices transitioned from analog to digital. Big Data analytics allows the decision maker to develop trends and interactively drill down to answer further questions as they become apparent. Procedural and performance metrics can be tracked to improve patient care and to reduce costs. Examples include resource utilization by scanner, time variances between scheduled and actual time, evidence-based outcomes, billing tracking, and quality outcomes measurements such as length of stay and charge costs (Woodside, 2014).

Big Data and access to information for patients, providers, payers, governments and others will allow joint responsibility for information and self-management of conditions that lead to improved levels of quality and care and of outcomes. Outcomes can be improved through measurements, surveillance, analysis, and interventions:

- Outcome measures can include patient assessments, health status, surveys, cost, providers, and other factors.
- Outcome surveillance is the monitoring of conditions or services for outcomes and can be used for environmental or communicable conditions. Also population-based approaches to health, such as cancer screenings, are useful.
- Outcome analysis converts the data into useful information.
- Outcome interventions are a result of the preceding analysis and are judged by effectiveness and efficiency and target improvements in population health.

Big Data is able to be collected from both structured and unstructured data sources. Each set of source systems should be identified, measured, and documented through data types, size, growth, and value (Woodside, 2013). According to a SAS survey, only 14% of respondents were likely to use big data sources, and according to McKinsey organizations use only 12% of the data available to them (Brown, 2012; Merrideth, 2013).

Big Data management solutions will benefit healthcare providers, payers, research, and government organizations. The solutions will also decrease variability, reduce costs, and improve quality by providers, while still delivering personalized care. Healthcare organizations are able to monitor metrics (such as service line, physician, payer, and patient category) to decrease costs per case, increase revenue, and increase utilization of facilities and staff. The capabilities include executive dashboards, reporting, and analysis, with data drill-down capabilities. Physicians can use predictive analytics for care treatment decisions and real-time decision-making and can incorporate both structured and unstructured sources to bring additional evidence. For patients, Big Data can help personalize care for patients based on genomic and test data, increase consistency of costs and treatment, and improve quality. A patient's social media data can be analyzed to improve lifestyle, care, and costs. In addition, historical trends can be reviewed and future trends predicted with greater confidence. Patients can be measured against larger comparison groups, patient outcomes predicted, and best pathways of care established. For providers, Big Data facilitates the review of unstructured and structured data,the identification of care patterns, and the identification of risk factors, all leading to improved outcomes through education, research, and care. For payers, Big Data can be used to improve patient wellness, compliance, fraud detection, incentive programs, and controlling costs. Researchers can integrate large disparate sets of information automatically in order to develop findings from previously siloed information, improving pattern detection and surveillance (Woodside, 2013).

Big Data and High-Performance Data Mining Model

Although SEMMA is a standard data mining methodology that is used for most data sets and projects, when using Big Data, a separate set of tools and methods can be used. A data mining method that is designed specifically for Big Data is BEMO (Business Opportunity, Exploration, Modeling, and Operationalization) and is a standard parsimonious process that was developed for conducting data mining projects in a reusable and repeatable fashion in a Big Data environment. The BEMO model is technology- and industry-agnostic and uses new high-performance data mining technologies such as SAS High-Performance Data Mining (HPDM). The principle of parsimony requires the abandonment of complex models and the use of simpler models that can generalize new problems. A parsimonious model is one that meets the necessary requirements while limiting factors. Occam's razor is a principle that states that, all things being equal, the simpler solution is preferred. For example, data mining decision trees are often pruned to develop simpler solutions and to prevent overfitting. During Business Opportunity, the goals (or objectives) are defined along with the problem being solved. During the Exploration step, data quality (including variable selection, outlier correction, duplication correction, and so on) is reviewed and updated. During the Modeling steps, the predictive models are used based on the combination of inputs and outputs. In the Operationalization step, the model is implemented and practical considerations of operations are incorporated (Woodside, 2016).

Figure 11.3: Big Data Mining Process

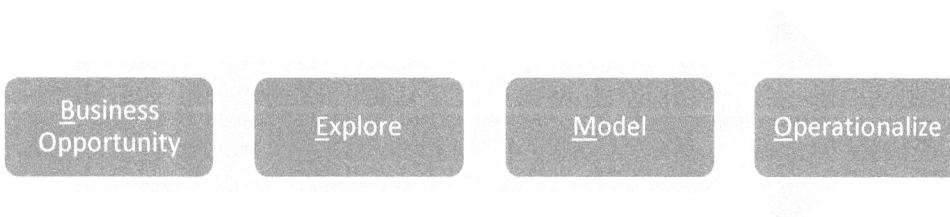

Process Step Overview

Business Understanding

During the Business Opportunity step, the goals (or objectives) are defined along with the problem being solved. This is a critical, although sometimes overlooked, component of a data mining project that ensures successful definition and outcomes in support of the business objectives. During this step, an executive summary is provided to communicate to key stakeholders from a customer, financial, employee, or operational perspective. An organizational and industry background is provided along with key objectives (or goals) that are expected as a result of the data mining project. The Business Opportunity section is kept intentionally brief and at a summary level for stakeholders who are conducting the data mining project.

Exploration

During the Exploration step, data quality is reviewed and updated. Data quality factors include variable selection, outlier correction, duplication correction, data reduction, data visualization, and data standardization (or normalization). During Exploration, variables must be assessed for type for future model use. With Big Data, traditional data processes, data management, and data quality components must be reassessed and improved.

Model

During the Modeling step, the predictive models are used based on the combination of inputs and targets. The models are similar to the ones that we have covered earlier, including regression, decision trees, and neural networks.

Operationalization

In the Operationalization step, the model is implemented, and practical considerations of operations are incorporated. Model advantages and disadvantages are discussed in the context of a business setting application. An overall project plan is developed to ensure that the project deliverable is on time, on budget, and on scope. The plan includes practical considerations of resources, timeline, and iterative phases. A formal assessment of the project is also made. Assessment items include return on investment (ROI), payback, net present value (NPV), and break-even analysis.

HPDM Tab Enterprise Miner Node Descriptions

High-Performance Data Mining (HPDM) capabilities have been developed in SAS in part due to the rise of Big Data and the ability to leverage technology such as parallel processing and distributed processing. The technology allows computationally intensive operations to be split across computer processors or a set of computers to improve overall speed of analysis. The HPDM capabilities are similar to those covered in previous chapters, including logistic regression, decision trees, and neural networks. The primary difference is the multi-processor and computer capabilities. For illustration and learning purposes, we are able to run the HPDM nodes on a single computer. However, to take full advantage of the capabilities, a distributed multi-computer environment can be set up. In one demonstration example from SAS, a data set with over 100 million records took over seven hours to run using traditional methods, although only one minute was used with the HPDM capabilities (SAS, 2016e). The distributed capability is valuable for computationally intensive research and can be used for more timely decision- making at the point-of-care by providers.

The HPDM nodes that are associated with the **HPDM** tab are not meant to replace existing nodes that are associated with the **SEMMA** tabs, though used based on big data processing requirements. Although a subset of nodes exists today, more nodes are planned for the future. For the most part, HPDM and other nodes from the associated **SEMMA** tabs cannot be combined. They are indented to indicate that they can be used separately. However, exceptions are the File Import and Model Comparison nodes. The High-Performance (HP) node icons match those for the non-HP SEMMA nodes. However, the HP nodes are identified by a red icon in the upper corner, and each node name contains the HP prefix. Here is an example:

Figure 11.4: HP Data Partition Node

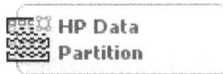

The HP Data Partition node splits the data set into training and validation.

Figure 11.5: HP Explore Node

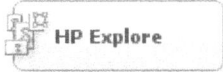

The HP Explore node generates a set of summary descriptive statistics.

Figure 11.6: HP Impute

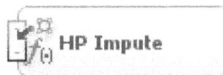

The HP Impute node generates values for missing values data sets.

Figure 11.7: HP Regression Node

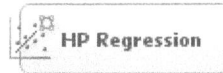

The HP Regression node generates high-performance linear and logistic regression models.

Figure 11.8: HP Tree Node

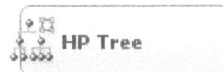

The HP Tree node generates high-performance decision tree models.

Figure 11.9: HP Neural Node

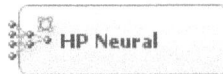

The HP Neural node generates high-performance neural network models. Now that you are familiar with the major trends, including Big Data health anamatics and High-Performance Data Mining (HPDM), we'll continue with an experiential learning application that deals with SIDS.

Experiential Learning Application: SIDS

In the U.S., 1 of 1000 children under one year of age die each year from unexplained causes. These unexplained deaths are referred to as Sudden Infant Death Syndrome (SIDS). One preventation method that was started in the 1990s, was to start putting babies to sleep on their backs, instead of their stomachs or sides. The SIDS prevention method and associated 'Back to Sleep' educational campaigns, led to a decrease in the number of deaths, and rate has remained consistent ever since. SIDS is often a combination of factors during a critical development time and most often occurs during the second month of life. Research studies in the past have often used a few hundred cases at a time to examine the problem. In a new effort led by Microsoft and Seattle Hospital, the aim is to examine and mine large data sets from the CDC to find patterns. By attempting to find the relationships between every single variable and SIDS, the researchers hope to find patterns that were not previously identified in small samples of data. Researchers have used similar large data sets in developing treatments for cancer and other diseases. The CDC collects 90 variables for each child born in the U.S., and between 2004 and 2010, that collection represents nearly 30 million records. Variables include medical care of the mother, race, education, income, along with other factors such as cause of death. One discovery that has already been made by the data scientists is that mothers who receive prenatal care within their first trimester have a lower risk of having a baby with SIDS, with the risk increasing up to 40% when starting prenatal care later. Other findings include the optimal number of physician visits, which can help policy makers determine best practices for prenatal care and physician appointments. Because the data shows how the risk of smoking while pregnant increases the SIDS rate each day, the recommendation to quit smoking earlier in pregnancy is an easy sell (Phillips et al., 2010; Bass, 2017; CDC, 2017h; Mayo Clinic, 2017c).

Although SIDS can impact any infant, a set of factors has been identified by researchers that can lead to higher risk. The factors include male gender, infants 2-3 months of age, race, family history, second-hand smoke, premature birth, and maternal factors such as age under 20, smoking, drugs, alcohol, and prenatal care (Mayo Clinic, 2017c). Premature births also have a similar set of risk factors. In a study in the U.S., the preterm birth (one that occurs before 37 weeks of gestation or pregnancy time) rate has been increasing and stands at nearly 10% of births. The study found that the U.S. has the highest rate of newborn deaths in the developed world, ranking below 130 countries. Factors that can lead to pre-term birth include lack of prenatal care, obesity, tobacco use, age, ethnicity, and accessibility of care (Fox, 2017).

The CDC tracks SIDS and unknown causes using ICD-9 and ICD-10 codes. SIDS is represented by 798.0 (or R95), accidental suffocation represented by E913.0 (or W75), and unknown causes represented by 799.0 (or R99), respectively, with the label Sudden Unexpected Infant Death (SUID). The CDC also monitors SUID within sixteen states, covering 30% of SUID cases within the U.S. Using the CDC SUID cases, the Michigan Public Health Institute (MPHI) identified an increase in SUID with families that also receive child protective services. In response, training and policies were implemented for all child protection workers, and a bill was passed to require hospitals to provide educational materials to new parents (CDC, 2017i).

Other researchers have examined parental behavior to link with SIDS. In one U.S. study, researchers reviewed alcohol consumption in relation to SIDS. The findings demonstrated the greatest spikes in alcohol consumption and SIDS cases occurred on New Year's Eve between 8 p.m. and 3:59 a.m. on New Year's Day. Additional increases in alcohol consumption and SIDS cases occurred on weekends. Although alcohol can be a risk factor in combination with other factors that lead to SIDS, the authors hypothesize that an alcohol-impaired caregiver might be less likely to have an infant sleep in the correct position and less likely to adequately monitor the infant during New Year's and weekends (Phillips et al., 2010). Another study in Croatia examined factors such as socioeconomic status, smoking, age of mother, age of father, and birth weight, among others. Their finding showed a greater relationship between SIDS and smoking, sleeping position, age of infant between 2-4 months, time of death, parental age above 30, and male gender. Socioeconomic factors were not found to have a significant relationship with SIDS (Rozman et al., 2014).

A sample has been provided for you to get started. The SIDS data set includes 13 variables and 21,000 records.

Objective: Determine the factors linked with SIDS.

Data Set File: 11_EL1_SIDS.xlsx

Variables:

- PatientID, unique identifier
- Gender, (F=Female, M=Male)
- Age_Months, in months
- Race, (Alaska Native, American Indian, Black, White)
- FamilyHistory, indicator (1=True, 0=False)
- SecondhandSmoke, indicator (1=True, 0=False)
- Premature, indicator (1=True, 0=False)
- MotherAge, in years
- MotherSmoke, indicator (1=True, 0=False)
- MotherAlcohol, indicator (1=True, 0=False)
- MotherPrenatalCare, indicator (1=True, 0=False)
- DiagnosisCode, ICD code
- SIDS, indicator (1=True, 0=False)

Step 1: Sign in to SAS OnDemand for Academics.

Step 2. Open SAS Enterprise Miner (Click the SAS Enterprise Miner link).

Step 3. Create a New Enterprise Miner project (Click **New Project**).

Step 4: Use the default SAS Server, and click **Next**.

Step 5: Add the project name, and click **Next**.

Figure 11.10: Create New Project

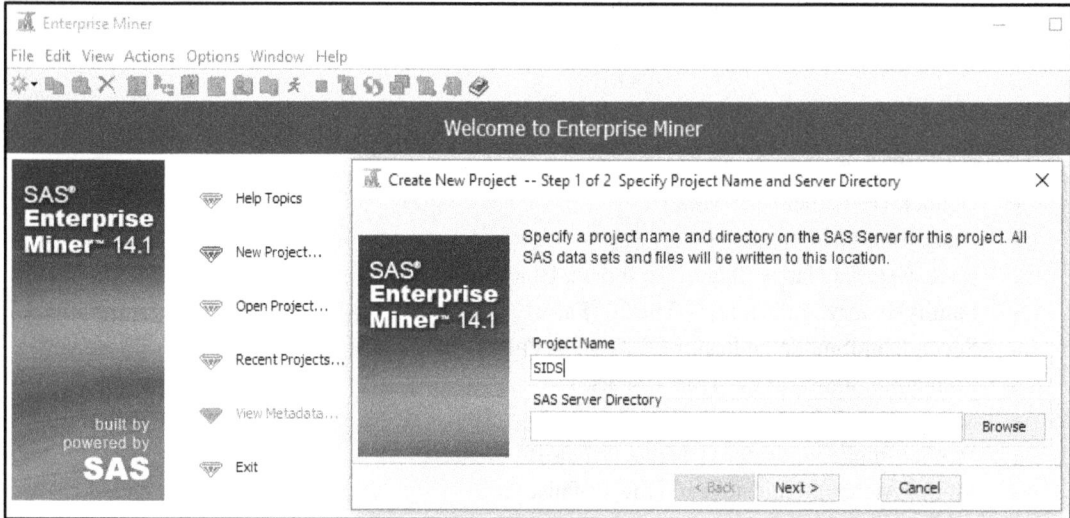

Step 6: SAS will automatically select your user folder directory (If using the desktop version, choose your folder directory), and click **Next**.

Step 7: Create a new diagram (Right-click **Diagram**).

Figure 11.11: Create New Diagram and Add Name

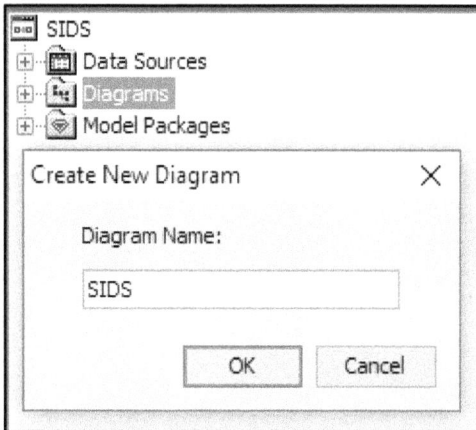

Step 8: Add a File Import node (Click the **Sample** tab, and drag into the diagram workspace).

Step 9: Click the File Import node, and review the property panel on the bottom left of the screen.

Step 10: Click **Import File** and navigate to the *11_EL1_SIDS.xlsx* Excel file.

Figure 11.12: File Import

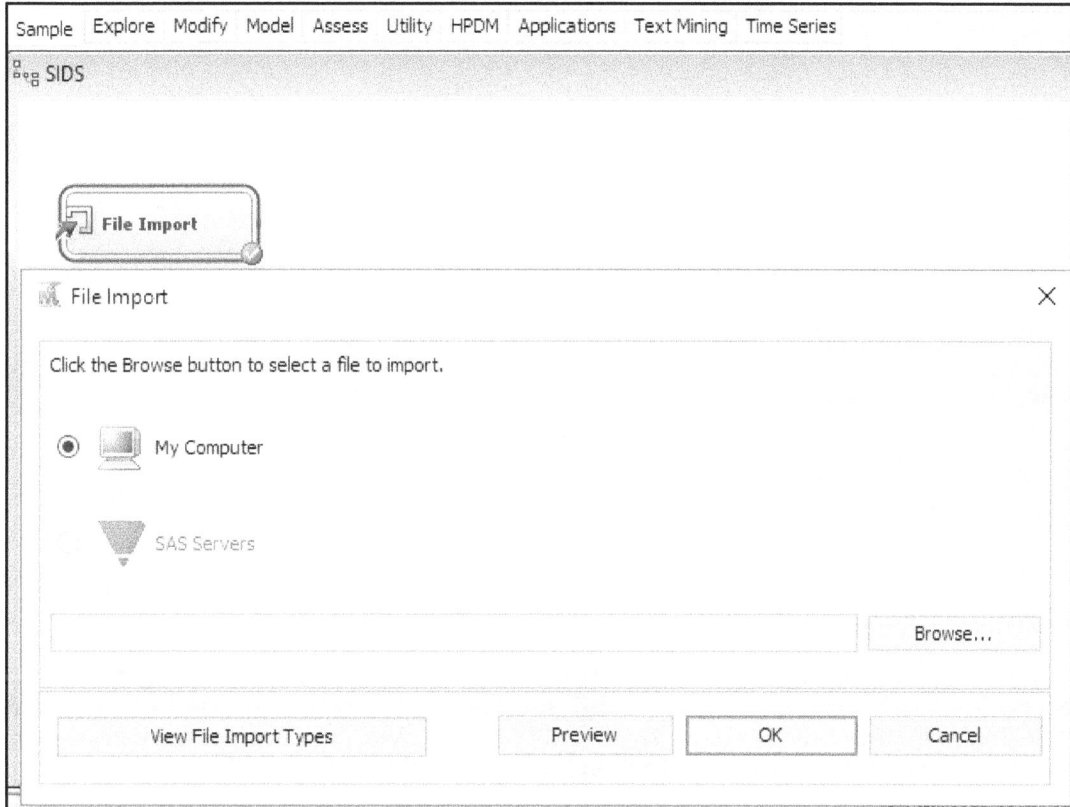

Step 11: Click **Preview** to ensure that the data set was selected successfully, and click **OK**.

Step 12: Right-click the File Import node and click **Edit Variables**.

Step 13: Set SIDS to the Target variable role, set PatientID to the ID role, set the DiagnosisCode to Rejected, and all other variables to the Input role. Explore and set the remaining variables according to their nominal, interval, or binary levels. To review an individual variable in order to verify its role and level assignment, click the variable name and then click **Explore**. After you have finished setting all variables, click **OK**.

Figure 11.13: Edit Variables

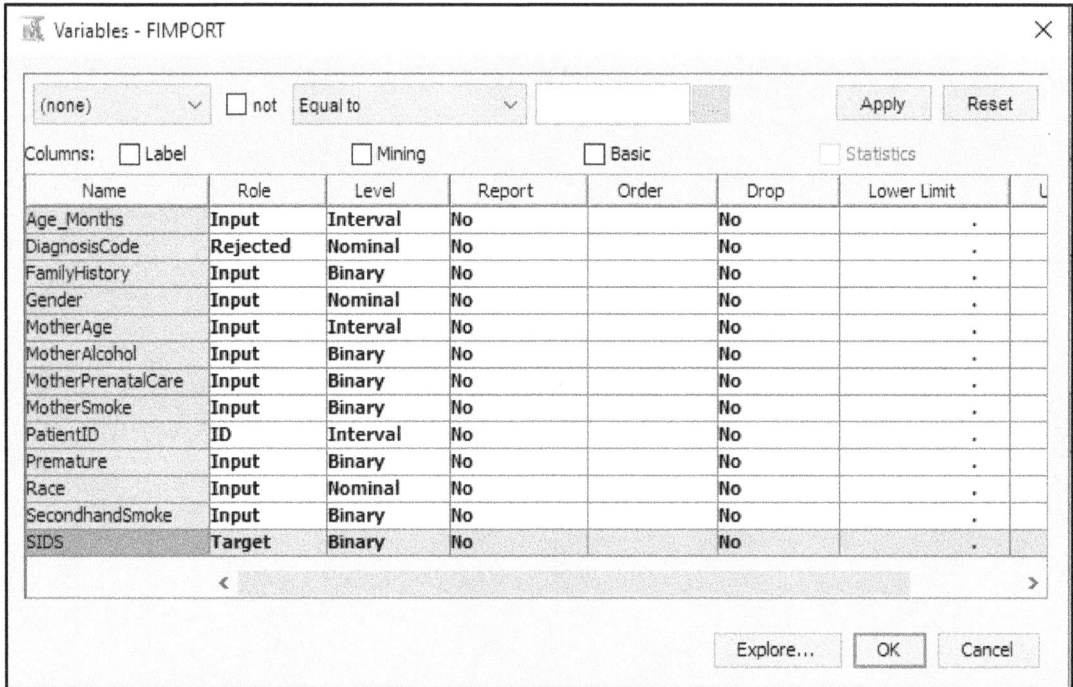

Name	Role	Level	Report	Order	Drop	Lower Limit	U
Age_Months	Input	Interval	No		No	.	
DiagnosisCode	Rejected	Nominal	No		No	.	
FamilyHistory	Input	Binary	No		No	.	
Gender	Input	Nominal	No		No	.	
MotherAge	Input	Interval	No		No	.	
MotherAlcohol	Input	Binary	No		No	.	
MotherPrenatalCare	Input	Binary	No		No	.	
MotherSmoke	Input	Binary	No		No	.	
PatientID	ID	Interval	No		No	.	
Premature	Input	Binary	No		No	.	
Race	Input	Nominal	No		No	.	
SecondhandSmoke	Input	Binary	No		No	.	
SIDS	Target	Binary	No		No	.	

Step 14: Add a Data Partition node (Click the **Sample** tab, and drag the node onto the diagram workspace). Set the Data Partition Property Data Set Allocations to 60.0 for Training, 40.0 for Validation, and 0.0 for Test. Review the data Partition Results. For non-HPDM applications, this would be similar to the process steps for selecting Data Partition from the **Sample** tab.

Figure 11.14: Data Partition Node

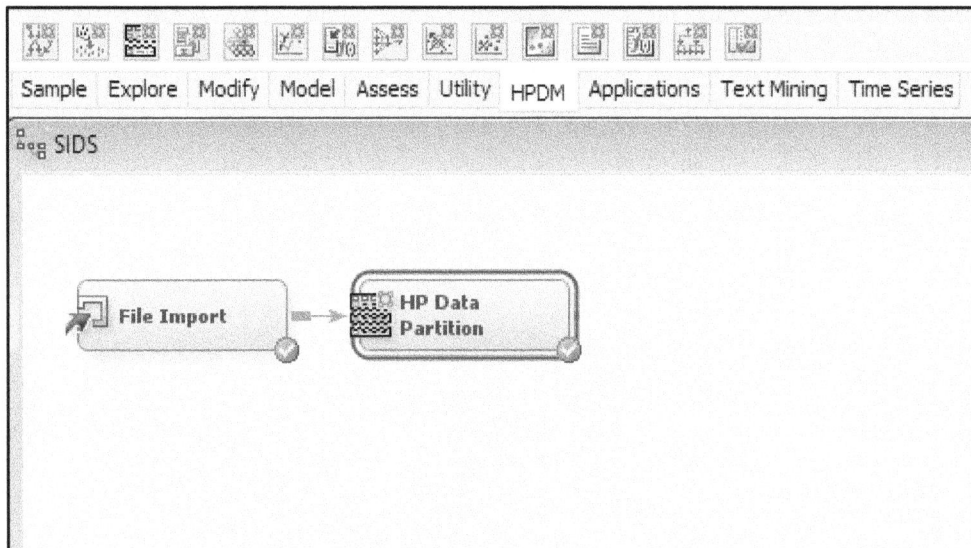

Step 15: Add an HP Explore node (Click the **HPDM** tab, and drag the node onto the diagram workspace). For non-HPDM applications, this would be similar to the process steps for selecting StatExplore, Graph Explore, and MultiPlot from the **Explore** tab.

Figure 11.15: HP Explore Node

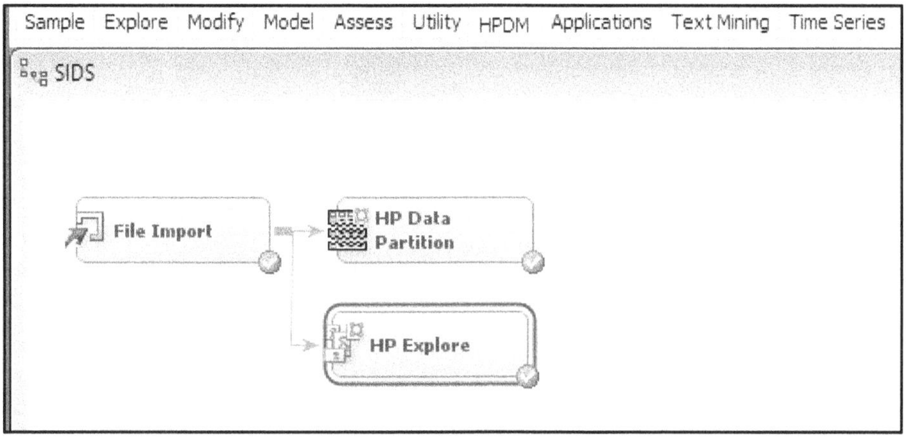

Step 16: Review results. From the HP Explore results, we find that there are a total of 21,000 records, with no missing data across variables and an initial indication of good data quality. There are two interval variables, Age_Months and MotherAge, as shown by the **Scale** VAR. The remaining variables are of **Scale** CLASS, indicating a binary variable or a nominal variable, or a class or category variable. The skewness and kurtosis for all variables is within acceptable ranges of -2 to 2, indicating an absence of extreme outlier values. Since neural network might be affected by missing or outlier values, we still want to add the Impute and Transform nodes to our final model diagram as a best practice, since future unknown data sets might contain these issues and we want to reuse the same model.

Figure 11.16: HP Explore Results

Statistics Table

Variable Name ▲	Scale	Missing	Percent Missing	Non Missing	Minimum	Mean	Maximum	Standard Deviation	Skewness	Kurtosis	Coefficient of Variation	Number of Levels
Age_Months	VAR	0	0	21000	1	5.965762	12	3.378558	0.345926	-1.23775	0.566325	
FamilyHistory	CLASS	0	0	21000								2
Gender	CLASS	0	0	21000								2
MotherAge	VAR	0	0	21000	15	24.86105	45	8.780112	0.798564	-0.80177	0.353167	
MotherAlcohol	CLASS	0	0	21000								2
MotherPrenatalCare	CLASS	0	0	21000								2
MotherSmoke	CLASS	0	0	21000								2
Premature	CLASS	0	0	21000								2
Race	CLASS	0	0	21000								4
SecondhandSmoke	CLASS	0	0	21000								2
SIDS	CLASS	0	0	21000								2

Step 17: Review results. Note that the execution mode of a single machine and number of threads. The threads are the cores, in this case. A quad-core processor was used on a single machine. A thread is the set of processing instructions that are sent through a single computer core. In this case, four simultaneous instructions were able to pass through four central processing units or cores, improving the run-time efficiency.

Figure 11.17: HP Explore Results

```
 Output
  28      The HPDMDB Procedure
  29
  30           Performance Information
  31
  32      Execution Mode        Single-Machine
  33      Number of Threads     4
  34
  35
  36              Data Access Information
  37
  38      Data                Engine     Role       Path
  39
  40      WORK._DMDBCLASS      V9         Output     On Client
  41      WORK._DMDBVAR        V9         Output     On Client
  42      EMWS1.FIMPORT_TRAIN  V9         Input      On Client
```

Step 18: Add an HP Impute node (Click the **HPDM** tab, and drag the node onto the diagram workspace). Verify that the Impute Property is set to Count for Class variables and Mean for Interval variables.

Figure 11.18: Impute Node

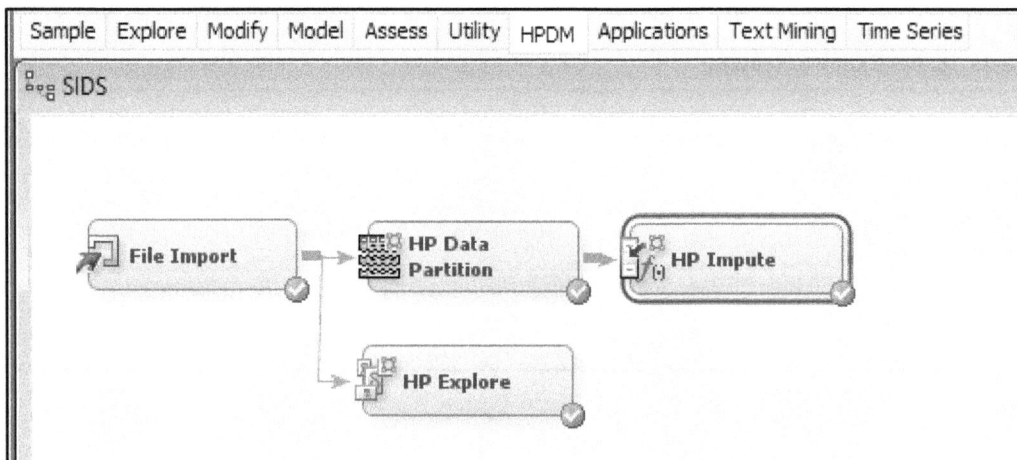

Step 19: Add an HP Regression node, an HP Decision Tree node, and an HP Neural Network node (Click the **HPDM** tab, and drag the nodes onto the diagram workspace).

Figure 11.19: Model Nodes

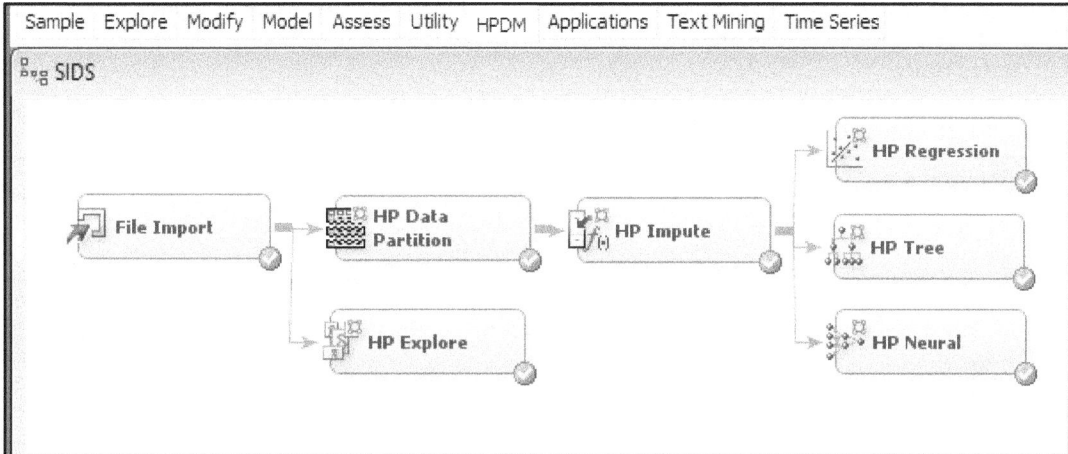

Step 20: Right-click the Regression node, Decision Tree, and HP Neural nodes, and click **Run**.

Step 21: Expand the Output Window Results for each model. Review the Fit Statistics for each model, including the misclassification rates and cumulative lift.

Step 22: Add a Model Comparison node (Click the **Assess** tab, and drag the node onto the diagram workspace). Instead of having to manually compare each of the three model outputs, the Model Comparison node will compare the three models and select a model based on the criteria in the Model Comparison node properties. By default, we will use the validation misclassification rate to select the best model, or we will use the model with the lowest event classification error percentage from our validation data set. Right-click the Model Comparison node, and click **Run**. Note that although the Model Comparison node is not present on the **HPDM** tab, it is compatible with the HPDM nodes.

Figure 11.20: Model Comparison Node

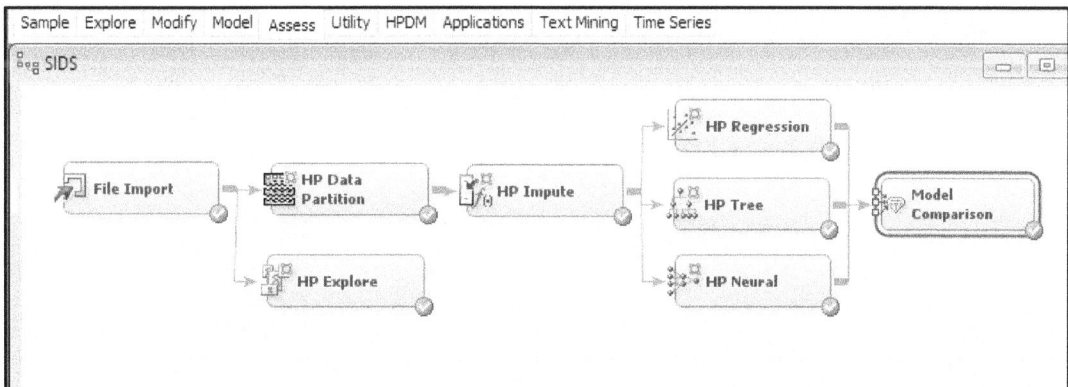

Step 23: Expand the Output Window Results for each model. Review the Fit Statistics for each model, including the misclassification rates and cumulative lift. With the results of the Model Comparison node, all three model results are included at once. In the Cumulative Lift chart, the Decision, Tree, Regression and Neural Network model perform similarly. For the validation misclassification rate, again all three models have a matching misclassification rate of 0.19%. The Decision Tree model is selected based on the slightly lower error rate. The decision tree can also prove to be a more valuable model because it can be interpreted easily.

Figure 11.21: Model Comparison Nodes Cumulative Lift

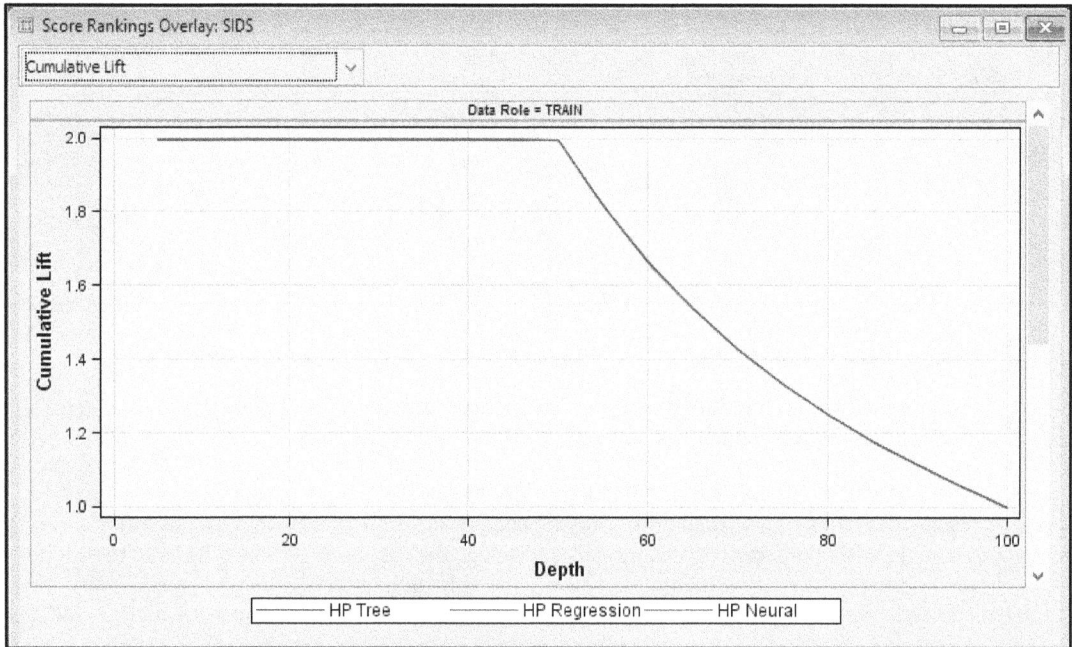

Figure 11.22: Model Comparison Nodes Fit Statistics

Selected Model ▼	Model Description	Target Variable	Selection Criterion: Valid: Misclassification Rate	Valid: Average Squared Error	Valid: Sum of Squared Errors
Y	HP Tree	SIDS	0.001905	0.001902	31.94537
	HP Neural	SIDS	0.001905	0.001916	32.18698
	HP Regres...	SIDS	0.001905	0.001919	32.24167

Step 24: Click the Decision Tree node in the model diagram, and expand the output window results. The first split of the tree is based on the Premature variable. The first split is to Node Id 1, Premature = 1 (or True), which contains the majority of the records. Following Node Id 3, the next split is MotherSmoke. The displayed English Rule for Node Id 3, where Premature = 1 and MotherSmoke = 1, shows that the infant is highly susceptible to SIDS. Using these IF…THEN rules can help reduce occurrence of SIDS, for example, through patient education that premature birth, smoking, and family history might increase the risk of SIDS.

Figure 11.23: Decision Tree Output

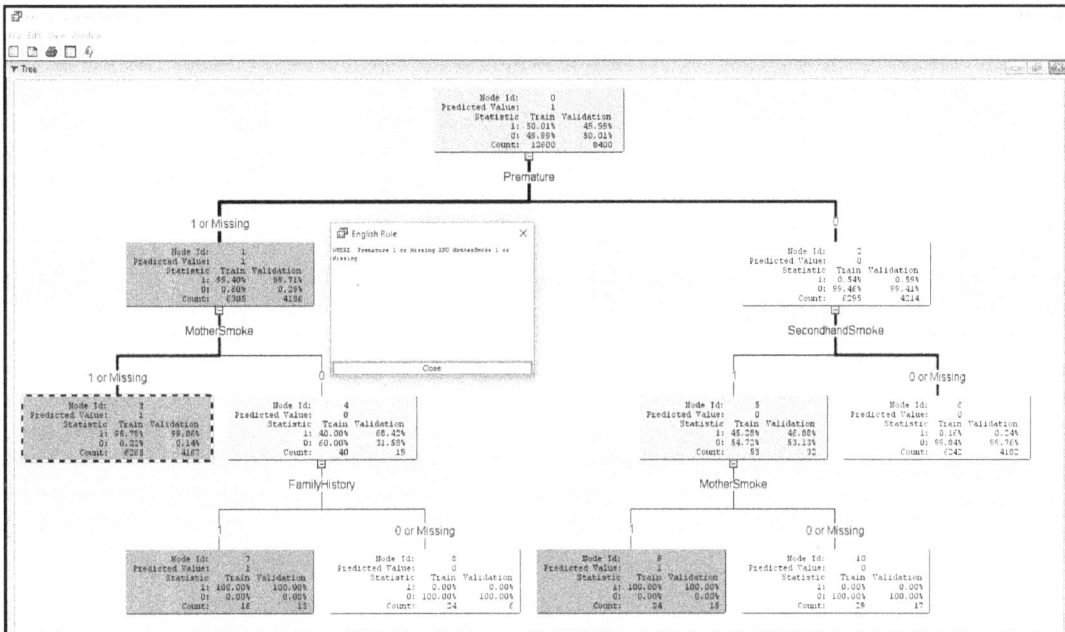

Model Summary

In summary, High-Performance Data Mining (HPDM) capabilities have been developed in SAS in part due to the rise of Big Data and the ability to leverage technology such as parallel processing and distributed processing.

- Input: Interval, Binary, or Nominal
- Target: Interval, Binary, or Nominal
- Model: HP Regression node, HP Tree node, and HP Neural node
- Evaluation: Assess Model Comparison node, or individual nodes evaluation: R^2, Adjusted R^2, Error, Lift, p-value, Odds Ratio, English Rules, Misclassification Rate

Healthcare Digital Transformation

In the context of future trends, digital disruption and digital transformation have perhaps the greatest potential to change markets more quickly than any other historical force. In Australia, an estimated one-third of companies are at immediate risk for disruption. Sectors such as finance, media, and information and communications technology are likely to experience the most immediate impact. As an advantage, sectors such as healthcare, which will also experience changes, have a longer timeline to react and to potentially learn about and more adequately prepare for the arriving change (Deloitte, 2017). In the U.S., Aetna recently announced plans to change their headquarters from Hartford, Connecticut to New York City. Aetna, a large health insurer, was founded in Hartford in 1853 and has been headquartered there for over 160 years. As part of the major change, Aetna's CEO indicated that the relocation was necessary due to their need to increase innovation through younger knowledge workers and technologists (Zacks, 2017).

Digital change within an industry can be measured along a continuum of no change, minor change, some core change, mainstream, predominately digital, and fully digital. Currently, most industries, including healthcare systems and services, fall into some core change area, indicating the future potential for change (Bughin et al., 2017). Despite the warning of impending change, in a Cisco Global Center for Digital Business Transformation survey, nearly half of companies have not acknowledged or addressed the threat of digital disruption, with only 1 in 4 taking a proactive approach to transform themselves before others do (Bradley, 2015). Based on an Accenture survey, by 2020 up to 40% of incumbent businesses are expected to be displaced by digital disruption, despite the fact that the digital economy will still represent only 25% of the overall global economy with even more room to grow (Adonis, 2018, Knickrehm et al., 2016). Several decades ago, the healthcare industry began its digital transformation through the introduction of electronic health records (EHRs). Today, increasing technologies (such as gene sequencing, consumer tracking products, robotics, and analytics) promise to make major shifts (Hewlett Packard, 2016). These capabilities will enable the next phase of digital transformation in the digital experience era, which enables digitally empowered consumers (or patients) to integrate experiences across all channels, improving content, engagement, and personalization (Seebacher, 2015).

McQuivey (2013) defines digital disruption as the new and extended opportunity for more people than ever before to directly interact with customers at the lowest costs possible. Whereas people, along with infrastructure, equate to disruption, innovators, along with digital infrastructure, equate to digital disruption, with massive digital disruption to occur at a scale and pace that is largely unprepared for by organizations. Using technology enablers, the digital disruptors can directly reach the end consumers and bypass all the traditional obstructions. In this newly created environment, continuously finding ways to adapt to the customers' requirements and to offer a better and more affordable solution is the ultimate goal and the essential key for successful disruption. Disruption creates the dilemma that companies need to face and decide upon: whether to create new and more affordable products, which might not always meet the existing customers' quality expectations, or to continue improving on the existing products for generating better profits. As disruption takes place at a continuously increasing pace, it is anticipated to create new trends in customers' expectations and reactions. Digital disruption increases the rate of competition and facilitates creativity with the cumulative effect of devastation to a company that operates under the norms of the previous century (McQuivey, 2013).

In healthcare, examples (such as FitNow, which is the producer of Lose It!, a personal fitness application) have proven that neither the size nor the tradition of a company matter when competing with a digital disruptor. Competing with traditional businesses like Weight Watchers and Jenny Craig, a team of eight employees and their new product Lose It!, attracted 10 million customers (Jan van Nouhuys, 2016). Digital disruption in healthcare represents nearly $9 trillion in potential value, and digital disruption is already occurring in several avenues of healthcare. However, barriers are present, including regulatory requirements, user adoption, and privacy and security (Newman, 2017).

Figure 11.24: Eras of Transformation in Healthcare

1.0 Era: EHR	2.0 Era: SMACITR	3.0 Era: DX
2000s	2010s	2020s
Provider Driven	Technology Driven	Data Driven
Value: Millions	Value: Billions	Value: Trillions

The first wave of digital technology change in healthcare was the adoption of electronic health records (EHRs) (Steger, 2014; Singh, 2017). This is considered the 1.0 era, where providers primarily in the 2000s, were adopting EHRs to improve quality and cost of care resulting in value of millions of dollars. The second wave includes SMACITR (social, mobile, analytics, cloud, IoT, and robotics and AI) technologies, and the shift to value-based care models, primarily in the 2010s. These digital technologies that make up SMACITR have great opportunity and potential within healthcare to reduce costs, improve efficiencies, and improve quality and healthcare value. Although valuable individually, when integrated, they can create additional benefits. Estimates in the U.S. health system from implementation of these technologies ranges from tens of billions to hundreds of billions of dollars in potential savings (Tiwari, 2016; ReportLinker, 2017; Collier et al., 2017; Ray, 2017; Deloitte, 2018). SMACITR technologies are an enabler and a transformer of the current business models to digitization, with the platform having the ability to increase the capabilities in a number of management areas. Transformations through SMACITR are already altering and changing the competitive environment. By 2020, there will be nearly 100 billion web-connected computing devices and nearly 2 zettabytes of data generated (Frank, 2013). IoT, for example, allows digital data exchange within a network of internet-enabled devices such as sensors. IoT is a key component of digital transformation within healthcare, as new business models are emerging, allowing improvements in patient outcomes and process improvements (Dimitrov, 2016; O'Brien, 2016; O'Brien, 2017). This is considered the 2.0 era, where stakeholders were adopting SMACITR technologies, resulting in a savings potential of billions of dollars.

The third wave includes the digital experience (DX) and transformation of healthcare delivery to include digital and non-digital capabilities primarily in the 2020s. This is considered the 3.0 era, with trillions of dollars in potential savings and capitalizing on the initial and ongoing investments from previous eras. The experience movement in healthcare emphasizes the human experience in which all beings care for others. It is embodied in every healthcare interaction in delivering the highest quality of care. The patient experience focuses on a core set of values, including accessibility, agility, innovation, inclusivity, and collaboration (Wolf, 2017). These values are also represented in the digital experience in which organizations are able to personalize digital experience solution to empower and engage patients with on-demand accessible information to improve patient outcomes.

Although nearly all of organizations claim to have a digital disruption strategy in place, the reality is that the exponential pace of change has created a capabilities gap within their organization. To improve capabilities, organizations can bring in new employees to act as catalysts for change, embed digital within the organizational culture, and be committed to digital. An example of a commitment is the appointment of a digital officer or dedicated staff members (Grossman, 2016). To address digital disruption, organizations might respond to minimize impact through three primary methods: 1) Improving legacy costs 2) Developing new revenue streams 3) Positioning corporate strategies (Deloitte, 2017). These methods provide a mechanism for increased agility within a changing environment and increased dynamic capabilities with the organization (Teece, et al., 2003, Deloitte, 2017). To be successful, organizations must capitalize on era 1.0 EHR and 2.0 SMACITR technologies to transform their business, to work directly with other organizations and customers, and to identify the processes where technologies can offer new strategic alternatives. Instead of focusing on technology first, organizations must focus on processes first, with the technology architecture following within the integrated process (Frank, 2013). Healthcare might arguably be the most complex industry for digital transformation (Sullivan, 2017). In order to enable digital transformation to improve the patient experience, healthcare organizations are recommended to include top leadership support, blending digital and physical experiences, using data and information assets for decision-making, and leveraging digital technologies for stakeholders (Sullivan, 2017). As a result of digital transformation, there is an ability to advance healthcare and improve patient outcomes and experiences. Similar to the 1990s, the 2020s are anticipated to be the beginning of another era of integration and significant change as a result of digital transformation in healthcare (Swartz, 2018).

Experiential Learning Application: Lifelogs

In Chapter 5, we introduced the concept of wearable technology, and, throughout the textbook, you have learned a variety of modeling techniques. In keeping with the learning approach of this book, for the final chapter experiential learning application, you will connect your knowledge from the prior chapters and complete an application on lifelogs.

An individual's lifelog is the set of digital information gathered from his or her life through smart devices, sensors, social network information, and medical data. This information tracked over a lifetime is the encapsulation of Big Data and is used to find patterns in the complex lifelog of individuals and used for real-time healthcare services. Using smart devices, individuals can be measured for personal health information (such as exercise, sleep, stress, heart rate via electrocardiogram, and blood pressure) and are provided with personalized healthcare services based on the resulting analysis. Researchers have developed frameworks for using lifelog data for detecting emergency events through smart devices (Choi et al., 2016).

Wearable technology along with augmented reality are two key technologies that have been increasingly combined and used over the last several years. Surgeons have started to use wearable technology during operations and, combined with augmented reality, can help improve interaction between providers and patients. The technology would allow surgeons to view a virtual 3-D map of a patient's internal organs without the need for an incision (Reynolds, 2017). There are many additional applications for wearable technology, such as in telemedicine and virtual examinations, as well as in education for healthcare professionals and patients through 3-D training tools (Govette, 2014; Govette, 2017). Shipments of wearable technology such as smart watches and glasses are forecasted to reach over 500 million devices with nearly $100 billion in revenue by 2021 (Wade, 2017). The first true wearable computer was invented and tested at a Las Vegas casino in the 1960s to help win at roulette. By using computers, the wearers were able to have superhuman capabilities as long as the technology remained hidden. A few years later, wearable technologies started to show up on people's faces, and these wearables were designed to enhance the wearer's ability to remember, record, and retrieve information. Recent wearable technologies that have generated significant interest and attention is Google Glass, an eyewear with connection to a smartphone, GPS, voice activation, and camera and video recording. Snapchat also has released Spectacles, an eyewear technology product that is similar to glasses, which allows users to take short videos or pictures, and to share them with friends or post online. These devices use demonstrated technology such as Wi-Fi, Bluetooth, Near Field Communication (NFC), and Global Positioning Systems (GPS) (Woodside and Reinhold, 2017).

A next generation of smart wearables involves smart clothing, which generates Big Data through tracking many physiological indicators. Health monitoring has a close relationship with Big Data due to ongoing monitoring capabilities, and assistance with reducing costs, and improved quality of healthcare. Application of smart clothing includes diagnosis, electrocardiographs, and emergency response. Given the future trends of population growth and aging, research indicates that wearable technology and smart computing as a potential solution to address the growing global shortage of medical facilities and personnel. Smart clothing and monitoring has applications for individuals living alone, children with autism, athletes, patients with depression, and pilots (Chen et al., 2016).

Smart devices can be used to monitor physiological signals and to alert medical professionals and medical centers of any emergency indicators. An emerging technology that is included within smart devices is the Internet of Things (IoT), which is the idea of an array of interconnected devices throughout one's environment. For example, hospitals can use IoT for location tracking of patients, staff, and medical devices. The IoT devices can also be used with environmental monitoring (such as verifying temperatures of refrigerators and IT centers). Companies such as Stanley and the open-source Kaa project seek to improve IoT integration and connectivity (Govette, 2017). Smart devices and lifelog analysis are an option to address rising chronic conditions. Smart devices follow closely with Big Data from the lifelog data collection and the use of analytics on cloud-based data storage. Today, most healthcare systems are set up for acute disease care, but are not prepared for prevention of chronic disease. Based on a World Health Organization study, current health systems can provide a solution for only 10% of medical issues. As a result, in the U.S., over half of the population has one or more chronic conditions, which account for 80% of health spending or nearly $3 trillion annually on chronic diseases alone. Nearly 50% of medical issues are dependent on individual lifestyle habits, 20% is a result of the environment, and 20% is the result of heredity factors. Therefore, smart devices and surveillance are useful strategies for improving existing healthcare systems and for addressing medical issues. The collection of large amounts of data can be used to improve efficiency, to increase quality of care, and to prevent chronic diseases (Chen, et al., 2016).

A sample has been provided for you to get started. The lifelog data set includes 18 variables and 20,012 records.

Objective: Determine the physiological factors that cause an emergency indicator from smart wearables.

Data Set File: 11_EL2_Life_Logs.xlsx

Variables:

- Person_ID, unique identifier
- Person_Gender, (F=Female, M=Male)
- Person_Weight, in pounds
- Person_BloodType, (A, B, AB, O)
- Person_Height, in inches

- Time_Hour, 24-hour time
- Blood_pressure_systolic, mmHg millimeters of mercury
- Blood_pressure_diastolic, mmHg millimeters of mercury
- Pulse, beats per minute
- DeviceID, unique identifier
- Device Type, (Smart Cloths, Smart Watch, Smart Phone)
- Body_Temp, in degrees Fahrenheit
- Diabetes_YN, indicator (Y=Yes/True, N=No/False)
- Alzheimer_YN, indicator (Y=Yes/True, N=No/False)
- CardiovascularDisease_YN, indicator (Y=Yes/True, N=No/False)
- Obesity_YN, indicator (Y=Yes/True, N=No/False)
- Hyperlipidemia_Cholesterol_YN, indicator (Y=Yes/True, N=No/False)
- Emergency_Y/N, indicator (Y=Yes/True, N=No/False)

Title	Lifelogs
Introduction	Provide a summary of the business problem or opportunity and the key objective(s) or goal(s). Create a new SAS Enterprise Miner project. Create a new Diagram.
Sample	Data (sources for exploration and model insights) Identify the variables data types, the input and target variable during exploration. Add a FILE IMPORT Provide a results overview following file import: Input / Target Variables Generate an HP DATA PARTITION
Exploration	Provide a results overview following data exploration Add an HP EXPLORE Summary statistics (average, standard deviation, min, max, and so on.) Descriptive Statistics Missing Data Outliers
Modify	Provide a results overview following modification Add an HP IMPUTE

Title	Lifelogs
Model	Discovery (prototype and test analytical models)
	Apply a regression, decision tree, and neural network model and provide a results overview following modeling.
	Add an HP REGRESSION
	Add an HP TREE
	Add an HP NEURAL Model description
	Analytics steps
	Model results (Lift, Error, Misclassification Rate)
	Selection Model
Assess and Reflection	ADD a MODEL COMPARSION
	Provide overall recommendations to business
	Model advantages / disadvantages
	Performance evaluation
	Model recommendation
	Summary analytics recommendations
	Summary informatics recommendations
	Summary business recommendations
	Summary clinical recommendations
	Deployment (operationalization plan: timeline, resources, scope, phases, project plan)
	Value (return on investment, healthcare outcomes)

Learning Journal Reflection

Review, reflect, and retrieve the following key chapter topics only from memory and add them to your learning journal. For each topic, list a one sentence description/definition. Connect these ideas to something you might already know from your experience, other coursework, or a current event. This follows our three-phase learning approach of 1) Capture, 2) Communicate, and 3) Connect. After completing, verify your results against your learning journal and update as needed.

Key Ideas – Capture	Key Terms – Communicate	Key Areas - Connect
Future Trends		
Digital Disruption and Digital Transformation		
Population and Consumer Changes		
Robotics and Automation		

Key Ideas – Capture	Key Terms – Communicate	Key Areas - Connect
Healthcare Globalization and Government		
Public Health		
Big Data Anamatics		
High-Performance Data Mining Nodes		
BEMO Process Steps		
SIDS Application		
LifeLogs Application		

Experiential Learning Application: Health Anamatics Project

From this textbook, I have shared my learning in hopes that these best practices will provide you with a competitive learning advantage for health anamatics in your career. Learning from others improves the efficiency of learning, and improving your current knowledge while connecting and expanding that knowledge to new knowledge is the key to long-term storage and lifelong learning. In Chapter 2 we discussed health anamatics and broccoli, now that you've completed this book, hopefully you have a renewed fondness of broccoli. Also consider if you have another 10, 20, 30, 40, or 50 years of career to go, how much learning you can gain, or helpings of broccoli that you can consume! As an even more appealing option, possibly consider the combination of health anamatics the same as combining your favorite foods like broccoli fries or broccoli ice cream!

You are now ready to apply and connect your learning and knowledge to new project areas in health anamatics. Below is a template to get started on a selected career project area.

Health Anamatics Project

Part I: Topic Overview
- 1-2 pages, single spaced
- Provide an executive summary on the topic explored
- Include 3 current industry technologies
- Include 3 current industry examples
- Develop a discussion question on the topic explored
- Post the topic executive summary and discussion question to discussion board.

Title:

Health Anamatics Project	
Executive Summary	Topic outline
	Big Picture Overview
	Key messages/objectives/goals
	Technologies/Analytics (3)
	Industry examples (3)
	Outside sources (3)
	Discussion Question:

Part II: Future Trends
- 1-2 pages, single spaced
- Develop a discussion question on the topic explored
- Post summary and 1 discussion question to discussion board

Future Trends	Next 5, 10, 25 years
	Future technology/business/legislative changes
	Impact/opportunity
	Outside sources (3)
	Discussion Question:

Part III: Final Project Document
- 7-10 pages, single-spaced (includes Parts I and II)
- Develop a health anamatics project for a selected organization and topic area

Executive Summary	Part I (use information completed in part I topic overview)
Organization	Name
	Industry
	General Environment
	Company Background
	Company Location
Healthcare-Business Problem/Opportunity Event	Who is the Decision Maker?
	What is the Decision to be Made? Business Question?
	When is the Decision to be Made?
	Why is the Decision important?
Decision Point	Problem-Specific Information (and alternative options)
	Technology/Analytics Options
Discussion/Decision Questions	Discussion/Decision Question 1:
	Discussion/Decision Question 2:
	Discussion/Decision Question 3:

Health Anamatics Project	
Data Set	Data description and definitions Data Resources • http://guides.lib.berkeley.edu/publichealth/healthstatistics/rawdata • www.pewinternet.org/datasets/ • https://www.cdc.gov/nchs/data_access/Vitalstatsonline.htm • https://archive.ics.uci.edu/ml/datasets.html • https://phpartners.org/health_stats.html
Data Mining	Data mining process (SEMMA, BEMO)
Recent Developments	What has occurred since the decision point to present
Implications	How should managers address this scenario? Best practices and recommendations
Future Trends	Part II (use information completed in part II future trends)
References	APA or MLA format Outside sources (10)
Exhibits	Financials Market Data

References

AAC&U. 2017. Integrative Learning. Available at: https://www.aacu.org/resources/integrative-learning.

AACSB. 2017. Eligibility Application for Business Accreditation. Available at: https://www.aacsb.edu/-/media/aacsb/docs/accreditation/eligibility%20application%20-%20process/business-eligibility-application-april2017.ashx?la=en.

AAPC. 2017. What is Medical Coding? Available at: https://www.aapc.com/medical-coding/medical-coding.aspx.

Abdous, M., and W. He. 2011. Using text mining to uncover students' technology-related problems in live video streaming. *British Journal of Educational Technology* 42(1):40–49.

Adonis, D. 2018. Digital Disruption - Cause and Effect. Whispir. Available at: http://www.whispir.com/news/digital-disruption-cause-and-effect.

AHA. 2017. Social Media Policy. American Hospital Association. Available at: http://www.aha.org/content/13/socialmediapolicy.pdf.

AHIMA. 2011. AHIMA Code of Ethics. Available at: http://bok.ahima.org/doc?oid=105098.

AHIMA. 2005. e-HIM Personal Health Record Work Group. Defining the personal health record. *Journal of AHIMA* 76(6):24–25.

AHIMA. 2017a. Get to Know AHIMA. The American Health Information Management Association. Available at: http://www.ahima.org/about/aboutahima.

AHIMA. 2017b. What is Health Information? Available at: http://www.ahima.org/careers/healthinfo.

Aikin, A. 2010. Case study: Social networking at the CDC. Washington, DC: CDC.

Almberg, M. January 21, 2016. Government funds nearly two-thirds of U.S. health care costs: American Journal of Public Health study. PNHP. Available at: http://www.pnhp.org/news/2016/january/government-funds-nearly-two-thirds-of-us-health-care-costs-american-journal-of-pub.

Alonso-Zaldivar, R., Gillum, J. 2015. Big Privacy Concerns Surround Government's Health Care Website. Inc. Available at: https://www.inc.com/associated-press/new-privacy-concerns-over-government-health-care-website.html.

American Dental Association (ADA). 2017. Code on Dental Procedures and Nomenclature (CDT Code). American Dental Association. Available at: http://www.ada.org/en/publications/cdt.

American Heart Association. 2016. Target Heart Rates. Available at: http://www.heart.org/HEARTORG/HealthyLiving/PhysicalActivity/FitnessBasics/Target-Heart-Rates_UCM_434341_Article.jsp#.V-KkZzXtoQM.

AMIA. 2017. The Science of Informatics. Available at: https://www.amia.org/about-amia/science-informatics.

Anderson, D.R., D.J. Sweeney, and T.A Williams. 2002. *Statistics for Business and Economics*, 8th ed. Cincinnati, OH South-Western Thomson Learning.

Anderson, K., Smith, L., and Garrett, D. 2012. Social media "likes" healthcare. PwC.

Andone, D. 2017. 10 hospitalized with botulism tied to nacho cheese sauce. CNN. Available at: http://www.cnn.com/2017/05/21/health/california-botulism-nacho-cheese/.

Andrews, J. 2011. Google Health shutdown spurs debate over PHR viability. Healthcare IT News. Available at: http://www.healthcareitnews.com/news/google-health-shutdown-spurs-debate-over-phr-viability.

Angeles, R., C.L. Corritoreb, S.C. Basuc, and R. Nath. 2001. Success factors for domestic and international electronic data interchange (EDI) implementation for US firms. *International Journal of Information Management* 330(21):329–347.

Angar, S. 2016. SAS Analytics in Action. SAS. https://www.sas.com/content/dam/SAS/de_de/doc/events/sas-partnertag/sas-partnertag-analytics-in-action.pdf

Apple. 2016. Your heart rate What it means, and where on Apple Watch you'll find it. https://support.apple.com/en-us/HT204666. Apple. 2018. iPhone X - Technical Specifications. Available at: https://support.apple.com/kb/SP770?locale=en_US

Archer, D., and T. Marmor. 2012. Medicare and Commercial Health Insurance: The Fundamental Difference. Health Affairs Blog. Available at: http://healthaffairs.org/blog/2012/02/15/medicare-and-commercial-health-insurance-the-fundamental-difference/.

Armstrong, C.M., V. Monteith, and D. Biel. 2010. ICD-10 Implementation for Health Care Providers: The business imperative for compliance. Deloitte Development.

Arts, D. G. T., N.F. de Keizer, and G.-J. Scheffer. 2002. Defining and Improving Data Quality in Medical Registries: A Literature Review, Case Study, and Generic Framework. *Journal of the American Medical Informatics Association: JAMIA* 9(6):600–611.

ASQ. 2016 What is Six Sigma? Available at: https://asq.org/learn-about-quality/six-sigma/overview/overview.html.

Associated Press. 2017. Maker of Nathan's hot dogs issues recall over metal concerns. The Washington Post. Available at: https://www.washingtonpost.com/business/maker-of-nathans-hot-dogs-issues-recall-over-metal-concerns/2017/05/21/748b36c0-3e4b-11e7-b29f-f40ffced2ddb_story.html?utm_term=.abda435d0569.

Auger, N., B.J. Potter, A. Smargiassi, M. Bilodeau-Bertrand, C. Paris, and T. Kosatsky. 2017. Association Between quantity and duration of snowfall and risk of myocardial infarction. *CMAJ* 189(6):E235–42.

Bakalar, N., K. Barrow, J. Huang, and D. Parker. 2012. Milestones in Medical Technology. The New York Times. Available at: http://www.nytimes.com/interactive/2012/10/05/health/digital-doctor.html?_r=0#/#time15_328.

Baltzan, P. 2015. *Business Driven Technology*, 6th ed. New York: McGraw-Hill Education.

Bass, D. 2017. A Bereaved Father and His Team of Microsoft Data Scientists Combat Infant Deaths. Bloomberg. Available at: https://www.bloomberg.com/news/articles/2017-06-07/a-bereaved-father-and-his-team-of-microsoft-data-scientists-combat-infant-deaths

Beaver, K., Herold, R. 2004. *The Practical Guide to HIPAA Privacy and Security Compliance*. Boca Raton, FL: Auerbach Publications.

Begley, S. 2017. House Republicans would let employers demand workers' genetic test results. STAT News. Available at: https://www.yahoo.com/news/house-republicans-let-employers-demand-100034946.html.

Bensusan, H. 2014. God doesn't always shave with Occam's razor – Learning When and How to Prune. University of Sussex, School of Cognitive and Computing Sciences.

Bertolucci, J. 2013. Big Data Analytics: Descriptive Vs. Predictive Vs. Prescriptive. *InformationWeek*.

Besler. 2018. How to calculate the LACE risk score. Available at: https://www.besler.com/lace-risk-score/.

Biernat, A. 2015. Healthcare costs unsustainable in advanced economies without reform. Organization for Economic Co-operation and Development. Available at: http://www.oecd.org/health/healthcarecostsunsustainableinadvancedeconomieswithoutreform.htm.

Bishaw, A., Macartney, S. 2010. Poverty: 2008 and 2009. Census.gov. Available at: https://www.census.gov/prod/2010pubs/acsbr09-1.pdf.

Bloom, N., R. Sadun, J. Van Reenen. 2014. Does Management Matter in Healthcare? Centre for Economic Performance.

BLS. 2016a. Healthcare Occupations. Bureau of Labor Statistics. Available at: https://www.bls.gov/ooh/healthcare/home.htm.

BLS. 2016b. Employment Projections: 2014-24 Summary. Bureau of Labor Statistics 2015. Available at: https://www.bls.gov/news.release/ecopro.nr0.htm.

BLS. 2017. Number of Jobs, Labor Market Experience, and Earnings Growth Among Americans at 50: Results from a Longitudinal Survey. Bureau of Labor Statistics. Available at: https://www.bls.gov/news.release/pdf/nlsoy.pdf

Blue Cross Blue Shield of Michigan (BCSBM). 2017. How to take your health assessment. Available at: http://www.bcbsm.com/index/health-insurance-help/faqs/topics/understanding-benefits/your-health-assessment.html.

Bostan, C. et al. 2014. Biological health or lived health: which predicts self- reported general health better? *BMC Public Health* 14:189.

Bradley, J. et al. 2015. Digital Vortex. Global Center for Digital Business Transformation. Available at: https://www.cisco.com/c/dam/en/us/solutions/collateral/industry-solutions/digital-vortex-report.pdf.

Brady, J., K. Ho, E. Kelley, and C.M. Clancy. 2007. AHRQs National Healthcare Quality and Disparities Reports: An Ever-Expanding Road Map for Improvement. *Health Services Research* 42(3):ix–xxi.

Bresnick, J. 2015. Four Use Cases for Healthcare Predictive Analytics, Big Data. HealthITAnalytics. Available at: https://healthitanalytics.com/news/four-use-cases-for-healthcare-predictive-analytics-big-data.

Brown, B., Sikes, J. 2012. Minding your digital business: McKinsey Global Survey results. McKinsey.

Brown, P.C., H.L. Roediger, and M.A. McDaniel. 2014. *Make it Stick: The Science of Successful Learning*. Cambridge, MA: The Belknap Press of Harvard University Press.

Bryan, B. 2017. Some of the most powerful people in the US are talking about a massive change to healthcare. Business Insider. Available at: http://www.businessinsider.com/warren-buffett-munger-aetna-ceo-trump-on-single-payer-healthcare-2017-5.

Bughin, J., L. LaBerge, and A. Mellbye. 2017. The case for digital reinvention. McKinsey Quarterly. Available at: https://www.mckinsey.com/business-functions/digital-mckinsey/our-insights/the-case-for-digital-reinvention.

Burke, L., and B. Weill. 2013. *Information Technology for the Health Professions,* 4th ed. Upper Saddle River, NJ: Pearson Education.

Byrne, C.M., L.M. Mercincavage, E.C. Pan, A.G., Vincent, D.S. Johnston, B. Middleton. 2010. The value from investments in health information technology at the U.S. Department of Veterans Affairs. *Health Affairs* 29(4):629–638.

Caldwell, L. A., and V. Hillyard. July 25, 2017. Senate Opens Debate on Health Care, Votes Down Repeal and Replace. NBC News. Available at: https://www.nbcnews.com/politics/congress/senate-sets-sights-skinny-repeal-obamacare-tuesday-s-voting-n786296.

Carson Tahoe Health. 2018. New technology taking health care to the next level at Carson Tahoe Health. Available at: https://www.nevadaappeal.com/news/local/new-technology-taking-health-care-to-the-next-level-at-carson-tahoe-health/.

CDC. 2011. Text Messaging. Centers for Disease Control and Prevention. Available at: https://www.cdc.gov/socialmedia/Data/Briefs/mobiletext.pdf.

CDC. 2013. Behavioral Risk Factor Surveillance System. Available at: https://www.cdc.gov/brfss/acbs/2013_documentation.html.

CDC. 2014. Text Messaging Guidelines & Best Practices. Centers for Disease Control and Prevention. Available at: https://www.cdc.gov/socialmedia/tools/guidelines/textmessaging.html.

CDC. 2015. CDC Health Disparities & Inequalities Report (CHDIR). Available at: https://www.cdc.gov/minorityhealth/chdireport.html.

CDC. 2016a. National Center for Chronic Disease Prevention and Health Promotion, Division of Population Health. Chronic Disease Indicators (CDI) Data. Available at: https://nccd.cdc.gov/cdi.

CDC. 2016b. Multiple Chronic Conditions. Available at: https://www.cdc.gov/chronicdisease/about/multiple-chronic.htm.

CDC. 2016c. Cholera General Information. https://www.cdc.gov/cholera/general/index.html.

CDC. 2016d. Mobile Telephone Text Messaging for Medication Adherence in Chronic Disease: A Meta-analysis. Centers for Disease Control and Prevention. Available at: https://www.cdc.gov/dhdsp/pubs/docs/sib_april2016.pdf.

CDC. 2017a. Heart Attack. Available at: https://www.cdc.gov/heartdisease/heart_attack.htm.

CDC. 2017b. Recommended Immunization Schedule for Children and Adolescents Aged 18 Years or Younger. Available at: https://www.cdc.gov/vaccines/schedules/hcp/child-adolescent.html.

CDC. 2017c. Recommended Immunization Schedules for Adults. Available at: https://www.cdc.gov/vaccines/schedules/hcp/adult.html.

CDC. 2017d. Behavioral Risk Factor Surveillance System (BRFSS) Prevalence Data, Table of Immunization. Available at: https://chronicdata.cdc.gov/Behavioral-Risk-Factors/BRFSS-Table-of-Immunization/ehiz-zk8j.

CDC. 2017e. Mission, Role and Pledge. Available at: https://www.cdc.gov/about/organization/mission.htm.

CDC. 2017f. CDC Strategic Framework. Available at: https://www.cdc.gov/about/organization/strategic-framework/index.html.

CDC. 2017g. New CDC report: More than 100 million Americans have diabetes or prediabetes. Available at: https://www.cdc.gov/media/releases/2017/p0718-diabetes-report.html.

CDC. 2017h. Sudden Unexpected Infant Death and Sudden Infant Death Syndrome. Centers for Disease Control. Available at: https://www.cdc.gov/sids/data.htm.

CDC. 2017i. SUID and SDY Case Registries. Available at: https://www.cdc.gov/sids/caseregistry.htm.

Chae, Y. M., H. S. Kim, C. K. Tark, J. H. Park, and S. H. Ho. 2003. Analysis of healthcare quality indicator using data mining and decision support system. *Expert Systems with Applications* 24(2):167–172.

Chai, M., J. Manyika, and M. Miremadi. 2016. Where machines could replace humans—and where they can't (yet). McKinsey. Available at: http://www.mckinsey.com/business-functions/digital-mckinsey/our-insights/where-machines-could-replace-humans-and-where-they-cant-yet.

Change Healthcare. 2017. Inspiring a Better Healthcare System. Available at: http://www.changehealthcare.com/about-us/company-overview.

Chapman, J., S. Schetzsle, and R. Wahlers. 2016. An innovative, experiential-learning project for sales management and professional selling students. *Marketing Education Review* 26(1):45–50.

Cheek, K. 2014. "An overview of analytics in healthcare payers", in *Analytics in Healthcare and the Life Sciences: Strategies, Implementation Methods, and Best Practices*, ed. D. McNeill. Upper Saddle River, NJ: Editor. Pearson.

Chen, P.W. January 7, 2010. Are Doctors Ready for Virtual Visits? The New York Times. Available at: https://www.nytimes.com/2010/01/07/health/07chen.html?_r=0.

Chen, M., Y. Ma, J. Song, C. Lai, and B. Hu. 2016. Smart clothing: Connecting human with clouds and big data for sustainable health monitoring. Mobile Network Applications.

Choi, J., C. Choi, H. Ko, and P. Kim. 2016. Intelligent healthcare service using health lifelog analysis. *Journal of Medical Systems* 40(8):1–10.

Choi, E., A. Schuetz, W. F. Stewart, and J. Sun. March 2017. Using recurrent neural network models for early detection of heart failure onset. *Journal of The American Medical Informatics Association* 24(2):361–370.

Christensen, C. 2017. Key Concepts. Available at: http://www.claytonchristensen.com/key-concepts/.

Chui, M., J. Manyika, and M. Miremadi. 2016. Where Machines Could Replace Humans - and Where They Can't (Yet). McKinsey Quarterly. Available at: https://www.mckinsey.com/business-functions/digital-mckinsey/our-insights/where-machines-could-replace-humans-and-where-they-cant-yet.

Clancy, C. M., and K. Cronin. 2005. Evidence-based decision making: Global evidence, local decisions. *Health Affairs* 24(1):151–162.

Cleveland Clinic. 2017. Social Media Policy. Available at: https://my.clevelandclinic.org/about/website/social-media.

CMS. 2011. Healthcare Common Procedure Coding System (HCPCS) Level II Coding Procedures. Available at: https://www.cms.gov/Medicare/Coding/MedHCPCSGenInfo/Downloads/HCPCSLevelIICodingProcedures7-2011.pdf.

CMS. 2012a. The ICD-10 Transition: An Introduction.

CMS. 2012b. Chronic Conditions among Medicare Beneficiaries, Chartbook: 2012 Edition. Available at: https://www.cms.gov/Research-Statistics-Data-and-Systems/Statistics-Trends-and-Reports/Chronic-Conditions/Downloads/2012Chartbook.pdf.

CMS. 2013. HCPCS Level II Coding Process & Criteria. Available at: https://www.cms.gov/Medicare/Coding/MedHCPCSGenInfo/HCPCSCODINGPROCESS.html.

CMS. 2014. Institutional paper claim form (CMS-1450). Available at: https://www.cms.gov/Medicare/Billing/ElectronicBillingEDITrans/15_1450.html.

CMS. 2016a. The National Health Expenditure Accounts. U.S. Department of Commerce, Bureau of Economic Analysis; and U.S. Bureau of the Census. Available at: https://www.cms.gov/Research-Statistics-Data-and-Systems/Statistics-Trends-and-Reports/NationalHealthExpendData/NationalHealthAccountsHistorical.html.

CMS. 2016b. Professional Paper Claim Form (CMS-1500). Available at: https://www.cms.gov/Medicare/Billing/ElectronicBillingEDITrans/16_1500.html.

CMS. 2016c. Readmissions Reduction Program (HRRP). CMS.gov. Available at: https://www.cms.gov/medicare/medicare-fee-for-service-payment/acuteinpatientpps/readmissions-reduction-program.html.

CMS. 2016a. the 80/20 rule increases value for consumers for fifth year in a row. Available at: https://www.cms.gov/CCIIO/Resources/Forms-Reports-and-Other-Resources/Downloads/Medical-Loss-Ratio-Annual-Report-2016-11-18-FINAL.pdf.

CMS. 2016. Code Sets Overview. Available at: https://www.cms.gov/Regulations-and-Guidance/Administrative-Simplification/Code-Sets/index.html.

CMS. 2016b. Health Expenditures by Age and Gender. Centers for Medicare & Medicaid Services. Available at: https://www.cms.gov/Research-Statistics-Data-and-Systems/Statistics-Trends-and-Reports/NationalHealthExpendData/Age-and-Gender.html.

CMS. 2017a. Eligible Professional Meaningful Use Table of Contents Core and Menu Set Objectives. Available at: https://www.cms.gov/Regulations-and-Guidance/Legislation/EHRIncentivePrograms/downloads/EP-MU-TOC.pdf.

CMS. 2017b. Payment-State. Available at: https://data.medicare.gov/Hospital-Compare/Payment-State/98ix-2iqy/about.

CMS. 2017c. About CMS. Available at: https://www.cms.gov/About-CMS/Agency-Information/CMS-Strategy/Downloads/CMS-Strategy.pdf.

CMS. 2017d. History. Available at: https://www.cms.gov/About-CMS/Agency-Information/History/index.html.

CMS. 2017e. NHE Fact Sheet. CMS.gov. Available at: https://www.cms.gov/research-statistics-data-and-systems/statistics-trends-and-reports/nationalhealthexpenddata/nhe-fact-sheet.html.

CMS. 2017f. How We Use Your Data. Healthcare.gov. Available at: https://www.healthcare.gov/how-we-use-your-data/.

CMS. 2017g. Glossary. Available at: https://www.cms.gov/apps/glossary/default.asp?Letter=ALL&Audience=7.

CMS. 2017h. ICD-10. Available at: https://www.cms.gov/Medicare/Coding/ICD10/.

CMS. 2017i. CMS President Speeches. Available at: https://www.cms.gov/About-CMS/Agency-Information/History/downloads/CMSPresidentsSpeeches.pdf.

Cognizant. 2014. Cognizant to Acquire TriZetto, Creating a Fully-Integrated Healthcare Technology and Operations Leader. Available at: https://investors.cognizant.com/2014-09-15-Cognizant-to-Acquire-TriZetto-Creating-a-Fully-Integrated-Healthcare-Technology-and-Operations-Leader.

Collier, M., Fu, R., Yin, L. 2017. Artificial intelligence: Healthcare's new nervous system. Accenture. Available at: https://www.accenture.com/us-en/insight-artificial-intelligence-healthcare.

Columbus, C. 2017. Could drones help save people in cardiac arrest? Available at: http://www.npr.org/sections/health-shots/2017/06/13/532639836/could-drones-help-save-people-in-cardiac-arrest.

Commins, J. 2015. Anthem Data Breach: Potential Game Changer for Healthcare. HealthLeaders Media. Available at: https://www.medpagetoday.com/PublicHealthPolicy/PublicHealth/49914.

Connor, P.E. 1997. Total quality management: A selective commentary on its human dimensions, with special reference to its downside. *Public Administration Review* 57(6):501–509.

Cottrell, Q. 2017. Bottled water is now more popular than soda — why you should avoid both. MarketWatch. Available at: http://www.marketwatch.com/story/bottled-water-overtakes-soda-as-americas-no-1-drink-why-you-should-avoid-both-2017-03-10?siteid=yhoof2&yptr=yahoo.

Court, E. 2017a. Zocdoc CEO Oliver Kharraz on why 2017's Health-care Uncertainty is a Good Thing. MarketWatch. Available at: https://www.marketwatch.com/story/zocdoc-ceo-oliver-kharraz-on-why-2017s-health-care-uncertainty-is-a-good-thing-2017-01-04.

Court, E. 2017. UnitedHealth Shares Plunge 4%, but Wall Street Analysts Respond with a Shrug. Available at: https://www.marketwatch.com/story/lawsuit-sends-unitedhealth-shares-plunging-4-but-wall-street-analysts-respond-with-a-shrug-2017-02-17.

Cox, Jr. L.A. 2013. Caveats for causal interpretations of linear regression coefficients for fine particulate (PM2.5) air pollution health effects. *Risk Analysis* 33(12):2111–2125.

Cronin, R.M., et al. 2015. National Veterans Health Administration inpatient risk stratification models for hospital-acquired acute kidney injury. *Journal of the American Medical Informatics Association* 22(5):1054–1071.

Cumming, R.B., D. Knutson, B.A. Cameron, and B. Derrick. 2002. A Comparative Analysis of Claims-based Methods of Health Risk Assessment for Commercial Populations. Society of Actuaries.

Davenport, T.H. 2013. Analytics 3.0. Harvard Business Review.

Davenport, T.H., and J.D. Miller. 2014. "An Overview of Analytics in Healthcare Providers". In: *Analytics in Healthcare and the Life Sciences: Strategies, Implementation Methods, and Best Practices*, ed. D. McNeill. Upper Saddle River, NJ: Pearson.

Davis, J.L. 2018. Women's Heart Attacks: How They Differ. WebMD. Available at: https://www.webmd.com/heart-disease/features/womens-heart-attacks-how-they-differ.

DEA. 2017a. DEA Mission Statement. Drug Enforcement Agency. Available at: https://www.dea.gov/about/mission.shtml.

DEA. 2017b. AlphaBay, the Largest Online 'Dark Market', Shut Down. Drug Enforcement Agency. Available at: https://www.dea.gov/divisions/hq/2017/hq072017.shtml.

DeGaspari, J. 2010. Healthcare Data Experiencing Explosive Growth. Healthcare Informatics.

Deloitte. 2016. 2016 Global health care outlook: Battling costs while improving care.

Deloitte. 2017. Digital disruption - Harnessing the 'bang'. Deloitte. Available at: http://www.deloitte.com/view/en_AU/au/news-research/luckycountry/digital-disruption/index.htm.

Deloitte. 2018. Improving health care efficiency with social, mobile, analytics, and cloud. Deloitte. Available at: https://www2.deloitte.com/us/en/pages/life-sciences-and-health-care/articles/social-mobile-analytics-cloud.html.

Desai, M., J. Chabra, and A.P. Ganpule. 2015. Robotic surgery is ready for prime time in India: For the motion. *Journal of Minimal Access Surgery* 11(1):2–4.

Dietshce, E. 2017. Car crashes into a pole, bringing down Epic EHR at Jefferson Healthcare. Becker's Hospital Review. Available at: http://www.beckershospitalreview.com/healthcare-information-technology/car-crashes-into-a-pole-bringing-down-epic-ehr-at-jefferson-healthcare.html.

Dimitrov, D.V. 2016. Medical internet of things and big data in healthcare. *Healthcare Informatics Research* 22(3):156–163.

Disney. 2017. Inside Out. Disney-Pixar. Available at: https://movies.disney.com/inside-out.

Dornhelm, E. 2017. US Average FICO Score Hits 700: A Milestone for Consumers. FICO Blog. Available at: http://www.fico.com/en/blogs/risk-compliance/us-average-fico-score-hits-700-a-milestone-for-consumers/.

Elizalde, E. 2016. Accepting Facebook friend requests may lengthen users' lives, study says. New York Daily News. Available at: http://www.nydailynews.com/life-style/health/accepting-facebook-friend-requests-leads-longer-life-report-article-1.2859772.

Epic Systems Corporation. 2017. Managed Care. Available at: https://www.epic.com/software#ManagedCare.

Evariant. 2017. About Evariant. Evariant, Inc. Available at: https://www.evariant.com/company/about-evariant/.

Falotico, R., C. Liberati, and P. Zappa. 2015. Identifying oncological patient information needs to improve e-health communication: A preliminary text-mining analysis. *Quality and Reliability Engineering International* 1115:doi:10.1002/qre.1853.

FDA. 2016a. Products and Medical Procedures. Available at: https://www.fda.gov/MedicalDevices/ProductsandMedicalProcedures/default.htm.

FDA. 2016b. Medical Devices Basics. Available at: http://www.fda.gov/AboutFDA/Transparency/Basics/ucm193731.htm.FDA. 2017a. What We Do. Available at: https://www.fda.gov/AboutFDA/WhatWeDo/default.htm.

FDA. 2017b. What Does FDA Regulate? Available at: https://www.fda.gov/AboutFDA/Transparency/Basics/ucm194879.htm.

FDA. 2017c. National Drug Code Directory. US Food & Drug Administration. Available at: https://www.fda.gov/Drugs/InformationOnDrugs/ucm142438.htm.

FDA. 2018. Digital Health. U.S. Food and Drug Administration. Available at: https://www.fda.gov/medicaldevices/digitalhealth/.

Fickenscher, K.M. 2005. The new frontier of data mining. *Health Management Technology* 26(10):26–30.

Field, A., J. Miles, and Z. Field. 2012. *Discovering Statistics Using R*. Thousand Oaks, CA: Sage.

Fisher, C. 2016. LinkedIn Unveils the Top Skills that can get you Hired in 2017, Offers Free Courses for a Week. LinkedIn Blog. Available at: https://blog.linkedin.com/2016/10/20/top-skills-2016-week-of-learning-linkedin.

FitzGerald, B. 2016. Telemedicine Adoption Continues Growth in 2016 and Beyond. HIMSS Analytics.

Fleming, J.H., C. Coffman, and J. K. Harter. 2005. Manage your human sigma. *Harvard Business Review* 83(7/8):106–114.

Florida Blue. 2015. Make This Your Healthiest Year Yet. Available at: https://www.floridablue.com/members/health-and-wellness/make-your-healthiest-year-yet.

Foley, K.E. 2017. For the first time, more than half of Americans are getting the recommended amount of exercise. Quartz. Available at: https://qz.com/989773/for-the-first-time-more-than-half-of-americans-are-getting-the-recommended-amount-of-exercise/

Ford, E., Carroll, J. A., H.E. Smith, D. Scott, and J. Cassell. 2016. Extracting information from the text of electronic medical records to improve case detection: a systematic review. *Journal of the American Medical Informatics Association* 23(5):1007–1015.

Fox, M. 2017. Preterm Birth Rates Have Increased in the U.S. NBC. Available at: http://www.nbcnews.com/health/health-news/preterm-birth-rates-have-increased-u-s-n778576.

Frank, M. 2013. Don't Get SMACked: How Social, Mobile, Analytics and Cloud Technologies are Reshaping the Enterprise. Cognizant.

Frazier, M. 2010. What We Mortals Can Learn From the 4-Minute Mile. NoMeatAthlete.

Friedman, G. 2017. Italy Is the Mother of All Systemic Threats. Forbes. Available at: https://www.forbes.com/sites/johnmauldin/2016/09/13/italy-is-the-mother-of-all-systemic-threats/.

Gartee, R. 2011. *Health Information Technology and Management*. Upper Saddle River, NJ: Pearson.

Gebrehiwot, T. G., M.S. Sebastian, K. Edin, and I. Goicolea. 2015. The health extension program and its association with change in utilization of selected maternal health services in Tigray region, Ethiopia: A segmented linear regression analysis. PLoS One.

Geller, J.S. 2017. "Exclusive: Upcoming Apple Watch to include game-changing health features". BGR. Available at: https://bgr.com/2017/05/15/apple-watch-fitness-glucose-monitoring/.

Glynn, T. 2014. The future of the doctor-patient relationship - A critique of current thinking. *Healthcare and Informatics Review Online* Available at: http://c.ymcdn.com/sites/www.hinz.org.nz/resource/collection/0F09C2E4-7A05-49FB-8324-709F1AB2AA2F/P2_Glynn.pdf.

Goldstein, A. 2017. IRS won't withhold tax refunds if Americans ignore ACA insurance requirement. The Washington Post. Available at: https://www.washingtonpost.com/national/health-science/trump-health-officials-propose-rule-to-shore-up-affordable-care-act-marketplaces/2017/02/15/1f69bd7c-f307-11e6-b9c9-e83fce42fb61_story.html?utm_term=.4fe50836d5f4.

Govette, J. 2014. The 7 Biggest Innovations in Health Care Technology in 2014. referralMD. Available at: https://getreferralmd.com/2013/11/health-care-technology-innovations-2013-infographic/.

Govette, J. 2017. 17 Amazing Healthcare Technology Advances of 2017. referralMD. Available at: https://getreferralmd.com/2017/01/17-future-healthcare-technology-advances-of-2017-referralmd/.

Grossman, R. 2016. The Industries That Are Being Disrupted the Most by Digital. Harvard Business Review. Available at: https://hbr.org/2016/03/the-industries-that-are-being-disrupted-the-most-by-digital.

Haliasos, N., K. Rezajooi, K.S. O'neill, J. Van Dellen, A. Hudovsky, S. Nouraei. 2010. Financial and clinical governance implications of clinical coding accuracy in neurosurgery: A multidisciplinary audit. *British Journal of Neurosurgery* 24(2):191–195.

Hall, J.L., and D. McGraw. 2014. For Telehealth to succeed, privacy and security risks must be identified and addressed. *Health Affairs* 33(2):216–221.

Handelsman, D. 2014. Surveying the analytical landscape in life sciences organizations. In: *Analytics in Healthcare and Life Sciences*, pp. 31–38. Upper Saddle River, NJ: Pearson.

Hao, H., and K. Zhang. 2016. The voice of Chinese health consumers: A text mining approach to web-based physician reviews. *Journal of Medical Internet Research* 18(5):e108–e120.

Hassan, A. A. Pons., and D. Collins. 2003. Global e-commerce: a framework for understanding and overcoming the trust barrier. *Information Management & Computer Security* 11(3):130–138.

HealthIT.gov. 2011. Information Security Policy Template. Available at: https://www.healthit.gov/providers-professionals/implementation-resources/information-security-policy-template.

Healthcare.gov. 2018. Rate Review & the 80/20 Rule. Available at: https://www.healthcare.gov/health-care-law-protections/rate-review/.

HealthIT.gov. 2013a. What is a Personal Health Record? Available at: https://www.healthit.gov/providers-professionals/faqs/what-personal-health-record.

HealthIT.gov. 2013b. How to Implement EHRs. Available at: https://www.healthit.gov/providers-professionals/ehr-implementation-steps.

Health IT Analytics. 2015. Four Use Cases for Healthcare Predictive Analytics, Big Data. Health IT Analytics. Retrieved from http://healthitanalytics.com/news/four-use-cases-for-healthcare-predictive-analytics-big-data.

HealthIT.gov. 2016a. What Is an Electronic Medical Record (EMR)? Available at: http://www.healthit.gov/providers-professionals/electronic-medical-records-emr.

HealthIT.gov. 2016b. Hospital health IT developers. Available at: https://dashboard.healthit.gov/quickstats/pages/FIG-Vendors-of-EHRs-to-Participating-Hospitals.php.

HealthGrid. 2017. HealthGrid Honored as Recipient of the 2017 Microsoft Health Innovation Awards. Available at: http://www.prnewswire.com/news-releases/healthgrid-honored-as-recipient-of-the-2017-microsoft-health-innovation-awards-300410821.html.

HealthIT.gov. 2017a. EHR Incentive Payment Timeline. Available at: https://www.healthit.gov/providers-professionals/ehr-incentive-payment-timeline.

HealthIT.gov. 2017b. Meaningful Use Definition & Objectives. Available at: https://www.healthit.gov/providers-professionals/meaningful-use-definition-objectives.

Hebda, T. and Czar, P. 2013. *Handbook of Informatics for Nurses & Healthcare Professionals*, 5th ed. Boston: Pearson Education.

Heilbrunn, E. 2014. "Top Health Insurance Companies". US News & World Report. Available at: https://health.usnews.com/health-news/health-insurance/articles/2013/12/16/top-health-insurance-companies.

Hennick, C. 2007. Will your health plan cover you overseas? Budget Travel. Available at: http://www.budgettravel.com/bt-dyn/content/article/2007/03/26/AR2007032600711.html.

Henry, J., et al. 2016. Adoption of Electronic Health Record Systems among U.S. Non-Federal Acute Care Hospitals: 2008-2015. ONC Data Brief 35. Available at: https://dashboard.healthit.gov/evaluations/data-briefs/non-federal-acute-care-hospital-ehr-adoption-2008-2015.php.

Heudecker, N. 2014. Big Data Challenges Move from Tech to the Organization. Gartner.

Hewlett Packard Enterprise. 2016. Enterprise.nxt. HPE. Available at: https://h20195.www2.hpe.com/V2/GetPDF.aspx/4AA6-7153ENW.pdf.

HHS. 2013. Minimum Necessary Requirement. Office for Civil Rights. Available at: http://www.hhs.gov/ocr/privacy/hipaa/understanding/coveredentities/minimumnecessary.html.

HHS. 2015. Better, Smarter, Healthier: In historic announcement, HHS sets clear goals and timeline for shifting Medicare reimbursements from volume to value. Available at: https://www.hhs.gov/about/news/2015/01/26/better-smarter-healthier-in-historic-announcement-hhs-sets-clear-goals-and-timeline-for-shifting-medicare-reimbursements-from-volume-to-value.html.

HHS. 2016. Summary of the HIPAA Privacy Rule. Available at: https://www.hhs.gov/hipaa/for-professionals/privacy/laws-regulations/index.html?language=esv.

HHS. 2017a. HHS Agencies & Offices. Available at: https://www.hhs.gov/about/agencies/hhs-agencies-and-offices/index.html.

HHS. 2017b. HHS FY 2017 Budget in Brief. Available at: https://www.hhs.gov/about/budget/fy2017/budget-in-brief/index.html.

HHS. 2017c. President's HHS FY 2017 Budget Factsheet. Available at: https://www.hhs.gov/about/budget/fy2017/budget-factsheet/index.html.

HHS. 2017d. Other Administrative Simplification Rules. Office for Civil Rights. Available at: https://www.hhs.gov/hipaa/for-professionals/other-administration-simplification-rules/index.html.

Highmark. 2017. Ensuring Maximum Medicare Revenues. Available at: http://www.sas.com/en_us/customers/highmark-enterprise-miner.html.

Himmelstein, D.U. 2014. A Comparison of Hospital Administrative Costs in Eight Nations: U.S. Costs Exceed All Others by Far. Health Affairs. Available at: https://www.healthaffairs.org/doi/10.1377/hlthaff.2013.1327.

Himmelstein, D.U., M. Jun, R. Busse, et al. 2014. A comparison of hospital administrative costs in eight nations: US costs exceed all others by far. *Health Affairs* 33(9):1586–1594.

Himmelstein, D.U., and S. Woolhandler. 2016. The current and projected taxpayer shares of US health costs. *American Journal of Public Health* 106(3):449–452.

HIMSS. 2012. Definitions of mHealth. Available at: http://www.himss.org/ResourceLibrary/GenResourceDetail.aspx?ItemNumber=20221.

HIMSS. 2014. Health Informatics Defined. Available at: https://www.himss.org/health-informatics-defined.

HIQA. 2012. What you should know about Data Quality. Health Information and Quality Authority.

Hirsch, L. and A. Sherman. 2018. Walgreens and AmerisourceBergen deal talks have cooled as takeover looks unlikely. CNBC. Available at: https://www.cnbc.com/2018/02/27/walgreens-and-amerisourcebergen-deal-talks-of-cooled-.html.

Hora, M.T. Beyond the skills gap. National Association of Colleges and Employers. Available at: http://www.naceweb.org/career-readiness/trends-and-predictions/beyond-the-skills-gap/.

Hughes, G. 2011. How big is 'Big Data' in healthcare? A Shot in the Arm Blog. Cary, NC: SAS Institute, Inc.

Humer, C. 2017. Anthem to leave Ohio's Obamacare insurance market in 2018. Reuters. Available at: https://finance.yahoo.com/news/anthem-plans-leave-obamacare-market-170058765.html.

Humer, C., Finkle, J. September 24, 2014. Your medical record is worth more to hackers than your credit card. Reuters. Available at: http://news.yahoo.com/medical-record-worth-more-hackers-credit-card-182251915--finance.html.

Hurley, R. F., and H. Estelami. 2007. An exploratory study of employee turnover indicators as predictors of customer satisfaction. *Journal of Services Marketing* 21(3):186–199.

IBM. 2014. The Four V's of Big Data.

IBM. 2015. Cognitive Computing. IBM Research.

IDRE. 2017. How Do I Interpret Odds Ratios in Logistic Regression? UCLA: Statistical Consulting Group. Available at: https://stats.idre.ucla.edu/stata/faq/how-do-i-interpret-odds-ratios-in-logistic-regression/.

IRS. 2018. Individual Shared Responsibility Provision. Internal Revenue Service. Available at: https://www.irs.gov/affordable-care-act/individuals-and-families/individual-shared-responsibility-provision.

iSixSigma. 2018. What is Six Sigma? Available at: https://www.isixsigma.com/new-to-six-sigma/getting-started/what-six-sigma/.

Ithaca College. 2017. What is Integrative Learning? Available at: https://www.ithaca.edu/icc/what_is_it/

Jan van Nouhuys, R. 2016. Short fuse, big bang: 4 ways that digital disruption will change your industry. Whiteboard. Available at: http://www.whiteboardmag.com/short-fuse-big-bang-4-ways-that-digital-disruption-will-change-your-industry/.

Jargon, J. 2017. This is what Panera did to make its menu clean. MarketWatch. Available at: http://www.marketwatch.com/story/what-panera-had-to-change-to-make-its-menu-clean-2017-02-28.

Jayanthi, A. 2015. 7 celebrity data breaches: When employees snoop on high-profile patients. Becker's Hospital Review. Available at: https://www.beckershospitalreview.com/healthcare-information-technology/7-celebrity-data-breaches-when-employees-snoop-on-high-profile-patients.html.

Jha, A., L. Lin, and E. Savoia. 2016. The use of social media by state health departments in the US: Analyzing health communication through Facebook. *Journal of Community Health* 41(1):174–179.

Jiwani, A., D. Himmelstein, S. Woolhandler, and J.G. Kahn. 2014. Billing and insurance-related administrative costs in United States' health care: synthesis of micro-costing evidence. *BMC Health Services Research* 14:556.

Johnson, N.B. 2014. CDC National Health Report: Leading Causes of Morbidity and Mortality and Associated Behavioral Risk and Protective Factors—United States, 2005–2013. *Morbidity and Mortality Weekly Report (MMWR)* 63(4):3–27.

Johnson, A.M., J.M. Woodside, A. Johnson, and J.M. Pollack. 2016. Spatial patterns and neighborhood characteristics of overall suicide clusters in Florida from 2001 to 2010. *American Journal of Preventive Medicine* 52(1):1–7.

Jonnagaddala, J., S. Liaw, P. Ray, M. Kumar, N. Chang, and H. Dai. 2015. Coronary artery disease risk assessment from unstructured electronic health records using text mining. *Journal of Biomedical Informatics* 58:S203–S210.

Jost, T. 2016. CMS White Paper Examines the ACA Risk Adjustment Methodology (Update). Health Affairs Blog. Available at: http://healthaffairs.org/blog/2016/03/29/cms-white-paper-examines-the-aca-risk-adjustment-methodology/.

Jui-Long H. 2012. Trends of e-learning research from 2000 to 2008: Use of text mining and bibliometrics. *British Journal of Educational Technology* 43(1):5–16.

Kaiser Family Foundation. 2016. 2016 Employer Health Benefits Survey. Kaiser Family Foundation. Available at: https://www.kff.org/report-section/ehbs-2016-section-ten-plan-funding/.

Kaiser Family Foundation. 2017. Health Care Expenditures per Capita by State of Residence. Kaiser Family Foundation. Available at: http://www.kff.org/other/state-indicator/health-spending-per-capita/?activeTab=map&currentTimeframe=0&selectedDistributions=health-spending-per-capita&sortModel=%7B%22colId%22:%22Location%22,%22sort%22:%22asc%22%7.

Kaiser Permanente. 2018. About Us: Fast Facts. Available at: https://share.kaiserpermanente.org/article/fast-facts-about-kaiser-permanente/.

Kalhori, S.R.N., M. Nasehi, and X.-J. Zeng. 2010. A logistic regression model to predict high risk patients to fail in tuberculosis treatment course completion. *IAENG International Journal of Applied Mathematics* 40:2.

Kaplan, R.S., and M.E. Porter. 2011. The Big Idea: How to Solve the Cost Crisis in Health Care. Harvard Business Review. Available at: https://hbr.org/2011/09/how-to-solve-the-cost-crisis-in-health-care.

Kart, L., G. Herschel, A. Linden, and J. Hare. 2016. Magic Quadrant for Advanced Analytics Platforms.

Kendall, K.E., and J.E. Kendall. 2014. Systems Analysis and Design. 9th edition. Upper Saddle River, NJ: Pearson.

Kendrick, M.M., et al. 2017. The Better Care Reconciliation Act of 2017 vs. The American Health Care Act: Summary of Key Differences. Akin Gump.

Kim, M.I., and K.B. Johnson. 2002. Personal health records: Evaluation of functionality and utility. *Journal of the American Medical Informatics Association* 9(2):171–180.

Klimberg, R., and B.D. McCullough. 2013. *Fundamentals of Predictive Analytics with JMP*. Cary, NC: SAS Institute, Inc.

Knickrehm, M., B. Berthon, and P. Daugherty. 2016. Digital disruption: The growth multiplier. Accenture. Available at: https://www.accenture.com/au-en/insight-digital-disruption-growth-multiplier.

Koo, M., M.-C. Lu, and S.-C. Lin. October 2016. Predictors of Internet use for health information among male and female Internet users: Findings from the 2009 Taiwan National Health Interview Survey. *International Journal of Medical Informatics* 94:155–163.

Kovell, M. 2011. A woman's Facebook PHR saved her life. MyPHR. Available at: http://www.myphr.com/Stories/SuccessStory.aspx?Id=382.

Kroenke, D.M., and R.J. Boyle. 2017. *Experiencing MIS*, 7th Edition. Boston, MA: Pearson.

Kuehn, B.M. 2017. Twitter streams fuel big data approaches to health forecasting. *Journal of American Medical Association* 314(10): 2010–2012.

Kulkarni, P., L.D. Smith, and K.F. Woeltje. 2016. Assessing risk of hospital readmissions for improving medical practice. *Health Care Management Science* 19(3):291–299.

Kunte, M. 2016. Employee wellness practices - A study in selected organizations. *SIES Journal of Management* 12(1):9–14.

Kurtz, A. 2013. Job-hopping millennials no different than their parents. CNN Money. Available at: https://money.cnn.com/2013/04/09/news/economy/millennial-job-hopping/index.html.

Ladstatter, F., E. Garrosa, B. Moreno-Jimenez, V. Ponsoda, J.M. Reales Aviles, and J. Dai. 2016. Expanding the occupational health methodology: A concatenated artificial neural network approach to model the burnout process in Chinese nurses. *Ergonomics* 59(2):207–221.

Lake, A., and M. Chan. 2017. Statement from UNICEF Executive Director Anthony Lake and WHO Director-General Margaret Chan on the cholera outbreak in Yemen as suspected cases exceed 200,000. World Health Organization. Available at: http://www.who.int/mediacentre/news/statements/2017/Cholera-Yemen/en/.

LaMagna, M. 2016. The Secret to Staying Young is Broccoli and Cabbage. MarketWatch. Available at: https://www.marketwatch.com/story/the-secret-to-staying-young-is-broccoli-and-cabbage-2016-10-31.

LaMagna, M. 2017. These are the Best Industries for Job Seekers Right Now. MarketWatch. Available at: http://www.marketwatch.com/story/these-are-the-best-industries-for-job-seekers-right-now-2017-02-10.

Lattin, J.M., J.D. Carroll, and P.E. Green. 2003. *Analyzing Multivariate Data*. Pacific Grove, CA: Thomson Brooks/Cole.

Lauer, K.A., et al. 2017. Fourth Circuit Declines to Address Use of Statistical Sampling in False Claims Act Cases. Latham & Watkins. Available at: https://www.lw.com/thoughtLeadership/fourth-circuit-declines-address-use-statistical-sampling-false-claims-act.

Lhuer, X. 2016. The next acronym you need to know about: RPA (robotic process automation). McKinsey. Available at: http://www.mckinsey.com/business-functions/digital-mckinsey/our-insights/the-next-acronym-you-need-to-know-about-rpa.

Lim, L. 2006. Foreign health insurers seek clients in China, in Bloomberg News. Available at: http://www.iht.com/articles/2006/03/29/bloomberg/bxchina.php.

Limeade. 2016. Well-Being & Engagement Report. Limeade Institute. Available at: http://www.limeade.com/content/uploads/2016/11/QW-LimeadeWellBeingEngagementReport-final.pdf.

Liou, F.-M., Y.-C. Tang, and J.-Y. Chen. 2008. Detecting hospital fraud and claim abuse through diabetic outpatient services. *Health Care Management Science* 11(4):353–358.

Livingston, S. 2016. Providers say commercial payers are unwilling to share risk. Modern Healthcare. Available at: http://www.modernhealthcare.com/article/20161012/NEWS/161019970.

Losa-Iglesias, M.E., R. Becerro-de-Bengoa-Vallejo, and K.R. Becerro-de-Bengoa-Losa. 2016. Reliability and concurrent validity of a peripheral pulse oximeter and health–app system for the quantification of heart rate in healthy adults. *Health Informatics Journal* 22(2):151–159.

Low, L.L., K.H. Lee, M.E.H. Ong, et al. 2015. Predicting 30-day readmissions: Performance of the LACE index compared with a regression model among general medicine patients in Singapore. *BioMed Research International*. https://doi.org/10.1155/2015/169870.

Ma, Y-N., J. Wang, G-H. Dong, M.-M. Liu, D. Wang, et al. 2013. Predictive equations using regression analysis of pulmonary function for healthy children in northeast China. *PLoS ONE* 8(5).

Makary, M.A., and M. Daniel. 2016. Medical error—the third leading cause of death in the US. *BMJ* 353:i2139.

Maldonado, M., J. Dean, W. Czika, and S. Haller, S. 2014. Leveraging Ensemble Models in SAS Enterprise Miner. Proceedings of the SAS Global Forum 2014 Conference. Cary, NC: SAS Institute, Inc.

Maney, K. 2017. How Artificial Intelligence Will Cure America's Sick Health Care System. Newsweek. Available at: http://www.newsweek.com/2017/06/02/ai-cure-america-sick-health-care-system-614583.html.

MarketWatch. November 8, 2016. Healthcare Analytics Market Growing at 27% CAGR to be Worth $24.55 Billion by 2021 Dominated by North America. Available at: https://www.marketwatch.com/story/healthcare-analytics-market-growing-at-27-cagr-to-be-worth-2455-billion-by-2021-dominated-by-north-america-2016-11-08-5203114.

Mayo Clinic. 2017a. Test Setup Information. Available at: https://www.mayomedicallaboratories.com/test-catalog/appendix/loinc-codes.html.

Mayo Clinic. 2017b. Sharing Mayo Clinic. Available at: http://sharing.mayoclinic.org/guidelines/for-mayo-clinic-employees/tml.

Mayo Clinic. 2017c. Sudden infant death syndrome (SIDS). Mayo Clinic. Available at: http://www.mayoclinic.org/diseases-conditions/sudden-infant-death-syndrome/basics/risk-factors/con-20020269.

Mazerolle, M. J. 2004. APPENDIX 1: Making sense out of Akaike's Information Criterion (AIC): its use and interpretation in model selection and inference from ecological data. Available at: http://avesbiodiv.mncn.csic.es/estadistica/senseaic.pdf.

Mazzolini, C. 2013. EHR holdouts: Why some physicians refuse to plug in. Medical Economics. Available at: http://medicaleconomics.modernmedicine.com/medical-economics/content/tags/2013-salary-survey/ehr-holdouts-why-some-physicians-refuse-plug?page=full.

McCarthy, D., and K. Mueller. 2009. Kaiser Permanente: Bridging the Quality Divide with Integrated Practice, Group Accountability, and Health Information Technology. The Commonwealth Fund.

McDowell, W., et al. 2018. Cigna to Acquire Express Scripts for $67 billion. Available at: https://www.cigna.com/newsroom/news-releases/2018/cigna-to-acquire-express-scripts-for-67-billion.

McGlynn, E.A., et al. 2003. The quality of health care delivered to adults in the United States. *The New England Journal of Medicine* 348(26):2635–2645.

McKesson. 2017. Claims Management Technology. Available at: http://www.mckesson.com/health-plans/network-and-financial-management/clinical-claims-management/claimsxten/.

McNeill, D. 2014. Grasping the brass ring to improve healthcare through analytics: Implementation methods. In: *Analytics in Healthcare and the Life Sciences: Strategies, Implementation Methods, and Best Practices*, ed. D. McNeill. Upper Saddle River, NJ: Pearson.

McQuivey, J. 2013. Digital disruption: Unleashing the next wave of innovation (excerpt). Huffington Post. Available at: http://www.huffingtonpost.com/james-mcquivey/digital-disruption_b_2868789.html.

MedicalBillingandCoding. 2017. How Does Commercial Health Insurance Work? Available at: http://www.medicalbillingandcodingu.org/how-does-commercial-health-insurance-work/.

MedicineNet. 2016. Medical definition of patient. MedicineNet, Inc. Available at: https://www.medicinenet.com/script/main/art.asp?articlekey=39154.

Medina, V. 2017. Tyson Foods Inc. recalls ready-to-eat breaded chicken products due to misbranding and undeclared allergens. USDA. Available at: https://www.fsis.usda.gov/wps/portal/fsis/topics/recalls-and-public-health-alerts/recall-case-archive/archive/2017/recall-067-2017-releasesegway.

Mehta, M. March 12, 2015. The Wisdom of the PaaS Crowd. Forbes.

Merrideth, F. 2013. SAS survey signals big data disconnect; only 12 percent on board. Caey NC: SAS Institute, Inc.

Miliard, M. August 3, 2012. 'Deluge of data' has hospitals swimming upstream. Healthcare IT News.

Minitab. 2016. Odds ratios for Binary Logistic Regression. Minitab Express Support. Available at: http://support.minitab.com/en-us/minitab-express/1/help-and-how-to/modeling-statistics/regression/how-to/binary-logistic-regression/interpret-the-results/all-statistics-and-graphs/odds-ratios/.

Moeller, M. 2009. Manage Medical Advances with Automated Prior Authorization. Managed Healthcare Executive.

Monica, K. 2018. Practice Fusion No Longer Offering Free EHR System Software. Available at: https://ehrintelligence.com/news/practice-fusion-no-longer-offering-free-ehr-system-software.

mThink. June 30, 2003. Health Care Technology: A History of Clinical Care Innovation. Available at: https://mthink.com/health-care-technology-history-clinical-care-innovation/.

Mukherjee, S. 2017. Graham-Cassidy may be dead, but Obamacare repeal is very much alive. Fortune. Available at: http://fortune.com/2017/09/26/healthcare-bill-graham-cassidy/.

Mukungwa, T. 2015. Factors Associated with full Immunization Coverage amongst children aged 12– 23 months in Zimbabwe. *African Population Studies*29:1761–1774.

Murphy, B. 2016. CMS penalizes 2.6k hospitals for high readmissions: 5 statistics. Becker's Hospital Review. Available at: http://www.beckershospitalreview.com/quality/cms-penalizes-2-6k-hospitals-for-high-readmissions-5-statistics.html.

Murphy, T. 2017. Anthem gives up Cigna bid, vows to fight on over damages. Associated Press. Available at: https://finance.yahoo.com/news/anthem-gives-cigna-bid-vows-181247826.html.

Naidu, A. 2009. Factors affecting patient satisfaction and healthcare quality. *International Journal of Health Care Quality Assurance* 22(4):366–381.

NCQA. 2017a. About NCQA. Available at: http://www.ncqa.org/about-ncqa.

NCQA. 2017b. HEDIS Data Submission. Available at: http://www.ncqa.org/hedis-quality-measurement/hedis-data-submission.

NCQA. 2017c. HEDIS® & Performance Measurement. Available at: http://www.ncqa.org/hedis-quality-measurement.

NCQA. 2017d. IDSS Import Template and Materials. Available at: http://www.ncqa.org/hedis-quality-measurement/hedis-data-submission/idss-import-template-and-materials.

NCQA. 2017e. Summary Table of Measures, Product Lines and Changes. Available at: http://www.ncqa.org/Portals/0/HEDISQM/HEDIS2017/HEDIS%202017%20Volume%202%20List%20of%20Measures.pdf?ver=2016-06-27-135433-350.

NCQA. 2018. Disease Management Organizations Report Card. Available at: http://www.ncqa.org/report-cards/other-healthcare-organizations/disease-management-organizations.

NCSL. 2011. The Affordable Care Act: A Brief Summary. Available at: http://www.ncsl.org/portals/1/documents/health/hraca.pdf.

NCSL. 2017. Health Insurance: Premiums and Increases. National Conference of State Legislatures. Available at: http://www.ncsl.org/research/health/health-insurance-premiums.aspx.

Nemours. 2017. Online Video Doctor Visit. Available at: https://www.nemours.org/services/pediatric-online-doctor-visit.html.

NHANES. 2017. Available at: https://catalog.data.gov/dataset/national-health-and-nutrition-examination-survey-nhanes.

Nesson, M. 2016. Dying Preventable Deaths: The Importance of Immunization. *Harvard International Review* 37:18–20.

Neuberg, G.W. 2009. The Cost of End-of-Life Care. Cardiovascular Quality and Outcomes. Available at: http://circoutcomes.ahajournals.org/content/2/2/127.full.

Newman, P. 2017. Digital Disruption in Healthcare: The $8.7 trillion opportunity in digital health. Business Insider. Available at: http://www.businessinsider.com/digital-disruption-in-health-care-the-87-trillion-opportunity-in-digital-health-2017-5.

NHS. 2017. About NHS England. Available at: https://www.england.nhs.uk/about/.

NIH. 2017a. Budget. Available at: https://www.nih.gov/about-nih/what-we-do/budget.

NIH. 2017b. Mission and Goals. Available at: https://www.nih.gov/about-nih/what-we-do/mission-goals.

NIST/SEMATECH. 2017. e-Handbook of Statistical Methods. Measures of Skewness and Kurtosis. NIST/SEMATECH. Available at: http://www.itl.nist.gov/div898/handbook/eda/section3/eda35b.htm.

Nonaka, I. 1994. A dynamic theory of organizational knowledge creation. *Organization Science* 5:14–37.

O'Brien, B. 2016. Disrupting Healthcare Through the Internet of Things. Aria. Available at: https://www.ariasystems.com/blog/disrupting-healthcare-through-the-internet-of-things/.

O'Brien, S.A. 2017. Why disrupting health care can be tricky. CNN. Available at: https://money.cnn.com/2017/03/04/technology/health-care-collective-health/index.html.

OCR. 2008. Personal Health Records and the HIPAA Privacy Rule. Office for Civil Rights. https://www.hhs.gov/sites/default/files/ocr/privacy/hipaa/understanding/special/healthit/phrs.pdf. [SAS Library Note: should the author be OCR or Office for Civil Rights? Unable to verify the 2008 date]

OCR. 2017. About Us. Office for Civil Rights. Available at: https://www.hhs.gov/ocr/about-us/index.html.

OCR. 2017. Summary of the HIPAA Privacy Rule. HHS.gov. Available at: https://www.hhs.gov/hipaa/for-professionals/privacy/laws-regulations/index.html.

OECD. 2011. Why is Health Spending in the United States So High? Available at: https://www.oecd.org/unitedstates/49084355.pdf.

Okoye, H. 2015. Understanding Healthcare's Value Chain. Ausmed. Available at: https://www.ausmed.com/articles/healthcare-value-chain/.

Orcutt, R. 2009. Common Coding Errors and How to Prevent Them. Clinical-Insights.

Ornstein, C. 2015. Celebrities' Medical Records Tempt Hospital Workers to Snoop. ProPublica. Available at: https://www.propublica.org/article/clooney-to-kardashian-celebrities-medical-records-hospital-workers-snoop.

Parikh, R.B., Z. Obermeyer, and D.W. Bates. 2016. Making Predictive Analytics a Routine Part of Patient Care. Harvard Business Review. Available at: https://hbr.org/2016/04/making-predictive-analytics-a-routine-part-of-patient-care.

Patel, V.L., et al. 2009. Cognitive and learning sciences in biomedical and health instructional design: A review with lessons for biomedical informatics education. *Journal of Biomedical Informatics* 42: 176–197.

Paul, K. 2017. College students would give up their friends' privacy for free pizza. MarketWatch. Available at: https://www.marketwatch.com/story/college-students-would-give-up-their-friends-privacy-for-free-pizza-2017-06-13.

Pearl, R. 2015. Are You a Patient or A Healthcare Consumer? Forbes. Available at: https://www.forbes.com/sites/robertpearl/2015/10/15/are-you-a-patient-or-a-health-care-consumer-why-it-matters/#2a6891a12b4d.

Pearson, B. 2018. 5 Ways A Walmart-Humana Partnership Could Change Everyone's In-Store Experience. Forbes. Available at: https://www.forbes.com/sites/bryanpearson/2018/04/12/5-ways-a-walmart-humana-partnership-could-change-everyones-in-store-experience/#2cc84a7a58a7.

PHDSC. October 2007. Public Health Data Standards Consortium. Users Guide for Source of Payment Typology. Available at: http://www.phdsc.org/about/committees/pdfs/SourceofPaymentTypologyUsersGuideOct2007.pdf.

Phillips, D.P., K.M. Brewer, and P. Wadensweiler. 2010. Alcohol as a risk factor for sudden infant death syndrome (SIDS). *Addiction* 106(30):516–525.

Pirani, F. 2017. 7 things you need to know before you send your spit to 23andMe. The Atlanta Journal-Constitution. Available at: http://www.ajc.com/news/national/things-you-need-know-before-you-send-your-spit-23andme/QdYVKNKLIq44aICIY1KvPJ/.

Pittman, D. 2014. 5 Problems with Mobile Health App Security. MedPage Today. Available at: https://www.medpagetoday.com/PracticeManagement/InformationTechnology/44161.

PNHP. 2017. What is Single Payer? Physicians for a National Health Program. Available at: http://www.pnhp.org/facts/what-is-single-payer.

Porterfield, A., K. Engelbert, amd A. Coustasse. 2014. Electronic prescribing: Improving the efficiency and accuracy of prescribing in the ambulatory care setting. *Perspectives in Health Information Management* 11(Spring):1g.

Premark, R. 2016. The soda industry is on the verge of losing one of its biggest battles ever. The Washington Post.

Pruss-Ustun, A., et al. 2016. Preventing disease through healthy environments. World Health Organization.

Punk, H. 2016. The 49 hospitals facing the highest readmission penalties from Medicare. Becker's Hospital Review. Available at: http://www.beckershospitalreview.com/quality/the-49-hospitals-facing-the-highest-readmission-penalties-from-medicare.html.

PwC. 2017. Top health industry issues of 2017: A year of uncertainty and opportunity. Available at: http://www.pwc.com/us/en/health-industries/top-health-industry-issues.html.

Qualcomm. 2017. XPrize Overview. Available at: https://tricorder.xprize.org/about/overview.

Ramsey, L. 2018. CVS and Aetna's megamerger could get blocked, and we'll soon get a hint of what will happen next. Business Insider. Available at: https://www.businessinsider.com/cvs-health-aetna-merger-potential-antitrust-challenges-to-the-deal-2018-4.

Ran, C., Waitzkin, H., Merhy, E. 2004. Managed Care Goes Global, in Multinational Monitor. Available at: http://www.thirdworldtraveler.com/Health/Managed_Care_Global.html.

Rashidian, A., H. Joudaki, and T. Vian. 2012. No evidence of the effect of the interventions to combat health care fraud and abuse: A systematic review of literature. *PLoS ONE* 7(8)

Ravussin, E., L.M. Redman, J. Rochon, S.K. Das, L. Fontana, W.E. Kraus, S. B. Roberts. 2015. A 2-year randomized controlled trial of human caloric restriction: Feasibility and effects on predictors of health span and longevity. *The Journals of Gerontology. Series A, Biological Sciences and Medical Sciences* 70(9):1097–1104.

Ravenna, A., C. Truxillo, and C. Wells. 2015. SAS Course Notes: SAS® Visual Statistics: Interactive Model Building. Cary, NC: SAS Institute, Inc.

Ray, B. 2017. IoT for Healthcare—A $163 billion opportunity. Medium. Available at: https://medium.com/iotforall/iot-for-healthcare-a-163-billion-opportunity-98e2a59b9849.

Raymond, N. 2017. U.S. investigates four insurers over Medicare payments. Reuters. Available at: https://finance.yahoo.com/news/u-probing-insurers-beyond-unitedhealth-over-medicare-charges-171239693--finance.html.

Rector, T.S., et al. 2004. Specificity and sensitivity of claims-based algorithms for identifying members of Medicare+Choice health plans that have chronic medical conditions. *Health Services Research* 39(6p1):839–1858.

Regenstrief Institute. 2017. About LOINC. Available at: https://loinc.org/about/.

Reider, J.R., and R. Tagalicod. 2013. Progress on Adoption of Electronic Health Records. HealthIT.gov. Available at: http://www.healthit.gov/buzz-blog/electronic-health-and-medical-records/progress-adoption-electronic-health-records/.

ReportLinker. 2017. Healthcare Cloud Computing: Global Markets to 2022. Available at: https://www.reportlinker.com/p05251939/Healthcare-Cloud-Computing-Global-Markets-to.html.

ReWalk. 2017a. What is ReWalk? Available at: http://rewalk.com.

ReWalk. 2017b. Florida Blue Cross Blue Shield to Cover ReWalk Exoskeleton for Paralyzed Plan Member Following Court Decision Deeming the Device Medically Necessary. Available at: http://rewalk.com/florida-blue-cross-blue-shield-to-cover-rewalk-exoskeleton/.

Reynolds, M. 2017. Augmented reality goggles give surgeons X-ray vision. *New Scientist*. Available at: https://www.newscientist.com/article/2130678-augmented-reality-goggles-give-surgeons-x-ray-vision/.

Roberts, T.G. 2003. An interpretation of Dewey's experiential learning theory. Available at: https://eric.ed.gov/?id=ED481922.

Roe, D.P. 2017. H.R. 277 - American Health Care Reform Act of 2017. Congress.gov. Available at: https://www.congress.gov/bill/115th-congress/house-bill/277.

Roser, M., and E. Ortiz-Ospina. 2017. OurWorldInData.org. Available at: https://ourworldindata.org/world-population-growth/.

Rossman, S. 2018. Why the romaine lettuce E. coli outbreak affects mostly women. 2018. Available at: https://www.usatoday.com/story/news/nation-now/2018/04/21/romaine-lettuce-e-coli-outbreak-affects-mostly-women/539024002/.

Royal College of Physicians. 2009. Improving clinical records and clinical coding together, Audit Commission, Editor.

Rozman, A., D. Habek, S. Zaputovic Brajnovic, E. Horvatic, and I. Jurkovic. 2014. Sudden infant death syndrome and perinatal risk factors. *Central European Journal of Medicine* 9(2):217–222.

Salesforce. 2017. About Us. Salesforce.com, Inc. Available at: https://www.salesforce.com/company/about-us/.

Sarasohn-Kahn, J. 2008. The Wisdom of Patients: Health Care Meets Online Social Media. California Health Care Foundation.

SAS Institute, Inc. 2003. Data Mining Using SAS® Enterprise Miner: A Case Study Approach, Second Edition. Available at: https://support.sas.com/documentation/onlinedoc/miner/casestudy_59123.pdf.

SAS Institute, Inc. 2014. Truxillo, C., and C. Wells, C. Advanced Business Analytics. Cary, NC: SAS Institute, Inc.

SAS Institute, Inc. 2015a. Getting Started with SAS® Enterprise Miner™ 14.1. Cary, NC: SAS Institute, Inc.

SAS Institute, Inc. 2015b. SAS OnDemand for Academics. Cary NC: SAS Institute, Inc. Available at: https://www.sas.com/en_us/software/on-demand-for-academics.html#students.

SAS Institute, Inc. 2016a. Health Analytics. Available at: https://www.sas.com/en_us/insights/health-analytics.html.

SAS Institute, Inc. 2016b. About SAS, Giving You the Power to Know since 1976. Available at: https://www.sas.com/en_us/company-information.html.

SAS Institute, Inc. 2016c. SAS Analytics in Action. SAS. Available at: https://www.sas.com/content/dam/SAS/de_de/doc/events/sas-partnertag/sas-partnertag-analytics-in-action.pdf.

SAS Institute, Inc. 2016d. SAS Federation Server 3.2: Administrator's Guide. Available at: http://support.sas.com/documentation/cdl/en/fedsrvmgrag/65976/HTML/default/viewer.htm#p0b2qr4wqsqqd8n1l6owuylwmiou.htm.

SAS Institute, Inc. 2016e. SAS® Enterprise Miner 14.2 High-Performance Procedures. Cary, NC: SAS Institute Inc.

SAS Institute, Inc. 2017a. Customer Stories. Available at: http://www.sas.com/en_us/customers.html#health-care.

SAS Institute, Inc. 2017b. Usage Note 24205: Rare-event oversampling for model fitting in SAS Enterprise Miner. Available at: http://support.sas.com/kb/24/205.html.

SAS Institute, Inc. 2018a. "About Nodes", Getting Started with SAS(R) Enterprise Miner(TM) 14.1. Available at: https://support.sas.com/documentation/cdl/en/emgsj/67981/HTML/default/viewer.htm#n1cpd0rgpneqwqn16mfcxp4sbjsb.htm.

SAS Institute, Inc. 2018b. "Combining Models", SAS Enterprise Miner 7.1 Extension Nodes: Developer's Guide. Available at: http://support.sas.com/documentation/cdl/en/emxndg/64759/HTML/default/viewer.htm#n1n7194dn7rsjbn1dv2qi7tisak3.htm.

Schencker, L. 2016. Providers Urge Court to Forbid Statistical Sampling in False Claims Cases. Modern Healthcare. Available at: http://www.modernhealthcare.com/article/20160329/NEWS/160329870.

Schwab, K. 2016. The Fourth Industrial Revolution: What it means, how to respond. World Economic Forum. Available at: https://www.weforum.org/agenda/2016/01/the-fourth-industrial-revolution-what-it-means-and-how-to-respond/.

Scitovsky, A.A. 2005. The high cost of dying: What do the data show? *The Milbank Quarterly* 83:825–841.

SearchHealthIT. 2014. Computerized Physician Order Entry (CPOE). TechTarget. Available at: http://searchhealthit.techtarget.com/definition/computerized-physician-order-entry-CPOE.

Seebacher, N. 2015. A Beginner's Guide to Digital Experience. CMSWire. Available at: https://www.cmswire.com/digital-experience/a-beginners-guide-to-digital-experience/.

Seeking Alpha. 2015. IBM Cloud Services (Part I). Seeking Alpha.

Selvam, A. 2012. Rules of the road. Modern Healthcare. Available at:
    http://www.modernhealthcare.com/article/20120421/magazine/304219937.

Senthilingam, M. 2017a. South Korea will take lead in life expectancy by 2030. CNN. Available at:
    http://www.cnn.com/2017/02/21/health/life-expectancy-increase-globally-by-2030/.

Senthilingam, M. 2017b. Death rate from Alzheimer's disease in the US has risen by 55%, says CDC. CNN. Available at:
    http://www.cnn.com/2017/05/26/health/alzheimers-disease-deaths-us-increase/.

Sermo. 2017. What is SERMO. Available at: http://www.sermo.com/what-is-sermo/overview.

Sharda, R., D. Delen, and E. Turban. 2014. *Business Intelligence: A Managerial Perspective on Analytics*, 3rd ed. Boston: Pearson.

Sharples, A.P., D.C. Hughes, C.S. Deane, A. Saini, C. Selman, and C.E. Stewart. 2015. Longevity and skeletal muscle mass: the role of IGF signalling, the sirtuins, dietary restriction and protein intake. *Aging Cell* 14(4):511–523.

Shay, R. 2017. EHR adoption rates: 20 must-see stats. Practice Fusion. Available at: https://www.practicefusion.com/blog/ehr-adoption-rates/.

Shmueli, G., N.R. Patel, P.C. Bruce. 2010. Data Mining for Business Intelligence: Concepts, Techniques, and Applications in Microsoft Excel with XLMiner, 2nd ed. Hoboken, NJ: Wiley.

SIIA. 2015. Self-Insured Group Health Plans. Self-Insurance Institute of America. Available at:
    https://www.siia.org/i4a/pages/index.cfm?pageID=4546.

Sillup, G.P., and R.K. Klimberg. 2011. Health Plan Auditing: 100-Percent-of-Claims vs. Random-Sampling Audits. Truven Health. Available at: http://truvenhealth.com/portals/0/assets/EMP_11266_0712_100claimsauditing_WP_WEB.pdf.

Simpao, A.F., L. M. Ahumada, J.A. Galvez, and M.A. Rehman. 2014. A review of analytics and clinical informatics in health care. *Journal of Medical Systems* 38(4):45.

Singh, N. 2017. Weathering change and the promise of digital transformation in healthcare. HealthcareITNews. Available at:
    http://www.healthcareitnews.com/news/weathering-change-and-promise-digital-transformation-healthcare.

Singh, T.B., S.S. Rathore, T.A. Choudhury, V.K. Shukla, D.K. Singh, and J. Prakash. 2013. Hospital-acquired acute kidney injury in medical, surgical, and intensive care unit: A comparative study. *Indian Journal of Nephrology* 23:24–29.

Skowronski, J. 2015. What is a FICO score? Bankrate. https://www.bankrate.com/brm/news/debt/debtmanageguide/vantage-scores2.asp?caret=40.

Skyttberg, N., J. Vicente, R. Chen, H. Blomqvist, and S. Koch. 2016. How to improve vital sign data quality for use in clinical decision support systems? A qualitative study in nine Swedish emergency departments. *BMC Medical Informatics and Decision Making* 16:61.

Smith, S. 2017. This new robotic suit could help millions of stroke victims walk again. Yahoo Finance. Available at:
    https://finance.yahoo.com/news/new-robotic-suit-help-millions-stroke-victims-walk-145604723.html.

Snipes, C. 2017. N.C.'s Health Information Exchange adds state's largest health care systems. Triangle Business Journal. Available at:
    https://www.bizjournals.com/triangle/news/2017/07/19/n-c-s-health-information-exchange-adds-states.html.

Soffen, K., D. Cameron, and K. Uhrmacher. 2017. How the House voted to pass the GOP health-care bill. The Washington Post.
    Available at: https://www.washingtonpost.com/graphics/politics/ahca-house-vote/?utm_term=.d44b3045bd8a.

Song, Q., Y.-J. Zheng, Y. Xue, W.-G. Sheng, and M.-R. Zhao. 2017. An evolutionary deep neural network for predicting morbidity of gastrointestinal infections by food contamination. *Neurocomputing* 226:16–22

Squires, D., and C. Anderson. 2015. U.S. Health Care from a Global Perspective: Spending, Use of Services, Prices, and Health in 13 Countries. The Commonwealth Fund. Available at: http://www.commonwealthfund.org/publications/issue-briefs/2015/oct/us-health-care-from-a-global-perspective.

SSA. 2013. Actuarial Life Table. Social Security Administration. Available at: https://www.ssa.gov/oact/STATS/table4c6.html.

Steger, L. 2014. Electronic Health Records Could Save Healthcare Billions. HealthcareITNews. Available at:
    https://www.abeo.com/electronic-health-records-could-save-healthcare-billions/.

Sullivan, T. 2017. Healthcare organizations must understand these elements of digital transformation today to succeed at DX in the near future. HealthcareITNews. Available at: http://www.healthcareitnews.com/news/5-steps-transform-digital-experience-healthcare.

Sussmuth-Dyckerhoff, C., and J. Wang. 2010. China's Health Care Reforms. Health International, No. 10. McKinsey & Company.

Sutner, S. 2017. Wisconsin provider uses Salesforce for Healthcare Marketing Campaign. TechTarget. Available at:
    http://searchhealthit.techtarget.com/feature/Wisconsin-provider-uses-Salesforce-for-healthcare-marketing-campaign.

Swartz, J. 2018. Digital Transformation Will Spark a Frenzy in Tech M&A in 2018: Prediction. Barrons. Available at: https://www.barrons.com/articles/digital-transformation-will-spark-a-frenzy-in-tech-m-a-in-2018-prediction-1519166445?mod=mw_hpm.

Syal, R. 2013. Abandoned NHS IT system has cost £10bn so far. The Guardian. Available at: https://www.theguardian.com/society/2013/sep/18/nhs-records-system-10bn.

Szalay, J. 2014. Broccoli: Health Benefits, Risks & Nutrition Facts. Live Science. Available at: http://www.livescience.com/45408-broccoli-nutrition.html.

Teece D.J., and G. Pisano. 2003. *The Dynamic Capabilities of Firms*. Handbook on Knowledge Management. International Handbooks on Information Systems, vol 2. Berlin: Springer.

Teece D.J, G. Pisano, and A. Shuen. 1997. Dynamic Capabilities and Strategic Management. *Strategic Management Journal* 18(7):509–533.

Tenconi, A. 2016. How to Turn Your Big Data into Business Insights: Analytics in Action. Bright Data Blog. Cary, NC: SAS Institute, Inc. Available at: https://blogs.sas.com/content/brightdata/2016/07/07/how-to-turn-your-big-data-into-business-insights-analytics-in-action/.

The George Mateljan Foundation. 2016. Broccoli. The George Mateljan Foundation. Available at: http://www.whfoods.com/genpage.php?tname=foodspice&dbid=9.

Thomas, L. 2017. Shares of this surgical robot maker just jumped to an all-time high on better-than-expected earnings. CNBC.

Thomas, L., and J.M. Woodside. 2015. Social media maturity model. *International Journal of Healthcare Management* 9(1):67–73.

Tiwari, R. 2016. mHealth Market to Represent $370 Billion in Annual Healthcare Cost Savings Worldwide by 2017. RnR Market Research. Available at: https://www.prnewswire.com/news-releases/mhealth-market-to-represent-370-billion-in-annual-healthcare-cost-savings-worldwide-by-2017-600842931.html.

TM Floyd & Company. 2006. Health Claims Processing in the US. TM Floyd & Company.

Tracer, Z., D. D. McLaughlin, and A.M. Harris. 2017. After two megadeals blocked, health insurers plot next moves. Bloomberg. Available at: https://www.bloomberg.com/news/articles/2017-02-09/anthem-s-bid-for-cigna-blocked-by-judge-as-anticompetitive.

Tricare. 2017. About Us. Available at: https://www.tricare.mil/About.

Trizetto. 2017. Claims Processing. Available at: http://www.trizetto.com/Provider-Solutions/Health-Systems/Claims-Processing/?utm_source=facebook&utm_medium=social&date=091514.

Tseng, W.-J., et al. 2013. Hip fracture risk assessment: artificial neural network outperforms conditional logistic regression in an age- and sex-matched case control study. *BMC Musculoskeletal Disorders* 14:207.

Twitter. 2015. Using Twitter Data to Study the World's Health. Twitter @Elaine. Available at: https://blog.twitter.com/2015/twitter-data-public-health.

Twitter. 2018. #EHRbacklash - Twitter Search. Available at: https://twitter.com/search?q=EHRbacklash.

UK Politics. 2013. NHS IT system one of 'worst fiascos ever', say MPs. BBC.

U.S. Census Bureau. 2016a. World Population. Available at: https://www.census.gov/population/international/data/worldpop/table_population.php.

U.S. Census Bureau. 2016b. World Population: 1950-2050. International Data Base. Available at: https://www.census.gov/population/international/data/idb/worldpopgraph.php.

U.S. Census Bureau. 2017. Historical Poverty Tables: People and Families - 1959 to 2016. Available at: https://www.census.gov/data/tables/time-series/demo/income-poverty/historical-poverty-people.html.

U.S. Senate. 2010. The Patient Protection and Affordable Care Act. Available at: https://www.dpc.senate.gov/healthreformbill/healthbill04.pdf.

UCMC. 2010. HIPAA Background. The University of Chicago Medical Center. Available at: http://hipaa.bsd.uchicago.edu/background.html.

Udwadia, T. E. 2015. Robotic surgery is ready for prime time in India: Against the motion. *Journal of Minimal Access Surgery* 11(1):5–9.

United Nations Department of Economic and Social Affairs. 2015. World Population Prospects: The 2015 Revision, Key Findings and Advance Tables. Available at: https://esa.un.org/unpd/wpp/publications/files/key_findings_wpp_2015.pdf.

UnitedHealthcare. 2017. Virtual Visits. Available at: https://www.uhc.com/individual-and-family/member-resources/health-care-tools/virtual-visits.

UPMC. 2017. UPMC Anywhere Care. Available at: https://myupmc.upmc.com/anywhere-care/.

UW Medicine. 2013. Protected Health Information, Limited Data Set, and De-Identification of Protected Health Information. Available at: https://depts.washington.edu/comply/docs/PP_19.pdf.

VA.gov. 2017. Veterans Health Administration. Available at: https://www.va.gov/health/.

Vandekerckhove, J., D. Matzke, and E.-J. Wagenmakers. 2014. Model Comparison and the Principle of Parsimony. CIDLab. Available at: http://www.cidlab.com/prints/vandekerckhove2014model.pdf.

Vandenbosch, B., and C.A. Higgins. 1995. Executive support systems and learning: A model and empirical test. *Journal of Management Information Systems* 12(2):99–130.

VFW. 2017. What is CHAMPVA and who is eligible? Available at: http://faq.vfw.org/faq/index.php?View=entry&EntryID=445.

Viaene, S., G. Dedene, R.A. Derrig. 2005. Auto claim fraud detection using Bayesian learning neural networks. *Expert Systems with Applications* 29(3):653–666.

Wade, J. 2017. Wearable Technology Statistics and Trends 2018. Smart Insights. Available at: https://www.smartinsights.com/digital-marketing-strategy/wearables-statistics-2017/.

Wagner, M. 2009. CDC Readies Internet Barrage to Combat Swine Flu. InformationWeek.

Walgreens. 2017. Flu Index. Available at: https://walgreens.maps.arcgis.com/apps/MapSeries/index.html?appid=40d0763cd3cc42428b26f85202108469.

Walgreens. 2018. Walgreens Flu Index™ Shows Flu Activity Gains in the Midwest. Available at: https://news.walgreens.com/press-releases/walgreens-flu-index-shows-flu-activity-gains-in-the-midwest.htm.

Walker, J. March 6, 2017. Alliance of Companies Announce Plans to Lower Their Health-care Costs. Available at: https://www.marketwatch.com/story/alliance-of-companies-announce-plans-to-lower-their-health-care-costs-2017-03-06.

Wang, P. 2012. Cutting the high cost of end-of-life care. Time. Available at: http://time.com/money/2793643/cutting-the-high-cost-of-end-of-life-care/.

Wang, R., G. Blackburn, M. Desai, et al. 2017. Accuracy of wrist-worn heart rate monitors. *JAMA Cardiology* 2(1):104–106.

Wang, S. 2017. How doctors are battling bad reviews on Yelp. The Wall Street Journal. Available at: http://www.marketwatch.com/story/how-doctors-are-battling-bad-reviews-on-yelp-2017-07-07.

Wang, W., Y. Jiang, and C.-H. Lee. 2016. Independent predictors of physical health in community-dwelling patients with coronary heart disease in Singapore. *Health and Quality of Life Outcomes* 14:113.

Weaver, C.A., and T. Hongsermeier. 2004. Measuring outcomes: bringing six sigma excellence to health care. In: *Healthcare Information Management Systems: Cases, Strategies, and Solutions*. 3rd ed., eds. M.J. Ball, C.A. Weaver, and J.M. Kiel. New York: Springer.

WebMD. 2017. Understanding Heart Attack: The Basics. Available at: https://www.webmd.com/heart-disease/understanding-heart-attack-basics#1.

Weiss, A.J., M.L. Barrett, and C.A. Steiner. 2014. Trends and Projections in Inpatient Hospital Costs and Utilization, 2003-2013. HCUP Statistical Brief #175. Agency for Healthcare Research and Quality.

White, S.K. 2015. Study reveals that most companies are failing at big data. CIO. Available at: https://www.cio.com/article/3003538/big-data/study-reveals-that-most-companies-are-failing-at-big-data.html.

Whitman, E. 2017. When a hospital patient is also a guest. Modern Healthcare. Available at: http://www.modernhealthcare.com/article/20170513/MAGAZINE/170519976.

WHO. 2017. Classification of Diseases. World Health Organization. Available at: http://www.who.int/classifications/icd/en/.

Wickell, J. 2016. How Your Credit Score is Calculated. The Balance. Available at: http://homebuying.about.com/cs/yourcreditrating/a/credit_score.htm.

Wilber, D.Q. 2017. Feds charge more than 400 with health-care fraud. The Wall Street Journal. Available at: http://www.marketwatch.com/story/feds-charge-more-than-400-with-health-care-fraud-2017-07-13.

Wiley, K. E., M. Steffens, N. Berry, and J. Leask. 2017. An audit of the quality of online immunisation information available to Australian parents. *BMC Public Health* doi:10.1186/s12889-016-3933-9.

Wisenberg Brin, D. 2016. Wellness Programs Raise Privacy Concerns over Health Data. Available at: https://www.shrm.org/resourcesandtools/hr-topics/technology/pages/wellness-programs-raise-privacy-concerns-over-health-data.aspx.

Witte, G., amd K. Adam. 2017. Hospitals across Britain paralyzed by cyberattack amid international assault. The Washington Post. Available at: https://www.washingtonpost.com/world/hospitals-across-england-report-it-failure-amid-suspected-major-cyber-attack/2017/05/12/84e3dc5e-3723-11e7-b373-418f6849a004_story.html?utm_term=.262584b57f25.

Wolf, J.A. 2017. Reaffirming the core values of patient experience. The Beryl Institute. Available at: http://www.theberylinstitute.org/blogpost/593434/267552/Reaffirming-the-Core-Values-of-Patient-Experience.

Wood, L. 2017. Global $189 Billion Mobile Health Market, 2025. PRNewswire. Available at: https://www.prnewswire.com/news-releases/global-189-billion-mobile-health-market-2025-300535596.html.

Woodside, J.M. 2007a. Economic externalities of health information technology: A game theoretic model for electronic health record adoption. *Journal of Healthcare Information Management* 21(4):25–31.

Woodside, J.M. 2007b. EDI and ERP: A real-time framework for healthcare data exchange. *Journal of Medical Systems* 31(3):178–184.

Woodside, J.M. 2008. Dynamic Healthcare Connectivity and Collaboration with Multi-Agent Systems. Proceedings of the Fourteenth Americas Conference on Information Systems.

Woodside, J.M. 2010a. Business intelligence and learning, drivers of quality and competitive performance. ETD Archive, paper 151.

Woodside, J.M. 2010b. A BI 2.0 Application Architecture for Healthcare Data Mining Services in the Cloud. The World Congress in Computer Science, Computer Engineering & Applied Computing – DMIN'10: The 2010 International Conference on Data Mining.

Woodside, J.M. 2011. Business Intelligence Best Practices for Success. International Conference on Information Management and Evaluation.

Woodside, J.M. 2012. Health Intelligence Model for Evaluation of Social Media Outcomes. International Conference on Information Technology- New Generations (ITNG).

Woodside, J.M. 2013a. Research Roadmap: Big Data in Healthcare. International Conference on Big Data.

Woodside, J.M. 2013b. Integrated care intelligence: A socio-technical approach for big data management. *Journal of Healthcare Information Management* 27(3):60–64.

Woodside, J.M. 2013c. Analytical Framework for Improving Organizational Quality. Institute for Healthcare Improvement Annual Forum.

Woodside, J.M. 2013d. sElection: Software Selection as a Presidential Election Mapping Process. Society for Information Technology & Teacher Education International Conference.

Woodside, J.M. 2013e. EDI and ERP: A real-time framework for healthcare data exchange. *Journal of Medical Systems* 31(3):178–184.

Woodside, J.M. 2014. Big Data Veracity in Healthcare. The 2014 International Conference on Advances in Big Data Analytics.

Woodside, J.M. 2015a. Advances in information, security, privacy & ethics: Use of cloud computing for education. In: *Handbook of Research on Security Considerations in Cloud Computing*, eds. S. Munir, M. S. Al-Mutairi, and A. L. Mohammed. Hershey, PA: Information Science Reference.

Woodside, J.M. 2016a. BEMO: A Parsimonious Big Data Mining Methodology. AJIT-e: *Online Academic Journal of Information Technology* 7(24):113–123.

Woodside, J.M. 2016b. Transforming Healthcare Provider and Patient Power Dynamics – Exploring the Impact of Mobile Healthcare. Americas Conference on Information Systems (AMCIS).

Woodside, J.M. 2018a. Real-world rigour: An integrative learning approach for industry and higher education. *Industry and Higher Education.*

Woodside, J.M. 2018b. Organizational health management through metaphor: a mission-based approach. *Journal of Health Organization and Management* 32(3):374–393.

Woodside, J.M., and A. Johnson. 2015. Developing curricular modules in context: mobile healthcare security. Proceedings of the 2015 NSF Workshop on Curricular Development for Computing in Context.

Woodside, J.M, amd I.U. Sikder. 2010. Space-time cluster analysis: Application of healthcare service data in epidemiological studies. *International Journal of Healthcare Information Systems and Informatics* 4(4):1842–1856.

Woodside, J.M., and S. Amiri. 2015. Theoretical Application of Business Decision Support. Proceedings of the 19th World Multi-Conference on Systemics, Cybernetics and Informatics: WMSCI 2015.

Woodside, J.M., and M. Florea. 2015. Security intelligence for healthcare mobile electronic commerce. Mobile Electronic Commerce: Foundations, Development, and Applications. June Wei [editor]. Boca Raton, FL: CRC Press.

Woodside, J.M., and R. Reinhold, 2018. Wearable Technology in Education. *Journal of Innovative Education Strategies* 6(1–2):39–52.

Woodside, J.M., and S. Amiri. 2018. Healthcare Hyperchain: Digital Transformation in the Healthcare Value Chain. Twenty-Fourth Americas Conference on Information Systems.

World Health Organization. 1996. European Health Care Reforms: Analysis of Current Strategies. Available at: http://www.euro.who.int/__data/assets/pdf_file/0005/111011/sumhecareform.pdf.

World Health Organization. 2010. Telemedicine: Opportunities and Developments in Member States: Report on the Second Global Survey on eHealth. Available at: http://www.who.int/goe/publications/goe_telemedicine_2010.pdf.

World Health Organization. 2012. The Bigger Picture for e-health. *Bulletin of the World Health Organization* 90(5):330–331.

World Health Organization. 2017. 10 Facts on Ageing and Health. Available at: http://www.who.int/features/factfiles/ageing/en/.Wright, J. 2013. Health Care's Unrivaled Job Gains and Where It Matters Most. Forbes. Available at: https://www.forbes.com/sites/emsi/2013/10/07/health-cares-unrivaled-job-gains-and-where-it-matters-most/#3a6acd6814a7.

Yang, F-C., A.J.T. Lee, and S.-C. Kuo. 2016. Mining health social media with sentiment analysis. *Journal of Medical Systems* 40(11):236.

Yang, W.-S., and S.-Y. Hwang. 2006. A process-mining framework for the detection of healthcare fraud and abuse. *Expert Systems with Applications* 31(1):56–68.

Yap, J., Y.C Chet, and M. Grover. 2015. A perspective of future healthcare landscape in ASEAN and Singapore. Deloitte. Available at: https://www2.deloitte.com/content/dam/Deloitte/sg/Documents/risk/sea-risk-future-healthcare-thought-leadership-noexp.pdf.

Yee, R. W. Y., A.C.L. Yeung, and T.C. Edwin. 2008. The impact of employee satisfaction on quality and profitability in high-contact service industries. *Journal of Operations Management* 26(5):651–668.

Yu, T-H., Y.-C. Hou, K.-C. Lin, and K.-P. Chung. 2014. Is it possible to identify cases of coronary artery bypass graft postoperative surgical site infection accurately from claims data? A multi-model comparison study over 2005–2008. *BMC Medical Informatics and Decision Making* 14:42.

Zacks. 2017. Aetna to Shift Headquarters, Aims Business Transformation. Zacks Equity Research. Available at: https://finance.yahoo.com/news/aetna-shift-headquarters-aims-business-123412373.html.

Zezza, M.A., A-M.J. Audet, and D. Hall. 2014. Incentives 2.0: A Synergistic Approach to Provider Incentives. The Commonwealth Fund Blog. Available at: http://www.commonwealthfund.org/publications/blog/2014/jul/synergistic-approach-provider-incentives.

Zwirn, E. 2017. Philly's soda tax is crushing the city's beverage business. New York Post.http://nypost.com/2017/03/05/phillys-soda-tax-is-crushing-the-citys-beverage-business/.

Index

Ready to take your SAS® and JMP® skills up a notch?

Be among the first to know about new books, special events, and exclusive discounts.
support.sas.com/newbooks

Share your expertise. Write a book with SAS.
support.sas.com/publish

sas.com/books
for additional books and resources.

SAS
THE POWER TO KNOW.

www.ingramcontent.com/pod-product-compliance
Lightning Source LLC
Chambersburg PA
CBHW081045220326
41598CB00038B/6987